Biological, Diagnostic and Therapeutic Advances in Alzheimer's Disease

Ghulam Md Ashraf • Athanasios Alexiou
Editors

Biological, Diagnostic and Therapeutic Advances in Alzheimer's Disease

Non-Pharmacological Therapies for Alzheimer's Disease

Springer

Editors
Ghulam Md Ashraf
King Fahd Medical Research Center
King Abdulaziz University
Jeddah, Saudi Arabia

Department of Medical Laboratory Technology, Faculty of Applied Medical Sciences
King Abdulaziz University
Jeddah, Saudi Arabia

Athanasios Alexiou
Novel Global Community Educational Foundation
Hebersham, NSW, Australia

AFNP Med
Wien, Austria

ISBN 978-981-13-9638-0 ISBN 978-981-13-9636-6 (eBook)
https://doi.org/10.1007/978-981-13-9636-6

This Springer imprint is published by the registered company Springer Nature Singapore Pte Ltd.
The registered company address is: 152 Beach Road, #21-01/04 Gateway East, Singapore 189721, Singapore

Preface

Alzheimer's disease (AD) is still a hardly curable disease with several symptoms of behavioral and cognitive impairments. Besides the altered protein levels and the side effects of oxidative stress, it seems that many crucial risk factors play vital roles, even though it is not yet clear what the etiology is and what the result is. The reasons why molecular mechanisms cause neurodegeneration in AD are not known. Alzheimer's disease is a progressive disorder that leads to dementia and affects approximately 10% of the population older than 65 years of age. Memory loss is the first sign of cognitive impairment followed by behavioral disturbances. These symptoms are associated with a rigorous neuronal decline and the appearance of two brain lesions, senile plaques and neurofibrillary tangles, which are mainly composed of Aβ and hyperphosphorylated tau protein, respectively.

While several attempts at reducing AD severity have been presented until now, targeting mainly the symptomatic treatment, it seems that the early diagnosis or even the prediction seems to be the most convincing approach. It is also crucial to mention that in AD, scientists often apply noninvasive therapeutic procedures or medications like cognitive-behavioral therapy and art therapy within enriched sensorimotor environments in order to engage attention, provide pleasure, and improve behavior and communication to the patients. Several published studies reveal the efficacy of applying environmental enrichment in order to increase the effects of enhanced sensory, cognitive, and motor stimulation on different brain areas of the patients, which can lead to improved neuronal activation, signaling, and plasticity between the brain regions.

This book aims to serve as a reference book for those who are interested in the healthy aging, the non-pharmacological methods on AD, and the importance of early diagnosis related to risk factors and biomarkers. The editors deeply acknowledge the excellent work of the authors on presenting the challenges against AD management, in a very unique, interesting, and innovative manner.

Tsagkaris et al. analyze the role of cognitive impairment as a predictive factor of AD, highly associated with the disease progression and the nonpharmaceutical management of the disease.

Bano et al. describe the latest studies on causes, symptoms, and preventing methods of AD, including risk factors and other related comorbidities.

Shah et al. analyze the way that dietary interventions can crucially affect the reduction of neurodegeneration and the various biochemical and genetic alterations and increase the healthy lifespan among the elderly.

Ashfaq et al. discuss carbon nanostructured material-based biosensors as an accurate biomarker for AD detection and identification.

Uddin et al. underlie the role of oxidative stress to AD progression in correlation to mitochondria dysfunction and Aβ- and tau-mediated neurotoxicity.

Alexiou et al. present a set of biologically inspired music algorithms to reveal the importance of art therapy methods on handling AD patients and increase social awareness.

Shah et al. investigate the crucial role of structural and functional neuroimaging on the accurate and efficient AD diagnosis and prognosis.

Ebada et al. demonstrate the role of human gut microbiota in the nervous system and the way that a variety of disturbances of the intestinal microbiota homeostasis may affect the gut-brain axis.

Verma et al. describe the approach of applying stem cell therapy for tissue regeneration in neurodegeneration and AD.

Srivastava et al. underlie the association and the effects of food, exercise, and nutrition over aging and AD progression in terms of age-associated cognitive decline, generation of stress, and neurological fitness.

Zubair presents the most common genetics and neuronal pathways that are associated with the AD leading to neuronal death, synaptic failure, and oxidative stress.

Ali et al. describe the most common biotechnological applications including the BCI and NGS methods, which are involved in the establishment of personalized tool medicine for gene identification and engineering.

Rizvi et al. present the molecular and cellular mechanisms which are correlated to the development and progression of AD and can serve as prospective therapeutic targets.

Hoque et al. reveal the importance of immunotherapy on targeting the senile plaques with limited side effects and toxicity.

Jeddah, Saudi Arabia — Ghulam Md Ashraf
Hebersham, NSW, Australia — Athanasios Alexiou

Keywords

Alzheimer's disease · Biomarkers · Brain imaging · Cognitive and behavioral neuroscience · Diet and nutrition · Early diagnosis · Immunotherapy · Nanostructures · Non-pharmacological therapies · Stem cells

Contents

About the Editors

Dr. Ghulam Md Ashraf currently working as an Associate Professor at King Fahd Medical Research Centre, King Abdulaziz University, Saudi Arabia, completed his PhD in Biochemistry at Aligarh Muslim University, Uttar Pradesh, India. His main research interests are in Proteomics and Neurology. He is also a Research Consultant/Supervisor at Glocal University, India, and an Honorary Faculty Member at the Novel Global Community Educational Foundation (NGCEF), Australia. He has acted as a Principal Investigator or Co-Investigator in various research projects funded by reputed funding agencies, for example, the Deanship of Scientific Research (DSR); King Abdulaziz University, Saudi Arabia; and Strategic Technologies Research Program, King Abdulaziz City for Science and Technology (KACST), Saudi Arabia. He is a Member of the Editorial Board of various journals, for example, *Scientific Reports* (Nature Publishing Group), *Journal of Proteomics* (Elsevier), *Frontiers in Aging Neuroscience*, etc. He is also affiliated with the Protein Society, USA, the Science Advisory Board, and Saudi Society for Genomics and Molecular Oncology, Saudi Arabia. He has published 178 articles in various national and international journals.

Dr. Athanasios Alexiou is a Mathematician with a PhD in Bioinformatics Algorithms and Mathematical Modeling in Biology and a 2-year Postdoctoral Research Fellowship in Neuroinformatics Applications. He currently serves as an Honorary Vice President and Faculty Member at the Novel Global Community Education Foundation (NGCEF), Australia, and as Head of the Scientific Advisory Board of the AFNP Med Company in Austria. His main research field is Biomedical Informatics, with a special interest in Medical Decision Support Systems, Algorithms in Neuroinformatics and Computational Biomedicine, Neurodegenerative Diseases, and Biomarker Modeling. He is an Associate Editor of the journals *Genetics*, *Pharmacology*, and *Sociology*, a Guest Associate Editor of the *Frontiers in Neuroscience* journal for the research topic "The Alzheimer's Disease Challenge," and the author/coauthor of several peer-reviewed scientific papers in various journals and books.

Cognitive Impairment and Rehabilitation in Alzheimer's Disease

1

Theodoros Angelopoulos, Dorothy Martha Scordilis, and Christos Tsagkaris

Abstract

A wealth of evidence emphasizes on the link between Alzheimer's disease (AD) and cognitive impairment (CI). CI is generally accepted as a decline in memory, learning capacity, concentration, or decision-making, leading to functional impairment. The role of mild cognitive impairment (MCI) as a predictive factor of AD and the CI linked to the establishment and progression of AD have been discussed in the context of epidemiology, genetics, pathophysiology, pathology, and clinical practice. AD-associated CI has been also addressed as a source of socioeconomic burden urging for the development of therapeutic interventions. Cognitive rehabilitation (CR) is a complementary non-pharmaceutical treatment for AD consisting of cognitive stimulation and training methods. Based on the neuronal plasticity concept, CR aims at increasing AD patients' functional status and retarding their mental decline. Although the method has encouraging results in several studies, there is still controversy as far as its efficacy is concerned and further research ought to be conducted before incorporating CR in AD's treatment plan globally. CR drawbacks should be addressed and counteracted in this context, while the current CR approach may be ameliorated through errorless learning techniques.

T. Angelopoulos · C. Tsagkaris (✉)
Faculty of Medicine, University of Crete, Heraklion, Greece
e-mail: med3618@edu.med.uoc.gr

D. M. Scordilis
Faculty of Medicine, Medical University of Sofia, Sofia, Bulgaria

G. M. Ashraf, A. Alexiou (eds.), *Biological, Diagnostic and Therapeutic Advances in Alzheimer's Disease*, https://doi.org/10.1007/978-981-13-9636-6_1

1.1 Introduction

Someone in the world develops dementia every 3 s. There were an estimated 46.8 million people globally suffering from dementia in 2015, and this number is believed to double every 20 years. Much of the growth is due to developing countries. At present, 58% of people with dementia live in low- and middle-income countries, but by 2050, this will increase to 68% (Baumgart et al. 2015). Alzheimer's disease (AD) is nowadays the leading cause of dementia, and it consists of progressive cognitive decline. Although researchers and health professionals focus on AD-associated memory deficits, in the recent years, it has become clear that AD is strongly linked to an overall cognitive decline and may also be predicted by mild cognitive impairment. A new therapeutic attitude known as cognitive treatment has been developing over the past three decades. With these facts, cognitive rehabilitation (CR) appears promising and is further investigated.

The purpose of this chapter is to discuss aspects of CI and CR in AD context. We first provide definitions of the important terms of this chapter, then we proceed with an overview of CI from a preclinical and clinical point of view, and finally we move to CR elaborating on its background, its efficacy, and practical aspects of its implementation.

1.2 Cognitive Impairment in AD

Alzheimer's disease (AD) is an irreversible, progressive brain disorder. It is the most common cause of dementia, a general term for memory loss. It is a type of dementia that causes problems with memory, thinking, and behavior. Dementia, a syndrome with many causes and types, is defined as an acquired deterioration in cognitive abilities that are serious enough to impair the successful performance of activities of daily life. While many people have trouble with memory, this does not mean they have Alzheimer's, taking into consideration the many different causes of memory loss (Larson et al. 1992; Mufson et al. 2012).

1.3 Cognition

Cognition refers to a brain activity such as awareness, perception, reasoning, and thinking. Cognitive impairment is when a person has trouble remembering, learning new things, concentrating, or making decisions that affect their everyday life. Experts divide the types of cognitive impairment into four categories: mild intellectual disability, moderate intellectual disability, severe intellectual disability, and profound intellectual disability, the most debilitating of the categories. Amnesia may coexist with CI or not (amnestic and non-amnestic type) (Takao 2012).

Although in the past years there was significant controversy concerning the characterization and the detection of CI's stages, nowadays it is widely known that mild cognitive impairment has been usually observed as a "prodromal" stage of AD development, whereas it is also deteriorated along with the progression of the disease. CI may alter the treatment options for AD.

1.4 Clinical Significance

Physicians must take a multidisciplinary approach when working with patients, family members, or caregivers of cognitive impairment (CI). Alzheimer's disease has shown to follow a characteristic pattern of events; however, in the early stages of AD, it may be difficult to diagnose. Physicians must not only be aware of the important risk factors and family history associated with the disease but must also have knowledge of the various initial signs and symptoms which may offer suspicion of AD. Early recognition of signs and symptoms offers substantial benefits to diagnosis, which can therefore lead to early-stage therapy and prevention, giving value to prognostic information and rehabilitation.

Another key consideration in the diagnosis of AD is patient education. AD is a disorder that is renowned worldwide, and the average population knows the common pathological features of it. The importance of patient education is another key factor in distinguishing early diagnosis. The more family members and spouses understand and accept the initial symptoms, the earlier a diagnostic intervention can ensue.

1.5 Cognitive Impairment (CI) and Alzheimer's Disease (AD)

AD is the most common cause of dementia, which is defined as a syndrome of global CI (https://www.ncbi.nlm.nih.gov/pubmed/1599598). The link between AD and CI is supported by (1) *epidemiology, genetics*, (2) *pathophysiology*, (3) *neuropathology*, and (4) *clinical evidence*. Some investigators claim that essentially all patients with MCI have AD neuropathologically, while others believe that many of these patients develop AD but that is not mandatory. At the same time, the amnestic subtype of mild cognitive impairment has a high risk of progression to Alzheimer's disease, and it could constitute a prodromal stage of this disorder, while the clinical presentation of AD is highly compatible with CI.

1.5.1 Epidemiology

The majority (>95%) of patients who develop AD disease are over 65 years of age (also known as late-onset AD), and up to 5% of all people with Alzheimer's have early-onset AD, normally appearing in the late four or early five decades of life. The two forms of AD cannot be easily clinically distinguished but they comply with different patterns of genetic epidemiology (de Souza-Talarico et al. 2016). The early-onset form displays more severe symptoms and shows higher progression rates than the late-onset form. The most important risk factor for Alzheimer's disease is advanced age. Every 5 years after the age of 65, the risk of suffering from the disease nearly doubles. Prevalence rates in less developed countries are reduced. In the USA, where the Hispanic population has a 30% decreased risk than the non-Hispanic white people, the chance of dying from Alzheimer's disease is 26% higher among the non-Hispanic white population than among the non-Hispanic black population (Jicha and Carr 2010; Takizawa et al. 2014).

1.5.2 Genetics and Pathophysiology of CI and AD

Cognitive impairment is often correlated with autosomal dominant inheritance. Several genes identified in AD pathogenesis are also supposed to play a role in MCI or CI. APOE genotype, as well as rare mutations in PS1, PS2, and APP, can cause familial forms of AD.

1.5.2.1 Early-Onset Alzheimer's Disease: Familial Alzheimer's Disease (FAD)

Three causative genes have been linked to autosomal dominant familial AD which include the APP itself and PSEN1 and PSEN2 which encode proteins required for APP breakdown and Aβ formation. Mutations in these three genes on chromosomes 21(APP), 14(PSEN1), and 1(PSEN2) are responsible for Aβ aggregation formation and early-onset disease. As a result, the dominant component of amyloid plaques in the brains of AD patients is amyloid-β (Aβ) (De Strooper et al. 1998). The mutations on APP spread Aβ levels (Aβ40, Aβ42, or both) increasing APP expression or activity of β-secretase and α-secretase inhibition.

Somatic variants in autosomal dominant genes like APP are a rare cause of AD. Mutations affect particularly the aβ production process, since they are most commonly located in or near the proteins coding exons (APP exons 16 and 17). Lastly, all three causal AD genes cause a common pathogenic AD pathway, directly affecting Aβ (Scheuner et al. 1996).

1.5.2.2 Secretases and Amyloid-Beta (Aβ) Plaques

Brain amyloid-beta (Aβ) plaques are a hallmark damage of patients with a clinical diagnosis of aMCI or AD. The distribution of Ab deposits changes in accordance with the evolution of the disease and reflects the spread of extracellular amyloid aggregates in the diseased brain. The distribution of amyloid deposits in MCI appears to be intermediate between the changes seen in the people with cognitive function (or NCI) and AD brain (Bagyinszky et al. 2016; Gouras et al. 2015).

Normally, in people not suffering from the AD disease, the dominant form of αβ is 40-amino-acid long and is called Aβ40. Aβ42 has two additional amino acid residues at the C-terminus and is lower than Aβ40. Increased proportion of Aβ42 appears to be responsible for EOAD. Aβ 42 is more easily oligomerized and forms fibrils than the Aβ40 peptide that is found in higher levels. The APP protein can be cleaved by three different secretases: α, β, or γ. Amyloid beta (Aβ or Abeta) denotes peptides of 36–43 amino acids that are derived from the amyloid precursor protein

(APP) after cleavage by beta secretase (first) and gamma secretase (second cleavage of the β-secretase product) to produce Aβ. The point of cleavage by γ-secretase determines the kind of aβ produced (Aβ40 or Aβ42). Both b- and γ-secretases are proteolytic proteins that increase in response to cellular stress such as oxidative stress, ischemia, and energy loss. β-Secretase1 requires the presence of glycosaminoglycans for effective cleavage. It is important to note that the function and the normal biological activity of αβ are not yet fully understood. Both the amyloidogenic and non-amyloidogenic pathways exist in healthy people, with AD presumably being caused through increased amyloidogenic cleavage or decreased Aβ turnover. Clots created in the presence of Aβ have an abnormal structure and are resistant to clearance (Bagyinszky et al. 2016). The amyloid plaques are extracellular deposits found in the brains of AD patients. The plaques are made of a hodgepodge of normally fibrillar aggregates known as amyloid fibers, a protein fold shared by other peptides, for example, the prions linked to protein misfolding diseases (Nicolas et al. 2018; Wu et al. 2012).

1.5.2.3 Late-Onset Alzheimer's Disease

Late-onset AD has been linked to genes that are not inherited in a Mendelian pattern. First-degree relatives of patients with late-onset AD are two times more likely to have the expected lifetime risk of this disease of individuals without an AD-affected first-degree relative. The fact that AD is more common in monozygotic than in dizygotic co-twins gives good reason to assume an important genetic contribution to this disorder (Jicha and Carr 2010; Reitz et al. 2011; Pierce et al. 2017).

1.5.2.4 ApoE

ApoE, one of the principal apolipoproteins in the brain that is expressed in humans as one of three common isoforms, transports both lipids and Aβ. The three ApoE variants, ApoE-ε2, ApoE-ε3, and ApoE-ε4, are encoded by three different alleles. The ApoE-ε2 allele is protective against AD, while ApoE-ε4 allele is the greatest risk factor. The *APOE* gene has been associated with late-onset AD and is located in chromosome 19, belonging in a cluster together with the genes encoding translocase of outer mitochondrial membrane 40 (TOMM40), apolipoprotein C1, and apolipoprotein C2. *APOE ε4* is responsible for as much as 20–30% of AD risk. The presence or absence of an *APOE ε4* allele determines the risk of AD and the age of AD onset by approximately 6 years for each allele (Myers et al. 1996; Kurz et al. 1996). The appearance of a single *APOE ε4* allele triples the risk of the disease, while the two copies are associated with a fivefold increase of it. In addition, the presence of this allele links up with memory impairment, MCI, and progression to dementia.

ApoE plays a significant role in many potential causes of AD, including Aβ plaque formation, τ-tangle formation, oxidative stress, lipid homeostasis deregulation, inflammation, synaptic plasticity loss, and cholinergic dysfunction. ApoE-ε4 allele has a low ability to remove Aβ plaques in contrast with ApoE-ε3 and ApoE-ε2 alleles. In patients with AD, ApoE is present in senile plaques (polymorphous beta-amyloid protein deposits), vascular amyloid, and neurofibrillary tangles and binds to APP (Liu et al. 2013; Corder et al. 1993; Bekris et al. 2010).

1.5.2.5 Further Risk Variants: SORL1

As it is already mentioned, the recycling of the amyloid precursor protein (APP) from the cell membrane (plasma membrane) through the endocytic pathways is crucial for the formation of amyloid β-peptide (Aβ) in Alzheimer's disease. SorL1 is one of the five type I transmembrane receptors often found in the CNS and contains a luminal, extracellular vacuolar protein sorting ten domains and is a specific receptor for APP holoprotein. SORL1 results in APP recycling preventing cleavage by beta-secretase. Dysfunction or absence of sortilin-related receptor SORL1 caused by inherited variants increases αβ production as the APP holoprotein can no longer be recycled. Aβ plaques accumulating on cortical and subcortical brain structures linked to cognitive functions lead to a dysfunction interpreted as CI (Nicolas et al. 2018; Rogaeva et al. 2007; Behrman et al. 2017).

1.5.3 Neuropathology

1.5.3.1 Mild Cognitive Impairment (MCI)

The pathologic and molecular features of patients with MCI are not entirely understandable. The neuropathological changes don't seem to follow a direct linear path. MCI is characterized by a complex background that includes plaque and tangle pathology and also significant morphological alterations of cells and molecules (Stephan et al. 2012). These changes are highly associated with cognitive deficit together with the compensatory responses to the development of the disease. The neuronal disconnection syndrome that is associated with MCI is variable indicating that there is no one and only unique event which precipitates this early stage of AD. It is possible that neuronal damage observed in MCI is deteriorated in AD enhancing a pattern of decline (Takao 2012; Petersen et al. 2014).

1.5.3.2 Characteristics of Brains in MCI

The cortical gyral and sulcal patterns have not shown any discernible difference between amnestic MCI (aMCI) and non-amnestic MCI brains. In aMCI and mild AD, the cerebrum is characterized by a widening of sulci, such as the ventral ramus of the lateral fissure along with a blunting of the anterior tip of the temporal pole in comparison with specimens from patients without MCI. As the disease evolves into AD, the morphological changes noted in aMCI are enlarged and extended to other cortical zones (Mufson et al. 2016; Larson et al. 1992).

1.5.3.3 Neuropathological Evidence of CI in AD Patients

Positive and negative neuropathological features occur on the brains of people with CI. Positive characteristics of the disease include neutrophil threads, plentiful amyloid plaques, and neurofibrillary tangles and dystrophic neurites involving hyperphosphorylated tau (p-tau) combined with additional astrogliosis and massive microglial irritation. Congophilic amyloid angiopathy is frequently an additional feature. Positive characteristics also include brain damage present in MCI and AD that mainly affects hippocampal formation. In the hippocampus, we can generally observe the major presence of Hirano bodies and granulovacuolar degeneration. As for the negative neuropathological lesions of MCI, these are characterized by significant losses of neurons, neutrophil, and synapses. All these damages have a unique distribution, for example, the plaques dominate the cortical mantle, and the tangles are mainly found in limbic and associated cortices. The early damages in the cortex and in particular in the entorhinal and the perirhinal area, then in the hippocampus proper, later in the association cortex, and finally in the primary neocortex represent a highly specific model of the neurofibrillary degeneration that occurs among brain regions (Jeong 2017).

All the above neuropathological characteristics are major diagnostic markers for AD. It needs to be added that the advancing degeneration of the basal forebrain, the limbic system, and the neocortical areas of the brain cannot be separated from the CI in Alzheimer's patients. More precisely, neuronal degeneration seems to be proportionate to CI. The pathological evolution of AD initiates with alterations to the synapses, proceeds with retrograde degeneration of the axons, and eventual results in atrophy of the dendritic tree and perikaryon. CI in patients with AD appears linked to neocortex and limbic system lesions. These brain areas are crucial components of the Papez circuit taking part in the genesis and consolidation of short-term memory processes, and hence their damage results in deficits identified in CI. The lesion's evolution inferiorly and posteriorly in the temporal lobes affects brain structures decisive for semantic memory. Anterior spread has the effect of involvement of subcortical areas including the cholinergic projections. Neuropsychiatric and behavioral characteristics of AD have resulted from the engagement of the anterior cingulate, amygdala, and other limbic structures in the basal forebrain. Eventually, AD as well as other neurodegenerative diseases starts focally and evolves outward, affecting gradually the entire brain in the end stage of disease and causing the typical phenotype in AD from normal cognition to end-stage disease (Stephan et al. 2012).

It is essential to highlight that the complete definition of anatomic disease progression is clearly in evolution. Clinicopathological correlation studies played a decisive role to develop a hypothetical model about the pathophysiology of the disease. The models are based on the existence of a continuum between normal aging and AD dementia and on the observation that the amyloid plaque formation is taking place principally before the onset of cognitive deficits, while neurofibrillary tangles, neuron loss, and particularly synaptic loss equally extend the progression of cognitive decline (Mufson et al. 2012).

1.5.3.4 Macroscopic Features

The visual examination of the AD brain has no diagnostic value. Nevertheless, most patients display certain clinical features as the considerable expansion of the lateral ventricles and specifically of their temporal horns. The excessive *accumulation* of CSF is a result of cortical atrophy affecting medial temporal lobes and primary sensory, motor, and visual cortices that characterize dementia. Additionally, the presence of several lacunar infarcts in the basal ganglia, cortical microinfarcts, and demyelination of the periventricular white matter is frequent due to the cerebrovascular disease that is more prevalent in older people and hence is correlated with cognitive impairment. It is also noteworthy that possible existence of a concurrent severe cerebral amyloid angiopathy would have a profound effect on brain morphology affecting especially the posterior parietal and occipital lobes and causing from cortical microbleeds to even substantial lobar hemorrhages. AD, dementia, and cognitive impairment are characterized by the death of dopaminergic neurons in substantia nigra. Only when Lewy bodies are present in the above cases, the substantia nigra has an abnormal coloration. Finally, the locus coeruleus is noticeably affected in the early stages of AD (Behrman et al. 2017).

1.6 Microscopic Features

1.6.1 Neurofibrillary Tangles: Tαu Protein

The neurofibrillary tangles (NFTs) are formed in the perikaryal region of pyramidal neurons. The NFTs are principally produced by paired helical filaments (PHFs) that are fibrils of $\approx$ 10 nm in diameter that outline pairs with a helical tridimensional conformation according to a standard repetitive pattern. A small number of fibrils in the NFTs remain unpaired and form straight filaments, a different kind of abnormal filament (Dos Santos Picanço et al. 2016). There are cases where filaments in NFTs include transition between a paired helical and a straight segment. Leaving structural units aside, the main component of NFTs is tau protein that has become defective. This protein normally stabilizes microtubules and is plentiful in neurons and less common in astrocytes and oligodendrocytes of CNS. Cognitive impairment, dementia, and AD are often characterized by microtubule instability secondary to tau protein deficiency. The abnormal tau protein originates from the pathological phosphorylation of tau and has the effect of the transformation of normal adult tau into PHF (paired helical filament)-tau and NFTs. Generally, the stabilization of the microtubules requires the effective interaction of tau protein isoforms with tubulin via phosphorylation (Jeong 2017). Lastly, the primarily responsible hyperphosphorylated tau isoforms are often caused by mutations affecting tau function and expression or even perhaps by increased protease action or interplay between polyanions and tau protein (Takahashi et al. 2017; Bagyinszky et al. 2016; Bloom 2014; Goedert 2015; Reitz et al. 2011).

Recent studies show that NFT formation is directly proportional to molecular and conformational changes in the tau protein before and during NFT development and maturation. The pathologic expansion of the NFTs from the MTL to the neocortex is defined by the Braaks. The topographic spread involves six stages according to the current location and the severity of the damaged neurons (transentorhinal stages I–II, clinically silent cases; limbic stages III–IV, incipient AD; neocortical stages V–VI, fully developed AD). Patients with CI as well as individuals without

CI demonstrate many potential Braak staging scores from stage 0 that shows total nonexistence of NFTs to stages V–VI indicating AD. The use of the Braak staging criteria seems not to clearly differentiate the normal and the CI brains. Nevertheless, a study suggests that Braak scores are very effective in categorizing amnestic MCI from non-amnestic MCI (ref) (Takahashi et al. 2017; Mufson et al. 2016; Behrman et al. 2017).

1.7 Protective Factors

1.7.1 Diet

Although diets higher in antioxidants and polyunsaturated fatty acids (PUFAs) have been shown in several studies to lead to decreases in the risks of dementia and AD, age-related cognitive decline, and MCI (more PUFA, less disease) (Yehuda et al. n.d.), other studies discovered no clear relationship between dietary PUFAs and cognitive impairment (Chiu et al. 2008). PUFAs are mostly found in vegetables, fish, and fruit. In some other studies, individuals with high levels of vitamins E and C were less susceptible to develop dementia than those with low levels of these vitamins. Antioxidants like vitamin C can prevent cell damage by oxidation, and vitamin C may help to revive other antioxidants, such as vitamin E (Engelhart et al. 2002; Morris et al. 2002). However, larger studies have not been able to determine such associations (Laurin et al. 2004).

Currently, clinical trials concerning dietary supplementation with omega-3 PUFAs have found no complete consequence on cognition in patients with MCI or AD. On the other hand, docosahexaenoic acid supplementation has a benign effect on cognitive function in patients harboring the *APOE ε4* allele and in the beginning of the AD.

A Mediterranean-type diet (MeDi) reduced the incidence of AD with the aim of reducing the risk of MCI and progressing from MCI to AD (Singh et al. 2014). A 2013 systematic review reached similar conclusions, and also found a negative association with the risk of progressing from mild cognitive impairment to Alzheimer's, but acknowledged that only a limited number of studies had been done on the topic. The principal aspects of this diet include proportionally high consumption of legumes; olive oil as the primary source of monounsaturated fat, unrefined cereals, fruits, and vegetables; moderate to high intake of fish; moderate consumption of dairy products (mostly as cheese and yogurt); moderate amount of wine consumption (mostly red); and conservative poor consumption of red non-fish meat goods. A following cohort study in France showed that Mediterranean-type diet doesn't affect the performance on the Isaacs Set Test, the Benton Visual Retention Test, or the Free and Cued Selective Reminding Test in contrast with high MMSE scores mentioned (Féart et al. 2009).

Several prospective studies exploring the effect of alcohol on dementia risk concluded that light to moderate alcohol consumption was associated with a reduction in the risk of AD and dementia (Xu et al. 2017).

1.8 Physical Activity

Researchers examining the link between objective measures of complete periodic physical activity and incidence of Alzheimer's disease are lacking. Recent experimental reports show that physical exercise possibly promotes brain health (Abbott et al. 2004).

Physical activity could affect cognition in many ways. Aerobic exercise (cardio) – a physical exercise of low to high intensity which – raises cerebral blood flow, oxygen extraction, and glucose utilization and activates growth factors that cause structural brain changes, such as an increase in capillary density. Furthermore, studies show that physical activity reduces the amyloid plaque formation (Fratiglioni et al. 2004; Féart et al. 2009).

1.8.1 Daily Intellectual Activity

Taking part in intellectual activities, such as reading, learning a new hobby, and playing board games and card games or even using the opposite hand when brushing your teeth, may delay or prevent dementia in older adults, even if these habits and practices take place in late life, according to the latest research. New researches suggest, for example, elderly people with greater levels of education had a lower incidence of dementia than those with no education. Cognitive activity was suggested to decrease the risk of cognitive decline by increasing cognitive reserve. However, it is important to point out that the exact effect of cognitive exercise on the risk of dementia remains unclear (Ball et al. 2002; Bidzan et al. 2016; Takizawa et al. 2014; Acevedo and Loewenstein 2007).

1.9 Economic and Social Burden

Dementia is one of the major causes of disability and dependency among the elderly. It is overwhelming not only for the people who have the syndrome but also for their caregivers and families. There is often a lack of awareness and understanding of dementia, which results in stigmatization and barriers to diagnosis and care. The impact of dementia on caregivers, family, and societies can be physical, psychological, social, and economic (Dementia 2017).

Before evaluating the cost of AD, cost terminology should be defined. The costs of dementia to society come from all goods and services that are given up to anticipate, diagnose, heal, or in other respects cope with dementia. Individuals, families, and carers are affected both economically and in terms of quality of life. Alzheimer's disease (AD) – the most common cause of dementia – is associated with a valuated health-care cost of US$172 billion annually (Alzheimer's Association 2010). The basic Alzheimer's disease (AD) symptoms including progressive cognitive, behavioral, and functional impairment have a direct influence on the patients, families, and the public health system. AD affects both indirect and direct costs. During the early stages of AD and for the community-dwelling patients, indirect costs (such as the serious adverse impact on patients and family members' financial situation) are the highest costs. The evolution of the disease leads to an increase of the direct costs (equally medical treatment and social services) when the patient is hospitalized or services of a caregiver are required. Although the drug therapies increase the direct costs they can reduce, other expenses are also involved. A number of studies have shown that there are total economic benefits to society while using drug therapies, and all related cost are considered, where results rely on specific patient and care setting characteristics. Indicatively, direct medical costs per year for mild AD ranged from €5476 in France to €27,380 in Spain. A good health system should take greater account of dementia, and governments must prepare a plan to tackle AD based on treatment options that are proportionate to the burden of the disease (Marešová and Zahálková 2016; Takizawa et al. 2014).

1.10 Diagnosis of Cognitive Impairment in AD

To standardize the diagnostic process, several major medical organizations have created diagnostic criteria for AD. In clinical practice, diagnosis of dementia (or neurocognitive disorder) is required prior to diagnosis of probable AD. The first one is the International Statistical Classification of Diseases and Related Health Problems (ICD), listed by the World Health Organization (WHO). The second one is the *Diagnostic and Statistical Manual of Mental Disorders* (DSM), published by the American Psychiatric Association (APA) (Sachdev et al. 2014). Both are used interchangeably worldwide, and preference of usage depends on the country, member states, and the physician's organization. The latter, for example, serves as the principal authority for psychiatric diagnoses in the USA. Another organization that has diagnostic criteria is the National Institute on Aging-Alzheimer's Association (NIA-AA) (Albert et al. 2011).

The range of disorders associated with CI is evident, highlighting that modern diagnostic procedures pay attention to cognitive deficits. For instance, DSM-5 diagnostic criteria for dementia have been renamed and split into "major neurocognitive disorder" (previously dementia) and "mild neurocognitive disorder," which is equivalent to mild cognitive impairment (MCI) (Wetterling et al. 1996). A short comparison between normal aging, MCI, and major neurocognitive disorder can be found in Table 1.1.

1.10.1 Diagnostics

1.10.1.1 Intro

When an individual has symptoms of dementia, a physician must conduct tests to identify the underlying brain disease or other condition that is causing these symptoms. Different types of dementia are associated with distinct symptom patterns and brain abnormalities (Langa and Levine 2014).

Table 1.1 Normal aging vs. mild and major neurocognitive disorder

Normal aging	Mild neurocognitive disorder	Major neurocognitive disorder
Primarily intact cognition	Inefficiency in daily activities	Needs help with daily activities
Subtle processing speed is decreased	Decline from lifelong abilities in one or more areas of thinking	Substantial decline in one or more cognitive abilities
Less efficient attention and executive reasoning	[a]ADLs progressively decreased and [b]IADLs decreased	ADLs severely decreased and IADLs not feasible

[a]Basic activities of daily living (ADLs): bathing, dressing, eating, transferring from bed to chair, continence, toileting
[b]Instrumental activities of daily living (IADLs): transportation, shopping, cooking, using the telephone, managing money, taking medications, housecleaning, laundry

CI changes of AD follow a characteristic pattern, beginning with memory impairment and progressing to language and visuospatial deficits. However, approximately 20% of patients with AD present with non-memory complaints such as word finding, organizational, or navigational difficulties. In other AD patients, upstream visual processing dysfunction (referred to as posterior cortical atrophy syndrome) and a progressive "logopenic" aphasia are the primary manifestations of AD for years before progressing to involve memory and other cognitive domains. Furthermore, other patients may present with an asymmetric akinetic–rigid–dystonic ("corticobasal") syndrome or a dysexecutive "frontal variant" of AD. Loscalzo MD, Phd, 2015 (Casper et al. 2015; Fratiglioni et al. 2004).

1.10.2 Approach to Diagnosis: Evaluation of the Patient

Physicians may have multiple objectives with alternating degrees of importance during their consultations with patients. These goals include, but are not limited to, the interpretation of signs and symptoms into diagnoses, assessing the stability or change in known conditions, providing information and counseling for future prevention, and the continuation or adjustment of therapeutic interventions. A general health check will increase the number of diagnoses for a patient, but it may not affect overall morbidity and mortality (Krogsboll et al. 2012).

The interaction between the patient and the physician represents not only a scientific encounter but also a social practice centered on the point of control and meeting each other's expectations. Patients expect that their health-care needs and concerns will be addressed efficiently. Physicians also have expectations: a need to feel that they have not missed something important in addressing diagnostic challenges, a need to put limits on the time available for each interaction, and a need to maintain open-mindedness so that their evaluations and recommendations are not clouded by any emotional feelings about the patient. The expertly performed rational clinical examination enhances the expected social practice and the likelihood of acquiring relevant data. It also optimizes the physician's ability to understand the patient's symptoms and concerns, as well as to facilitate the healing process (Verghese et al. 2011).

Table 1.2 Overview- approach to evaluating a patient with dementia: routine evaluation and optional diagnostic tests

Routine evaluation	Optional focus tests	Occasionally helpful tests
History	Psychometric testing	EEG
Physical examination	Chest X-ray	Parathyroid function
CT/MRI	Lumbar puncture	Adrenal function
Laboratory tests: thyroid function (TSH), vitamin B12, complete blood count, electrolytes	Other tests: liver function, renal function, urine toxin screen, HIV, apolipoprotein E, RPR or VRDL	Other tests: urine heavy metals, lab screen for autoantibodies, RBC sedimentation rate, angiogram, brain biopsy, SPECT, PET

Abbreviations: *CT* computed tomography, *EEG* electroencephalogram, *MRI* magnetic resonance imaging, *PET* positron emission tomography, *RBC* red blood cell, *RPR* rapid plasma reagin (test), *SPECT* single-photon emission computed tomography, *TSH* thyroid-stimulating hormone, *VDRL* Venereal Disease Research Laboratory (test for syphilis)
Source: Casper et al. (2015)

Table 1.3 Overview- approach to evaluating a patient with dementia: diagnostic categories

Reversible causes	Irreversible/degenerative dementias	Psychiatric disorders
Examples: Hypothyroidism Thiamine deficiency Vitamin B12 deficiency Normal pressure hydrocephalus Subdural hematoma Chronic infection Brain tumor Drug intoxication Autoimmune encephalopathy	Examples: Alzheimer's disease Frontotemporal dementia Huntington's Dementia with Lewy bodies Vascular dementia Leukoencephalopathies Parkinson's disease	Examples: Depression Schizophrenia Conversion reaction

Table 1.4 Overview-approach to evaluating a patient with dementia: associated treatable conditions

Depression	Agitation
Seizures	Caregiver "burnout"
Insomnia	Drug side effects

When it comes to patients with dementia, including patients suffering from CI in AD context, three major points should be kept in focus: (1) What is the best fit for a clinical diagnosis? (2) What component of the dementia syndrome is treatable or reversible? (3) Can the physician help alleviate the burden on caregivers? A broad overview of the approach to evaluating a patient with dementia including optional diagnostic tests (Table 1.2), diagnostic categories (Table 1.3), and associated treatable conditions (Table 1.4) can be found under Tables 1.2, 1.3, and 1.4.

The major degenerative dementias can usually be distinguished by the initial symptoms; neuropsychological, neuropsychiatric, and neurological findings; and neuroimaging features.

Table 1.5 Patient's medical history

Description of the patient: Age, gender, ethnic background, occupation, chief reason for seeking medical care, state the purpose of the evaluation (usually in the patient's words)
Other physicians involved in the patient's care: Include the clinician that the patient identifies as his/her primary provider or the physician who referred the patient. Record contact information for all physicians who should receive information about the visit
History of the reason for seeking medical care: In chronologic fashion, determine the evolution of the indication for the visit and then each major symptom. It is best to address the patient's reason for seeking care first rather than what the physician ultimately believes is most important
Be careful to avoid "premature closure," in which a diagnosis is assumed before all the information is collected
Past medical and surgical history: List other illnesses and previous surgeries not related to the current problem. List all prescribed and OTC medications with dosages. Remember to ask about vitamin and herbal supplements
Allergies and adverse reactions: List allergic reactions to medications and food. Record the specific reaction (e.g., hives). Distinguish allergies from adverse reactions or intolerance to medication (e.g., dyspepsia from NSAIDs)
Social, occupational, and military history: Describe patient's current family and a typical day for the patient. The occupational history should focus on current and past employment as it might relate to the current problem
Risk factors: Include history of tobacco use, alcoholism, illegal drug use, and risk factors for sexually transmitted diseases (e.g., HIV and hepatitis)
Family history: History of any diseases in first-degree relatives and a list of family members with any conditions that could be risk factors for the patient (e.g., CVD, known genetic disorders, malignancy)
Review of systems (see Table 1.6)

Abbreviations: *CVD* cardiovascular disease, *HIV* human immunodeficiency virus, *NSAIDs* nonsteroidal anti-inflammatory drugs, *OTC* over-the-counter

1.10.3 History

The history begins by asking the patients to describe, in their own words, the reason for seeking medical care (Table 1.5). Although patients may have many reasons for initiating a visit to the physician, they should be encouraged to select the most important or top two most important concerns they have and state them. The physician should reassure the patient that other concerns will not be ignored but that it's important to understand what is the most concerning to the patient at that moment.

The history should concentrate on the onset, duration, and tempo of progression when it comes to dementia and AD. For example, an acute or subacute onset of confusion may be due to a different cause like delirium and should trigger the search for intoxication, infection, or metabolic derangement. An elderly person with slowly progressive memory loss over several years is likely to suffer from AD. Nearly 75% of patients with AD begin with memory symptoms, but other early symptoms include difficulty with managing money, shopping, driving, following instructions, navigating, or finding words. The latter symptoms are really important in the frame of CI.

The history should be taken in a stepwise manner, and the spoken approach should also be taken into consideration. We should begin with acquiring the history of the present illness or concern at large. Open-ended questions produce a description of the patient's concerns in the patient's own words, whereas specific questions fill in gaps and help clarify significant points. These questions should be asked in an order lead by the story the patient tells and targeted to fit the individual problem. Important questions to consider include description of onset and chronology, location of symptoms, quality (character) of symptoms, intensity, and precipitating, aggravating, and relieving factors. Ask if this problem or similar problems occurred before and, if so, whether a diagnosis was established at the time.

After taking the history of the present illness, we should ask about past medical and surgical history, a list of current medications including prescriptions, herbal drugs, over-the-counter medications, and vitamins. Next, we may attempt to assess social and occupational history, along with risk factors. Social history can help us understand the patient's values and support systems. Information that can influence risk factors for disease should be gathered including occupational history, substance abuse, and sexual history. Marital status and the living situation of the patient are as important as risk factors for disease and help determine how to provide the best care for the patient.

An intelligent physician is aware that patients may not report all their problems because they may have forgotten or simply may not want to discuss them.

The family history is never diagnostic, but it allows risk stratification and it may indicate signs of CI which will not be mentioned by the patient (Goldman and Schafer 2012). Studies of family history say that if you have a close relative who has been diagnosed with AD, your risk increases by 30%. This is a relative risk increase, meaning a 30% rise in your existing risk. This means your risk may be higher, but not that much higher, if you consider the absolute numbers. The risk of being diagnosed with AD is 2% per year if you are over the age of 65. Family history then raises this 2% by 30%, to about 2.6% per year. Age raises the chances of AD more than family history considering that people above the age of 70 have a 5% chance of being diagnosed. Family history would again raise this by 30%, making 5% go to 6.5%. Nevertheless, a positive family history of AD should always be taken into consideration. The review of systems is the structural assessment of each of the major organ systems and can help elicit symptoms or signs that are not covered or may be overlooked in the history of the present illness (Table 1.6).

1.10.4 Screening

1.10.4.1 Cognitive and Neuropsychiatric Examination

Cognitive and neuropsychiatric examinations, or mental status testing, evaluate memory, ability to solve simple problems, and other thinking skills. Such tests give an overall sense of whether a person is aware of symptoms; knows the date, time, and where he or she is; or can remember a short list of words, follow instructions, and do simple calculations.

Table 1.6 Review of systems[a]

Focus all questions on a specific time frame (e.g., within the past "month" or "now") and on items not already addressed during the clinical examination
Change in weight or appetite, change in vision, change in hearing, new or changing skin lesions
Chest discomfort or sensation of skipped beats, shortness of breath, dyspnea on exertion
Abdominal discomfort, constipation, melena, hematochezia, diarrhea, difficulty with urination
Depression; joint or muscle discomfort; sensation of unsteadiness when walking, standing, or getting up from a chair; problems with sleep; difficulty with sexual function

[a]Clinicians may start with this basic list and adapt the items to their specific patient population by considering factors such as age, gender, medications, and the problems identified during the examination. The process is facilitated by developing a routine personal approach to these questions, typically going through the systems from "head to toe"

Brief screening tools such as the Mini-Mental State Examination (MMSE), Mini-Cog test, Montreal Cognitive Assessment (MOCA), and Cognistat can be used to capture dementia and follow progression. None of these tests is highly sensitive to early-stage dementia or discriminates between the different dementia syndromes. In most patients with MCI and some with clinically apparent AD, bedside screening tests may be normal and therefore may require a more comprehensive set of neuropsychological tests to be conducted.

The MMSE is a questionnaire designed to test a range of everyday mental skills. It is a 30-point test of cognitive function, with each correct answer being scored as 1 point, making a maximum MMSE score of 30 points. It includes tests in the areas of orientation (e.g., identify season/date/month/year/floor/hospital/city/state or region/country); registration (e.g., name and restate three objects); recall (e.g., remember the same above mentioned three objects 5 min later); and language (e.g., name pencil and watch; repeat "no if's, and's, or but's"; follow a three-step command; and write a sentence and copy a design). A score of 20–24 suggests mild dementia, 13–20 suggests moderate dementia, and less that 12 indicates severe dementia. On average, the MMSE score of a person with Alzheimer's declines about 2–4 points per year.

The Mini-Cog test is a 3-min instrument to screen for cognitive impairment in older adults. The Mini-Cog is scored in two parts: the first part uses a three-item recall test for memory and the second part is a simply scored clock-drawing test (CDT) (Fig. 1.1). These are then added together for a total score. The maximum score for the item recall test is 3, and the maximum score for the clock-drawing test is 2, with the complete total score being a maximum of 5. A total score of 3, 4, or 5 indicates lower likelihood of dementia but does not rule out some degree of cognitive impairment.

The three-item recall score is simply 1 point for each word recalled without cues, scoring 1, 2, or 3. The clock-drawing is scored as 2 points for a normal clock or 0 (zero) points for an abnormal clock-drawing. A normal clock must include all numbers (1–12), each only once, in the correct order and direction (clockwise). There

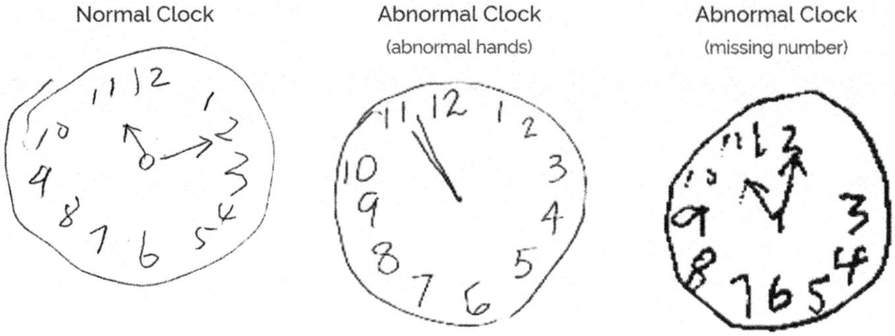

Fig. 1.1 Example of normal and abnormal clock-drawing test (CDT)

must also be two hands present, one pointing to eleven (11) and one pointing to two (2). Hand length is not scored in the Mini-Cog algorithm.

1.10.5 Physical Examination

1.10.5.1 Physical and Neurologic Examination

A comprehensive general and neurological examination is fundamental in diagnosing dementia, AD, and any signs of cognitive impairment. The physical and neurological examination is essential evidence when it comes to documenting the progression of cognitive decline. It is also useful for the discovery of other signs of nervous system involvement and to search for clues that may suggest an underlying cause that may be responsible for the cognitive disorder.

A classic medical workup consists of medical history, diet, nutrition, use of alcohol, review of medications, blood pressure check, temperature, pulse, auscultation of heart and lungs, and collection of blood or urine samples for laboratory testing. Information obtained from the physical examination and laboratory tests can help identify health issues that can cause symptoms of dementia, like conditions other than Alzheimer's that could cause confusion, trouble focusing, memory problems, and thinking problems which include anemia, infection, diabetes, hepato-renal disease, cardiovascular disease, lung disease, or vitamin and hormonal abnormalities.

A classic neurological examination consists of testing reflexes, coordination, muscle tone and strength, eye movement, speech, and sensation. The neurological examination may also include a brain imaging study depending on the outcome of the physical examination, symptoms, and patient concerns. The typical neurological examination is done in order to closely evaluate the patient for problems that may signal brain disorders other than Alzheimer's. It is important to look for signs of small or large strokes, Parkinson's disease, brain tumors, fluid accumulation in the brain, and other illnesses that may impair memory or thinking.

Typical AD spares the motor systems until later on in the course, whereas frontotemporal dementia (FTD) patients usually develop axial rigidity, supranuclear gaze palsy, or a motor neuron disease reminiscent of amyotrophic lateral sclerosis (ALS). In dementia with Lewy bodies (DLB), the initial symptoms may include the new onset of a parkinsonian syndrome (resting tremor, cogwheel rigidity, bradykinesia, festinating gait), but DLB most often starts with visual hallucinations or dementia. Creutzfeldt–Jakob disease (CJD) is advocated by the presence of an akinetic-mute state, diffuse rigidity, and prominent, startle-sensitive myoclonus.

1.10.6 Laboratory, Microscopic, and Imaging Findings

There are no definitive laboratory tests available that will positively diagnose AD during life. Currently, the only definite diagnosis of AD is to microscopically examine a section of the person's brain tissue after death. Histopathologists will look for senile plaques and neurofibrillary tangles which are characteristic of AD. Since plaque and tangle formation are also seen in the normal aging process, the sample must be compared to a control sample of normal, non-AD brain tissue from a person of the same age.

A physician may use a range of traditional laboratory tests to rule out other conditions and deficiencies that could be affecting the patient's memory. They may also look for overmedication and may use imaging tools such as computed tomography (CT) and magnetic resonance imaging (MRI) scans in order to look for evidence of tumors, trauma, or stroke that could cause dementia symptoms. Imaging can also help look for brain atrophy or shrinkage that may be seen later in the AD progression. The American Academy of Neurology recommends the routine measurement of a complete blood count, electrolytes, renal and thyroid function, a vitamin B12 level, and neuroimaging studies (CT or MRI).

1.11 Differential Diagnosis

Common causes of dementia include Alzheimer's disease, which is the number one cause. Vascular dementia, alcoholism, and drug/medication intoxication are other common causes. Less common causes of dementia include vitamin deficiencies, endocrine and other organ failure, other psychiatric disorders, degenerative disorders, chronic infections, head trauma and diffuse brain damage, intracranial hypotension, neoplasm, toxic disorders, and others.

Differential diagnosis should always be in mind when the physician is taking a history. Personality change, disinhibition, and weight gain or compulsive eating suggest frontotemporal dementia (FTD), for example, and not AD. Early visual hallucinations, parkinsonism, and rapid eye movement (REM) behavior disorder (RBD; the loss of skeletal muscle paralysis while dreaming) may suggest diagnosis of dementia with Lewy bodies (DLB). Rapid progression with motor rigidity and myoclonus suggests Creutzfeldt–Jakob disease (CJD). Seizures may indicate strokes or neoplasm but also occur in AD, particularly early-age-of-onset AD (Casper et al. 2015). Furthermore, a history of stroke with irregular progression suggests vascular dementia. Vascular dementia is also commonly seen in the setting of hypertension, atrial fibrillation, peripheral vascular disease, and diabetes. Therefore, in patients with cerebrovascular disease, it can be difficult to determine whether the dementia is due to AD, vascular disease, or a mixture of the two because many of the risk factors for vascular dementia, including diabetes, high cholesterol, and lack of exercise which are also presumptive risk factors for AD.

1.12 Treatment Options for AD Patients with CI

The treatment options for MCI and CI are either pharmacological or psychological. Currently, the FDA has not approved any medication for MCI treatment since AD medication has not been proved effective as far as preventing or delaying the deterioration of MCI is concerned. Several psychological or occupational interventions have shown encouraging results; however, most of them should be personalized, while their efficacy depends on the experience of the therapist, the compliance of the patient, and the support of the carers.

CI in AD patients is treated with pharmacological agents on the grounds of CI's pathophysiology knowledge. More specifically cholinesterase inhibitors (ChEIs) and N-methyl-D-aspartate (NMDA) antagonists are considered as the first-line therapy for global cognitive impairment. Several studies indicate the benefits of psychological approaches in combination with pharmaceutical treatment. In this frame several approaches have been developed. To name just a few, we may shortlist cognitive behavioral therapy, stimulation training, cognitive training, cognitive stimulation, and cognitive rehabilitation. The latter consists of both stimulation and training techniques and focuses on cognitive aspects of the patient's disorder.

1.13 Cognitive Rehabilitation in AD

1.13.1 Introduction

Cognitive rehabilitation (CR) has been defined as a comprehensive program aiming to cognitive enhancement. It includes several training approaches, and it has been developed as a method of rehabilitation for people with cognitive impairment of various etiologies. CR is an individualized treatment in accordance with the principles of biopsychosocial approach. CR has been studied as a part of the treatment of several conditions including traumatic brain injury (TBI), cerebral vascular accident, cerebral palsy, Down syndrome, Alzheimer's disease (AD), attention deficit hyperactivity disorder (ADHD), and developmental disorders such as autism, schizophrenia, and Parkinson's disease (Choi and Twamley 2013).

Due to the low efficacy of pharmacological approaches up to this day, CR is expected to play an important role in AD's treatment. In CR, all facets of neuropsychological deficits are approached in the context of behavior and social functioning. Practitioners pay special attention to cognitive stimulation and training. Clare and Woods published their work on cognitive enhancement in AD back in 2004 and grouped the various treatments into three broad categories: cognitive stimulation, cognitive training, and cognitive rehabilitation (Clare and Woods 2004).

1.13.2 Neuropsychological Background

The idea of cognitive treatments for AD is based on the concept of neuronal plasticity. A continuous loss of cerebral capacity including neuromodulation has been related to aging. Accumulating evidence recommends that the sensory system is capable of altering its structural and hence its functional pattern in accordance with various stimuli. To put it in plain words, this means that the mind has the capacity for rebuilding itself so as to adjust to changing conditions or novel stressors. This feature has been established through plasticity promoting studies in ordinary old individuals (Choi and Twamley 2013).

Training appears as a promising method of inducing brain plasticity since it involves individuals in stimulating cognitive, sensory, and psychomotor tasks. The rationale behind this statement has been established from the level of neurons to the level of cortical representations. Training represents a repeated stimulus which enhances signaling pathways and eventually gene expression in molecular and cellular levels (Limond and Leeke 2005).

On these grounds, we assume that neurons which undergo these procedures can act as an assembly enhancing memory patterns and networks. This applies to existing memory patterns, which are enhanced and may also be able to form new patterns. A very important aspect of this concept is that minority neuron circuits may be enhanced, and in functional level, this signifies that an individual who faces a working memory impairment could witness an improvement of his daily life through cognitive training (Winocur et al. 2000; Choi and Twamley 2013).

Several neuropsychological experiments or clinical studies support the efficacy of training. Most of these studies focus on visual stimuli and suggest that the main cerebral areas affected by training experiments are the right fusiform face area, right parahippocampal cortex, right temporal parietal junction, and right medial prefrontal cortex. The efficacy of training could be assessed through either higher or lower activation of neural circuits. Higher activation has been correlated with higher functionality, whereas in different studies, lower activation is considered an encouraging finding suggesting that a task can be accomplished easier (Ginarte-Arias n.d.).

More specifically, Heiss et al. conducted a PET study of 70 patients with mild AD comparing social support and pharmacological and/or cognitive treatment. Their results suggest that a combination of cognitive training and pharmacological treatment (phosphatidylserine or pyritinol) was associated with increased brain glucose metabolism in temporal–parietal brain areas during a task based on recognition of visual stimuli (Heiss et al. n.d).

Moreover, a single-blind randomized controlled trial consisting of CR and relaxation therapy versus no treatment in mild AD conducted by Clare et al. in 2010 found an increase in blood oxygen level-dependent signals in the CR group in areas forming part of the network for visual associative encoding and learning, whereas individuals in the no-treatment control group showed reduced BOLD activity over time (Clare et al. 2010).

Other studies showed a progressive decrease in the neuronal activity associated with the accomplishment of a task upon training. Haier et al.'s study included young individuals performing repeatedly a complex visuospatial/motor task. The individuals underwent brain PET scans before and after practice which revealed a decrease in regional subcortical glucose metabolic rate with practice may reflect changes in cognitive strategy that are a part of the learning process (Haier et al. 1992).

Such findings formulate the neuropsychological background of CR allowing respective studies to be conducted. They support CR's scientific rather than empirical character, they provide practitioners with constant updates to improve their practice, and they also establish a feedback network between practitioners and researchers. All in all, this accumulating evidence is an added value to cognitive rehabilitation and its role in AD treatment (Limond and Leeke 2005; Petersen et al. 2014).

1.13.3 Cognitive Rehabilitation in AD Methods

Cognitive rehabilitation (CR) is defined as a comprehensive cognitive enhancement program, which puts under the same roof cognitive stimulation (CS), cognitive training (CT), and other approaches. In this point, it would be useful to provide a short definition of cognitive stimulation and cognitive training given that these methods have been developed for patients suffering from AD of different stages. CS focuses on the engagement of the patient in discussions about familiar to him everyday affairs in order to stimulate mental activity. More impaired AD patients and mainly inpatients in proper facilities are usually treated with CS. On the other hand, CT consists of various tasks designed to exercise specific cognitive functions or to work on patients with a relatively fair cognitive status so as to support its impaired aspects. Consequently, patients who have enough cognitive resources for a therapist or a computer program to guide them are usually treated with CT (Kelly and Maria 2015).

As a combination of the aforementioned methods, CR functions as a model of treating the cognitive decline depending on the assessed behavioral and social disability rather than focusing on specific cognitive deficits. In this frame CR not only alleviates cognitive deficits working closely on them, but it also creates compensatory mnemonic pathways so as to restore the functionality of the patient to the greatest extent. This compensation ranges from training the patients to dealing with finances in such a way that the monthly utility bills are easier to remember to learning how to use virtual or paper aids to organize and recall important information (medication, appointments, etc.). Verbal instructions along with physical

Table 1.7 An overview of CR patterns

Guiding principles	Recall strategies	Specific interventions
Effortful processing dual	Mnemonics	Face-name recall
Cognitive support	Cueing	Number recall
Errorless learning	Chunking	Story recall
	Method of loci	List/object recall
	Spaced retrieval	Procedural memory
		Fluency training
		Semantic impairments

An overview of principles, strategies, and intervention in the frame of CR (Kelly and Maria 2015)

demonstration and support items are used to teach the patient how to develop methods applicable to its own cognitive deficits. The outcomes of CR can be observed and assessed during the interaction between the patient and his/her environment (Clare et al. 2010).

To this point, the exact methods of CR seem to be opaque. Some clarifying examples can be found in clinical studies on CR and in health associations' guidelines papers. Hence, we should briefly review the guidelines of the Alzheimer's Society of Ireland and a CR therapy study conducted by Loewenstein et al. back in 2004.

According to the guidelines, CR is a tripartite pattern consisting of guiding principles, recall strategies, and specific interventions (Table 1.7).

To our knowledge, CR is performed on an individual basis although many studies group AD patients with a similar impairment score. The majority of the CR approaches belong to three different categories: (a) techniques without external memory aids, (b) techniques with nonelectronic external memory aids, and (c) techniques using electronic technologies. Studies have recorded everyday items and electronic applications, notepads, alarm clocks, traditional computer games such as Tetris, or more sophisticated electronic applications used in cognitive rehabilitation settings. The selection of these supportive items varies depending on availability, personal experience of the health professionals practicing CR, and familiarity of the patient with any given item. Familiarity is also linked to the optimal treatment setting given that space is another important feature of CR interventions (Ginarte-Arias n.d; (Choi and Twamley 2013).

A comfortable setting such as the patient's home plays an important role concerning the efficacy of the intervention. Moreover, CR ought to be perceived as a dynamic procedure in which the patient and his family/carers are actively involved, assisted, and coordinated by the health professional who acts as therapist. The patient should be encouraged to practice the strategies he/she elaborated on during the session on his own and apply them to as many of his daily activities as possible. At the same time, the family/carers should be debriefed after each session and also be instructed with strategies for practice outside the CR sessions (Kelly and Maria 2015).

Evidence suggests that the outcomes of CR should be assessed in the beginning and in the end of each CR session, in the short and in the long term. Moreover, it is considered useful for the health professional and for the family/carers of the patient to take notes so as to observe better the capacity of the patient to deal with everyday tasks. Although scientific studies observe and evaluate systematically the efficacy of any tested approach, such an individual monitoring appears as an essential part of each treatment plan, and it seems to be helpful as far as the personalization of CR, the commitment of the patient, and the therapist–patient communication are concerned (Kelly and Maria 2015; Ginarte-Arias n.d)

1.14 Efficacy of CR in AD patients

The efficacy of CR in AD has been debated for a long time. Back in 1999, an NIH resolution pointed out that few studies are available, and even these studies are greatly diverse as far as their methods are concerned or present data from a small number of patients. Even though evidence has grown since then, the approaches remain diverse, and it was difficult to group all studies. We provide the main categorization system of these studies and present the results of three representative ones:

1. Studies targeting techniques without external memory aids
2. Studies targeting techniques with nonelectronic external memory aids
3. Studies focusing on the use of assistive electronic technologies

Germain et al. conducted an observational, prospective study in a sample of 52 patients. CR sessions were facilitated by experienced therapists once weekly during 3 months at home and once monthly contact for 9 months. Patients with mild AD dependence and caregiver's burden were evaluated immediately after the intervention and had follow-ups at 6 months and 1 year. The dependence of the patients with regard to adapted activity as well as the caregiver's burden was decreased at 3 months, 6 months, and 1 year. Moreover, global cognition slightly decreased in a 1-year interval. The results of this observational study are encouraging for patients with mild AD and their caregivers, given the fact that patients' dependence decreased and the global cognitive decline is limited (Germain et al. 2018).

Clare et al. conducted a single-blind *randomized controlled trial* (RCT) comparing CR supported by practical components with relaxation therapy and no treatment. Sixty-nine individuals from a community – outpatient setting (41 women and 28 men), mean age 77.78 years (standard deviation 6.32, range = 56–89), and who have been diagnosed with AD or mixed AD and *vascular dementia* and have a *Mini-Mental State Examination score* of 18 or above – were also included. All patients were treated pharmacologically (stable dose of acetylcholinesterase-inhibiting medication). Eight weekly individual sessions of CR were held, and goal performance and satisfaction were assessed. The assessment was based on the Canadian Occupational Performance Measure; questionnaires assessing mood, quality of life, and carer strain; and a brief neuropsychological test battery. CR was associated with significant improvement in ratings of goal performance and satisfaction. An imaging study (fMRI) supported the behavioral changes in a subset of the CR group. Hence, these findings suggest that the item which assisted CR may be effective against early-stage AD (Clare et al. 2010).

Loewenstein et al. conducted an RCT of CR against mental stimulation facilitated by computer puzzle games. According to their findings, CR ameliorated orientation, learning and recalling faces and names for at least 3 months. The CR group also made noticeable improvements on an untrained functional task at a level of statistical significance. However, only informant patients in the CR condition noted a significant improvement in memory function on the Informant Questionnaire of the Cognitive Decline in the Elderly scale. These results suggest that long-term outcomes may be achieved provided that CR intervention focuses on everyday life – real-world affairs – and the patient is aware of his condition and motivated to ameliorate it (Loewenstein et al. 2004).

All in all, although CR interventions have encouraging results, they are still considered somewhat controversial. Many researchers claim that CR is beneficial only for a short time and that the results are mainly reported in patients with mild- or early-stage AD. Moreover, a Cochrane review focusing on individuals with early-stage AD or vascular dementia assumes that the available evidence is not strong enough for the application of such interventions because limited randomized controlled trials (RCTs) are available and considerable methodological limitations have been identified. The efficacy of individual CR approach appears even more opaque because of a complete absence of relevant RCTs. Further and more comprehensive research ought to be conducted in order to properly assess the efficacy of CR. This fact has a negative impact on the future incorporation of CR in clinical practice for the forthcoming patients with AD (Winocur et al. 2000; Jacquemin 2009; Bahar-Fuchs et al. 2013a).

1.15 Benefits and Drawbacks of CR on Patients with AD

CR is considered beneficial for various reasons. First of all, it is a non-interventional, non-pharmaceutical treatment, and thus it does not share the side effects of these treatments. Moreover, CR is a personalized treatment adjusted to the mental status and the specific needs of the patients. In this concept it engages both the patient and the carers in the therapeutic procedure, and thus it enhances doctor's and patient's communication according to the principles of biopsychosocial healthcare. What is more a wide spectrum of health professionals can be trained and qualified to perform CR, and thus many patients in different settings will have access to CR. More specifically, CR may be performed by an occupational therapist, a physical therapist, a speech/language pathologist, a neuropsychologist or other psychologist, or a neuropsychiatrist, psychiatrist, or other physician (Bottino et al. 2005; Viola et al. 2011; Bahar-Fuchs et al. 2013a).

On the other hand, CR is due to several drawbacks with regard to the method in general and the application of the method to patients with AD in particular. In the first category, we may list the variable efficacy of the method depending on the experience of the therapist, the ambivalent efficacy – which discourages health professionals from considering it as a treatment for their patients – and the fact that carers should be engaged to achieve optimal compliance and efficacy (Kelly and Maria 2015).

In the second category, we may refer to indigenous factors of the disease. Factors such as the progressive mental status decline, the patient's low understanding, or even awareness of the illness and the highly prevalent depression among geriatric patients have been shown to alter the potential outcomes of CR. As a result of this, the purpose of CR in AD is to slow down this cognitive decline offering to the patient some months of independent or less dependent function (Hwang et al. 2015). Even though this is considered as a huge benefit in terms of quality of life and care-associated direct and indirect costs, it does not meet the expectations of the patients' carers, and this results in a lower level of carers' engagement, weakening the overall efficacy of the treatment. A more detailed overview of obstacles has been presented by Choi et al. (Choi and Twamley 2013).

1.16 Cost and Overall Implementation Overview

Not all the patients with AD are guaranteed access to CR. Reviewing many US-based insurance brands, it is evident that there is a consensus not to cover CR expenses in patients suffering from different conditions than traumatic brain injury (TBI) and cerebral vascular accident. Medicare documents also reveal a similar attitude of the US public sector toward CR in AD. The background of this attitude is an NIH resolution in 1999 which concluded that CR has not been proved effective in conditions such as cerebral palsy, Down syndrome, Alzheimer's disease (AD), attention deficit hyperactivity disorder (ADHD), and developmental disorders such as autism, schizophrenia and Parkinson's disease. Hence, it appears that influential health-care systems are not willing to fund CR interventions as a counteract to AD's cognitive impairment (Winocur et al. 2000; Limond and Leeke 2005; Bahar-Fuchs et al. 2013b).

The cost of CR interventions varies accordingly to the methods applied. Interventions based on computer programs cost more than interventions based on simple notepads. Nowadays studies investigate cost–benefit aspects of the low-cost CR. Their findings are encouraging; nevertheless, more evidence concerning AD in particular and the impact on CR on direct and indirect treatment cost will be published in due time.

In the future, in case CR is approved for insurance coverage, it seems that the criteria will be strict. Taking into account the procedure for patients with TBI and HIV-associated encephalitis, patients with AD should prove their impaired cognitive status on the grounds of:

1. Neuropsychological testing leading to neuropsychological results which will be used in treatment planning and directing rehabilitation strategies.

2. Neuropsychiatric and neuropsychological evaluation.
3. Ability of the patient to actively participate in a cognitive rehabilitation program (e.g., is not comatose or in a vegetative state).
4. An expected significant amelioration of the patient's cognitive status (*Cognitive Rehabilitation Therapy* | TRICARE n.d).

1.17 Toward an Optimal CR

Errorless learning approach has been suggested as a promising CR intervention by several researchers. Errorless training prioritizes avoidance of errors during sessions and consequently targets the participants it is supposed to study and immediately reproduces any piece of given information. It appears that this feature prevents participants from attempting to retrieve target information from long-term memory. In this frame, both everyday materials and sophisticated computer software may be effective. However, error elimination could be a significant pitfall leading to a gradual impairment of retrieval practice ability and thus confusing the patients further, discouraging them and their carers, and eventually weakening CR treatment plan. (Petersen et al. 2014; Germain et al. 2018). Errorless learning has been designed to make sure that the patient memorizes accurately and correctly in order to achieve an appropriate level of functionality.as long as possible. This approach appears as the most promising one and is expected to reform all existing kinds of CR (Middleton and Schwartz 2012).

1.18 Conclusion

All in all, the significance of CI in AD has been more and more acknowledged during the past years. This knowledge has been incorporated in contemporary research and diagnostics, while the CR treatment model has been developed to counteract AD's cognitive features (Hwang et al. 2015). Several factors from the current understanding of pathophysiology to the patients' compliance interfere with the investigative, diagnostic, and therapeutic procedures (Choi and Twamley 2013). In the future, further and more comprehensive research ought to be conducted in order to enhance the understanding of CI in AD context and to establish the most efficient treatment models.

References

Abbott RD et al (2004) Walking and dementia in physically capable elderly men. JAMA 292(12):1447–1453. https://doi.org/10.1001/jama.292.12.1447

Acevedo A, Loewenstein DA (2007) Nonpharmacological cognitive interventions in aging and dementia. J Geriatr Psychiatry Neurol 20(4):239–249. https://doi.org/10.1177/0891988707308808

Albert MS et al (2011) The diagnosis of mild cognitive impairment due to Alzheimer's disease: recommendations from the National Institute on Aging-Alzheimer's Association workgroups on diagnostic guidelines for Alzheimer's disease. Alzheimers Dement 7(3):270–279. https://doi.org/10.1016/j.jalz.2011.03.008

Alzheimer's Association (2010) 2010 Alzheimer's disease facts and figures. Alzheimers Dement 6(2):158–194. https://doi.org/10.1016/j.jalz.2010.01.009

Bagyinszky E et al (2016) Mutations, associated with early-onset Alzheimer's disease, discovered in Asian countries. Clin Interv Aging Volume 11:1467–1488. https://doi.org/10.2147/CIA.S116218

Bahar-Fuchs A, Clare L, Woods B (2013a) Cognitive training and cognitive rehabilitation for mild to moderate Alzheimer's disease and vascular dementia. Cochrane Database Syst Rev 6:CD003260. https://doi.org/10.1002/14651858.CD003260.pub2

Bahar-Fuchs A, Clare L, Woods B (2013b) Cognitive training and cognitive rehabilitation for persons with mild to moderate dementia of the Alzheimer's or vascular type: a review. Alzheimers Res Ther 5(4):35. https://doi.org/10.1186/alzrt189

Ball K et al (2002) Effects of cognitive training interventions with older adults: a randomized controlled trial. JAMA 288(18):2271–2281. Available at: http://www.ncbi.nlm.nih.gov/pubmed/12425704. Accessed 13 Dec2018

Baumgart M et al (2015) Summary of the evidence on modifiable risk factors for cognitive decline and dementia: a population-based perspective. Alzheimers Dement 11(6):718–726. https://doi.org/10.1016/j.jalz.2015.05.016

Behrman S, Valkanova V, Allan CL (2017) Diagnosing and managing mild cognitive impairment. Practitioner 261(1804):17–20. Available at: http://www.ncbi.nlm.nih.gov/pubmed/29120563. Accessed 13 Dec 2018

Bekris LM et al (2010) Genetics of Alzheimer disease. J Geriatr Psychiatry Neurol 23(4):213–227. https://doi.org/10.1177/0891988710383571

Bidzan L, Bidzan M, Pąchalska M (2016) The effects of intellectual, physical, and social activity on further prognosis in mild cognitive impairment. Med Sci Monit 22:2551–2560. Available at: http://www.ncbi.nlm.nih.gov/pubmed/27434501. Accessed 13 Dec 2018

Bloom GS (2014) Amyloid-β and Tau. JAMA Neurol 71(4):505. https://doi.org/10.1001/jamaneurol.2013.5847

Bottino CM et al (2005) Cognitive rehabilitation combined with drug treatment in Alzheimer's disease patients: a pilot study. Clin Rehab 19(8):861–869. https://doi.org/10.1191/0269215505cr911oa

Casper LD et al (2015) 19th edition Harrison's principles of internal medicine: volume II, 19th edn. McGraw-Hill Education, New York/Chicago/San Francisco/Athens/London/Madrid/Mexico City/Milan/New Delhi/Singapore/Sydney/Toronto

Chiu C-C et al (2008) The effects of omega-3 fatty acids monotherapy in Alzheimer's disease and mild cognitive impairment: a preliminary randomized double-blind placebo-controlled study. Prog Neuro-Psychopharmacol Biol Psychiatry 32(6):1538–1544. https://doi.org/10.1016/j.pnpbp.2008.05.015

Choi J, Twamley EW (2013) Cognitive rehabilitation therapies for Alzheimer's disease: a review of methods to improve treatment engagement and self-efficacy. Neuropsychol Rev 23(1):48–62. https://doi.org/10.1007/s11065-013-9227-4

Clare L, Woods RT (2004) Cognitive training and cognitive rehabilitation for people with early-stage Alzheimer's disease: a review. Neuropsychol Rehabil 14(4):385–401. https://doi.org/10.1080/09602010443000074

Clare L et al (2010) Goal-oriented cognitive rehabilitation for people with early-stage Alzheimer disease: a single-blind randomized controlled trial of clinical efficacy. Am J Geriatr Psychiatry 18(10):928–939. https://doi.org/10.1097/JGP.0B013E3181D5792A

Cognitive Rehabilitation Therapy | Tricare (n.d.). Available at: https://tricare.mil/CoveredServices/IsItCovered/CogRehabTherapy. Accessed 13 Dec 2018

Corder EH et al (1993) Gene dose of apolipoprotein E type 4 allele and the risk of Alzheimer's disease in late onset families. Science (New York, NY) 261(5123):921–923. Available at: http://www.ncbi.nlm.nih.gov/pubmed/8346443. Accessed 13 Dec 2018 de Souza-Talarico JN et al (2016) Dementia and cognitive impairment prevalence and associated factors in indigenous populations. Alzheimer Dis Assoc Disord 30(3):281–287. https://doi.org/10.1097/WAD.0000000000000140

De Strooper B et al (1998) Deficiency of presenilin-1 inhibits the normal cleavage of amyloid precursor protein. Nature 391(6665):387–390. https://doi.org/10.1038/34910

Dos Santos Picanço LC et al (2016) Alzheimer's disease: a review from the pathophysiology to diagnosis, new perspectives for pharmacological treatment. Curr Med Chem 25(26):3141–3159. Available at: http://www.ncbi.nlm.nih.gov/pubmed/27978805. Accessed 13 Dec 2018

Engelhart MJ et al (2002) Dietary intake of antioxidants and risk of Alzheimer disease. JAMA 287(24):3223–3229. Available at: http://www.ncbi.nlm.nih.gov/pubmed/12076218. Accessed 13 Dec 2018

Féart C et al (2009) Adherence to a Mediterranean diet, cognitive decline, and risk of dementia. JAMA 302(6):638. https://doi.org/10.1001/jama.2009.1146

Fratiglioni L, Paillard-Borg S, Winblad B (2004) An active and socially integrated lifestyle in late life might protect against dementia. Lancet Neurol 3(6):343–353. https://doi.org/10.1016/S1474-4422(04)00767-7

Germain S et al (2018) Efficacy of cognitive rehabilitation in Alzheimer disease: a 1-year Follow-Up study. J Geriatr Psychiatry Neurol 32(13):089198871881372. https://doi.org/10.1177/0891988718813724

Ginarte-Arias Y (n.d.) [Cognitive rehabilitation. Theoretical and methodological aspects]. Rev Neurol 35(9):870–876. Available at: http://www.ncbi.nlm.nih.gov/pubmed/12436386. Accessed 13 Dec 2018

Goedert M (2015) Alzheimer's and Parkinson's diseases: the prion concept in relation to assembled A, tau, and -synuclein. Science 349(6248):1255555. https://doi.org/10.1126/science.1255555

Goldman L, Schafer IA (2012) In: Goldman L, Schafer A (eds) Goldman-cecil medicine 25th edition volume I, 25th edn. Elsevier Seaunders, Philadelphia

Gouras GK, Olsson TT, Hansson O (2015) β-Amyloid peptides and amyloid plaques in Alzheimer's disease. Neurotherapeutics 12(1):3–11. https://doi.org/10.1007/s13311-014-0313-y

Haier RJ et al (1992) Regional glucose metabolic changes after learning a complex visuospatial/motor task: a positron emission tomographic study. Brain Res 570(1–2):134–143. Available at: http://www.ncbi.nlm.nih.gov/pubmed/1617405. Accessed 13 Dec 2018

Heiss WD et al (n.d.) Long-term effects of phosphatidylserine, pyritinol, and cognitive training in Alzheimer's disease. A neuropsychological, EEG, and PET investigation. Dementia (Basel, Switzerland) 5(2):88–98. Available at: http://www.ncbi.nlm.nih.gov/pubmed/8038871. Accessed 13 Dec 2018

Hwang J-H, Cha H-G, Cho H-S (2015) The effects of cognitive rehabilitation on Alzheimer's dementia patients' cognitive assessment reference diagnosis system performance based on level of cognitive functioning. J Phys Ther Sci 27(9):2875–2877. https://doi.org/10.1589/jpts.27.2875

Jacquemin A (2009) Strategy and methods of cognitive rehabilitation in patients with Alzheimer's disease or other dementia. Psychol Neuropsychiatr Vieil 7(4):265–273. https://doi.org/10.1684/pnv.2009.0182

Jeong S (2017) Molecular and cellular basis of neurodegeneration in Alzheimer's disease. Mol Cell 40(9):613–620. https://doi.org/10.14348/molcells.2017.0096

Jicha GA, Carr SA (2010) Conceptual evolution in Alzheimer's disease: implications for understanding the clinical phenotype of progressive neurodegenerative disease. J Alzheimers Dis 19(1):253–272. https://doi.org/10.3233/JAD-2010-1237

Kelly ME, Maria BO (2015) Strategies and techniques for cognitive rehabilitation manual for healthcare professionals working with individuals with cognitive impairment. Available at: https://www.alzheimer.ie/Alzheimer/media/SiteMedia/Services/Cognitive-Rehabilitation-Manual.pdf. Accessed 12 Dec 2018

Krogsboll L, Jorgensesn K, Larsen C (2012) General health checks in adults for reducing morbidity and mortality from disease: conchrane systematic review and meta-analysis. BMJ 345:e7191

Kurz A et al (1996) Apolipoprotein E type 4 allele and Alzheimer's disease: effect on age at onset and relative risk in different age groups. J Neurol 243(6):452–456. Available at: http://www.ncbi.nlm.nih.gov/pubmed/8803817. Accessed 13 Dec 2018

Langa KM, Levine DA (2014) The diagnosis and management of mild cognitive impairment. JAMA 312(23):2551. https://doi.org/10.1001/jama.2014.13806

Larson EB, Kukull WA, Katzman RL (1992) Cognitive impairment: dementia and Alzheimer's disease. Annu Rev Public Health 13(1):431–449. https://doi.org/10.1146/annurev.pu.13.050192.002243

Laurin D et al (2004) Midlife dietary intake of antioxidants and risk of late-life incident dementia: the Honolulu-Asia Aging Study. Am J Epidemiol 159(10):959–967. Available at: http://www.ncbi.nlm.nih.gov/pubmed/15128608. Accessed 13 Dec 2018

Limond J, Leeke R (2005) Practitioner review: cognitive rehabilitation for children with acquired brain injury. J Child Psychol Psychiatry 46(4):339–352. https://doi.org/10.1111/j.1469-7610.2004.00397.x

Liu C-C et al (2013) Apolipoprotein E and Alzheimer disease: risk, mechanisms and therapy. Nat Rev Neurol 9(2):106–118. https://doi.org/10.1038/nrneurol.2012.263

Loewenstein DA et al (2004) Cognitive rehabilitation of mildly impaired Alzheimer disease patients on cholinesterase inhibitors. Am J Geriatr Psychiatry 12(4):395–402. https://doi.org/10.1097/00019442-200407000-00007

Marešová P, Zahálková V (2016) The economic burden of the care and treatment for people with Alzheimer's disease: the outlook for the Czech Republic. Neurol Sci 37(12):1917–1922. https://doi.org/10.1007/s10072-016-2679-6

Middleton EL, Schwartz MF (2012) Errorless learning in cognitive rehabilitation: a critical review. Neuropsychol Rehabil 22(2):138–168. https://doi.org/10.1080/09602011.2011.639619

Morris MC et al (2002) Dietary intake of antioxidant nutrients and the risk of incident Alzheimer disease in a biracial community study. JAMA 287(24):3230–3237. Available at: http://www.ncbi.nlm.nih.gov/pubmed/12076219. Accessed 13 Dec 2018

Mufson EJ et al (2012) Mild cognitive impairment: pathology and mechanisms. Acta Neuropathol 123(1):13–30. https://doi.org/10.1007/s00401-011-0884-1

Mufson EJ et al (2016) Braak staging, plaque pathology, and APOE status in elderly persons without cognitive impairment. Neurobiol Aging 37:147–153. https://doi.org/10.1016/j.neurobiolaging.2015.10.012

Myers RH et al (1996) Apolipoprotein E epsilon4 association with dementia in a population-based study: the Framingham study. Neurology 46(3):673–677. Available at: http://www.ncbi.nlm.nih.gov/pubmed/8618665. Accessed 13 Dec2018

Nicolas G et al (2018) Somatic variants in autosomal dominant genes are a rare cause of sporadic Alzheimer's disease. Alzheimers Dement 14(12):1632–1639. https://doi.org/10.1016/j.jalz.2018.06.3056

Petersen RC et al (2014) Mild cognitive impairment: a concept in evolution. J Intern Med 275(3):214–228. https://doi.org/10.1111/joim.12190

Pierce AL, Bullain SS, Kawas CH (2017) Late onset Alzheimer disease. Neurol Clin 35(2):283–293. https://doi.org/10.1016/j.ncl.2017.01.006

Reitz C, Brayne C, Mayeux R (2011) Epidemiology of Alzheimer disease. Nat Rev Neurol 7(3):137–152. https://doi.org/10.1038/nrneurol.2011.2

Rogaeva E et al (2007) The neuronal sortilin-related receptor SORL1 is genetically associated with Alzheimer disease. Nat Genet 39(2):168–177. https://doi.org/10.1038/ng1943

Sachdev PS et al (2014) Classifying neurocognitive disorders: the DSM-5 approach. Nat Rev Neurol 10(11):634–642. https://doi.org/10.1038/nrneurol.2014.181

Scheuner D et al (1996) Secreted amyloid beta-protein similar to that in the senile plaques of Alzheimer's disease is increased in vivo by the presenilin 1 and 2 and APP mutations linked to familial Alzheimer's disease. Nat Med 2(8):864–870. Available at: http://www.ncbi.nlm.nih.gov/pubmed/8705854. Accessed 13 Dec 2018

Singh B et al (2014) Association of mediterranean diet with mild cognitive impairment and Alzheimer's disease: a systematic review and meta-analysis. J Alzheimers Dis 39(2):271–282. https://doi.org/10.3233/JAD-130830

Stephan BCM et al (2012) The neuropathological profile of mild cognitive impairment (MCI): a systematic review. Mol Psychiatry 17(11):1056–1076. https://doi.org/10.1038/mp.2011.147

Takahashi RH, Nagao T, Gouras GK (2017) Plaque formation and the intraneuronal accumulation of β-amyloid in Alzheimer's disease. Pathol Int 67(4):185–193. https://doi.org/10.1111/pin.12520

Takao M (2012) [Neuropathology of mild cognitive impairment]. Rinsho Shinkeigaku 52(11):851–854. Available at: http://www.ncbi.nlm.nih.gov/pubmed/23196441. Accessed 13 Dec 2018

Takizawa C et al (2014) Epidemiological and economic burden of Alzheimer's disease: a systematic literature review of data across Europe and the United States of America. J Alzheimers Dis 43(4):1271–1284. https://doi.org/10.3233/JAD-141134

Verghese A, Brady E, Kapur C (2011) The bedside evaluation: ritual and reason. Ann Intern Med 155:550–553

Viola LF et al (2011) Effects of a multidisciplinary cognitive rehabilitation program for patients with mild Alzheimer's disease. Clinics (Sao Paulo) 66(8):1395–1400. Available at: http://www.ncbi.nlm.nih.gov/pubmed/21915490. Accessed 13 Dec 2018

Wetterling T, Kanitz RD, Borgis KJ (1996) Comparison of different diagnostic criteria for vascular dementia (ADDTC, DSM-IV, ICD-10, NINDS-AIREN). Stroke 27(1):30–36. Available at: http://www.ncbi.nlm.nih.gov/pubmed/8553399. Accessed 13 Dec 2018

Winocur G et al (2000) Cognitive rehabilitation in clinical neuropsychology. Brain Cogn 42(1):120–123. https://doi.org/10.1006/brcg.1999.1179

Wu L et al (2012) Early-onset familial Alzheimer's disease (EOFAD). Can J Neurol Sci 39(4):436–445. Available at: http://www.ncbi.nlm.nih.gov/pubmed/22728850. Accessed 13 Dec 2018

Xu W et al (2017) Alcohol consumption and dementia risk: a dose–response meta-analysis of prospective studies. Eur J Epidemiol 32(1):31–42. https://doi.org/10.1007/s10654-017-0225-3

Yehuda S et al (n.d.) The role of polyunsaturated fatty acids in restoring the aging neuronal membrane. Neurobiol Aging 23(5):843–853. Available at: http://www.ncbi.nlm.nih.gov/pubmed/12392789. Accessed 13 Dec 2018

Alzheimer's: A Progressive Brain Disease: Causes, Symptoms, and Prevention

2

Fakhra Amin, Anas Shamsi, Muhammad Nadeem Asghar, Peerzada Shariq Shaheen Khaki, Mohd Shahnawaz Khan, Shams Tabrez, Syed Kashif Zaidi, Wajihullah Khan, and Bilqees Bano

Abstract

Alzheimer's disease (AD) is an irretrievable, mysterious, and devastating neurodegenerative disorder that leads to memory loss and impaired cognitive function. Specifically, AD patients suffer from poor thinking skills and lack potential to perform simple activities. Dementia is multifactorial disorder and considered as the main cause of AD. Alzheimer's disease is characterized by the presence of large quantities of two remarkable structures, i.e., amyloid plaques and neurofibrillary tangles which are the hallmarks of AD. Both of these structures are misfolded proteins and thought to play an important role in the degeneracy of neurons, ultimately leading to most of AD symptoms. Another common feature

F. Amin · W. Khan
Department of Zoology, Faculty of Life Sciences, Aligarh Muslim University, Aligarh, India

A. Shamsi · P. S. S. Khaki · B. Bano (✉)
Department of Biochemistry, Faculty of Life Sciences, Aligarh Muslim University, Aligarh, India

M. N. Asghar
Department of Medical Biology, University of Quebec, Trois Riveres, Canada

M. S. Khan
Protein Research Chair, Department of Biochemistry, College of Sciences, King Saud University, Riyadh, Saudi Arabia

S. Tabrez
King Fahd Medical Research Center, King Abdulaziz University, Jeddah, Saudi Arabia

Department of Medical Laboratory Technology, Faculty of Applied Medical Sciences, King Abdulaziz University, Jeddah, Saudi Arabia

S. K. Zaidi
Center of Excellence in Genomic Medicine Research, King Abdulaziz University, Jeddah, Saudi Arabia

G. M. Ashraf, A. Alexiou (eds.), *Biological, Diagnostic and Therapeutic Advances in Alzheimer's Disease*, https://doi.org/10.1007/978-981-13-9636-6_2

of AD is the wrecking of intracellular connections which results in reduced cell function and cell death. Neurofibrillary tangles are twisted fibers which are found inside the neurons and are produced by hyperphosphorylation of a microtubule-associated protein, i.e., tau. Smoking and obesity are the putative risk factors for dementia and cardiovascular disease. Deficiency of vitamins, especially thiamine (B1), riboflavin (B2), niacin (B3), pyridoxine (B6), folate (B9), and cobalamin (B12), is associated with cognitive dysfunction and AD. Type-2 diabetes (T2D) and prediabetic insulin resistance have also been suggested as the risk factors for cerebrovascular injury and cognitive decline, eventually leading to dementia. Cognitive impairment in T2D is caused by deposits of amylin, an amyloidogenic hormone synthesized and cosecreted with insulin by pancreatic β-cells. The menace of AD can be lowered down by certain anti-inflammatory drugs, proper intake of vitamin B, nurturing physical and leisure activities. In addition, AD can also be controlled by decreasing the candidate risk factors for cardiovascular disease. Anti-amyloid approaches such as vaccination may provide a promising tool to establish safer therapeutic approaches.

Keywords

Alzheimer's · Dementia · Amyloid plague · Tau proteins · Neurofibrillary tangles

2.1 Introduction

Alzheimer's disease (AD) is named after Dr. Alois Alzheimer. In (1906), Dr. Alzheimer observed changes in the brain tissue of a woman who had died of an abnormal mental illness (Alzheimer's Disease Education & Referral (ADEAR) Center 2012). With growing age, AD becomes more of a concern from medical, social, and public perspective. It is a disease causing relentless and irreversible loss of rational thinking toward any activity. AD patients tend to lose their entire memory gradually even that they do not recognize their own family members and forget all the previously learned skills. There is a clear effect on person's potential to memorize things, imagine clearly, and take correct decisions. In short period of time, AD patients become completely devoid of fundamental self-care, thus, commanding a big responsibility on their families and communities (Petra 2001). The symptoms appear after the age of 60 in majority of AD patients. AD has been shown to cause degeneration of large number of neurons in the brain (Jabir et al. 2015; U.S. Department of Health and Human Services 2011). According to a recent survey, 5 million Americans or 1 in 9 people over the age of 65 are living with this neurological disorder which is the most common cause of dementia (Igor 2014). The world's population is rapidly aging, and the number of people with dementia is expected to grow from 35 million as of today to 65 million by the year 2030.

2.2 Disorders Associated with Alzheimer: Types and Their Possible Causes

2.2.1 Dementia

There are several factors that are responsible for the progress of dementia. Neurodegenerative disorders such as AD and dementia lead to a complete loss of neurons and brain activity. At present, there are no remedies available to treat such disorders, however, other types of dementia can be ceased or still overturned with the treatment. Sometimes even infectious disorders and vascular problems (Ashraf et al. 2016) that are treatable can lead to dementia. There are certain drugs and vitamin deficiencies that can be a source of neurological deficits similar to dementia.

2.2.2 Tauopathies

Disorders associated with the pathological accumulation of tau are termed as tauopathies. There is a twisting of Tau protein in a way that it is aggregated and eventually results in bundle formation referred to as neurofibrillary tangles, inside the neurons. Morphologically, these tangles are the consequence of hyperphosphorylation of tau. There is another notable development in this disorder, i.e., the presence of unusual clumps (plaques) of amyloid in the spaces between brain cells. Both of these uncharacteristic structures are somehow responsible for the reduced function and nerve cell death. The reason underlying this relationship is still ambiguous. It is not clear that either these uncharacteristic structures are responsible for the disorder or presence of these uncharacteristic structures somehow leads to the other processes that are responsible for neuronal death during AD. In some cases of dementia, there is a clumping of tau protein within the nerve cells which in turn ceases the proper functioning of these cells and hence ultimately leads to the death.

2.2.3 Corticobasal Degeneration (CBD)

Also known as corticobasal ganglionic degeneration (CBGD), it is a dynamic neu rological disorder characterized by loss of nerve cell and atrophy (shrinkage) of specific region of the brain including the cerebral cortex and the basal ganglia. This usually occurs in the patients with age around 45–70 and according to a current estimate, nearly 2000–3000 Americans likely disesteem the frequency of this disorder. Initially, only one part of the body becomes affected; however, after a while both sides become impaired. A person may have obscurity in making use of hand or it may also attain an abnormal position. There can be an occurrence of visual-spatial problems that put together tough in analyzing visual information, say in finding the distance between objects. There are certain helpful therapies that can reduce the severity of certain symptoms but there is no cure for CBD. For instance, Botulinum toxin can help in controlling the muscle contractions. Similarly, speech therapy and physical therapy may also be helpful in performing routine activities.

2.2.4 Frontotemporal Disorders (FTD)

This is an umbrellalike term referred to as a group of uncommon disorders specifically aimed at frontal and temporal lobes of the brain, the region of brain responsible for personality, behavior, and language. There are certain forms of FTD that are referred to as tauopathies. There are some cases where FTD is associated with mutations in the tau gene (MAPT) and tau aggregates. Behavioral variant frontotemporal dementia (BvFTD) leads to an individual undergoing changes in behavior and traits. People suffering from this disorder have a tendency to become impulsive, start stealing, and behave rudely with other people. They may become repetitive in their behavior such as chanting, appreciation, or rumbling other person's dialogue.

2.2.5 Primary Progressive Aphasia (PPA)

This neurological disorder is described as the condition in which language capabilities become slowly and progressively impaired. An individual experiences problem in expressing one's thoughts or speak one's mind.

2.2.6 Progressive Supranuclear Palsy (PSP)

It is an uncommon brain disorder that destroys the upper brain stem together with the substantia nigra, which is responsible for controlling the movement in midbrain. This part of the brain is also affected in Parkinson's disease which somehow underlies the fact that some of the symptoms are common in both these disorders.

2.2.7 Argyrophilic Grain Disease

Argyrophilic grain disease is well thought out to be a progressive disorder that may or may not be connected with dementia and is characterized by tau deposits called argyrophilic grains in brain space involved in memory and emotion. There is a quite clear demarcation between its sign and symptoms from the late-onset AD. It is only after getting the autopsy done that one can be sure of this neurological condition.

2.2.8 Synucleinopathies

It is a group of neurodegenerative disorders in which there are fibrillary aggregates of alpha-synuclein protein in the cytoplasm of selective populations of neurons and glia. There is a mounting evidence that relates the development of abnormal filamentous aggregates to the onset and progression of clinical symptoms and the degeneration of affected brain regions during neurodegenerative disorders. The early symptoms may differ but as the disease progresses, the symptoms that develop are quite similar to the ones observed in the case of other neurodegenerative disorders, i.e., Parkinson's disease. This forms the basis of linkage of this disorder with this Parkinson and hence this area is under intense research to understand the molecular mechanisms underlying this connection.

2.2.9 Dementia with Lewy Bodies (DLB)

Lewy body dementia engages aggregates of protein known as Lewy bodies, balloon-like structures, making up the inner side of neurons. It is a widespread form of progressive dementia. Its symptoms include sleeping difficulties, smell loss, and other visual problems such as visual hallucination which is followed by movement related problems lasting for as long as 10 years.

2.2.10 Parkinson's Disease Dementia (PDD)

It is a medical condition interrelated to DLB arising in individuals with Parkinson's disease. In PDD, there is a clear effect on memory of a person along with difficulty in social judgment coupled with reasoning problems. Through autopsy experiments, it is found that individuals suffering from PDD have amyloid plaques and tau tangles parallel to those common in patients of AD; however, the reason underlying these similarities is still not understood.

2.2.11 Vascular Dementia

Vascular dementia refers to subsequent decline of remembrance and other cognitive functions. The symptoms of patients suffering from this disorder are quite similar to the ones suffering from AD. Interestingly, the cause behind the appearance of the similar symptoms is not AD pathology rather than due to persistent reduced blood flow in the brain, ultimately causing dementia. There are certain brain imaging studies that support the fact that this vascular disease somehow is involved in a patient's cognitive impairment (Ashraf et al. 2016). This condition is quite prevalent in older age people; however, because of its complicated diagnosis, many studies investigating its occurrence may be incorrect. In the USA and Western Europe, the ratio of VaD to AD is generally thought to be 1:5, and dementia next to stroke is thought to occur in one quarter to one third of the cases for stroke.

2.2.12 Mixed Dementia (MID)

Autopsy studies from older adults, i.e., 80 and above, are evident that they usually suffer from "mixed dementia" (MID), which is mustered by AD-related neurodegenerative processes in combination with vascular disease-related activities. In a person with mixed dementia, it may perhaps not be clear exactly how many of a person's symptoms are due to AD or another type of dementia. A recent study showed that approximately 40% of people who were thought to have AD also have some form of cerebrovascular disease which is only confirmed after autopsy (THE DEMENTIAS 2013). In this type of dementia, there is an occurrence of many small strokes that destroys brain cells. Some of these strokes can occur even without noticeable symptoms and are referred to as "silent strokes." In contrast to similar incidence of dementia in man and women, MID is higher in men, especially in adults from the age group ranging from 60 to 75 years. Because of the similarities in the symptoms of MID and AD, it is quite hard for a clinician to make an accurate diagnosis. Depending on the region of the brain that is influenced, there could be several outcomes such as impairment of language that may occur or a portion of the body may be disproportionally affected.

2.2.13 Normal Pressure Hydrocephalus (NPH)

The term "normal pressure hydrocephalus" was at first introduced by Hakim and Adams in 1965 (Hakim and Adams 1965). This type of dementia is characterized by an excessive accumulation of cerebrospinal fluid (CSF) in the ventricles of the brain which eventually leads to a variety of problems related to thinking and walking coupled with loss of bladder control. This is also known as "normal pressure" as in spite of surplus fluid, the cerebrospinal fluid pressure as calculated during a spinal tap is normal. With enlargement of brain ventricles because of excessive cerebrospinal fluid, there can be disruption of the surrounding brain tissue, thus giving rise to the symptoms of NPH. CSF dynamics is affected from two aspects in this, viz., production and absorption. The CSF moves out of the ventricular system, thereby passing into the subarachnoid space contiguous to the brain and in the region of the spinal cord. The CSF, in the subarachnoid space, provides a cushioning effect to the brain. Most of the CSF is absorbed into the arachnoid villi, which are permeable and allow the CSF to move out from the subarachnoid space into the venous sinuses (Verrees and Selman 2004; Hickey 2003). In this disorder, the arachnoid villi fail to maintain an adequate elimination of CSF (Byrd & RN, MSN, ANP, ACNP, CNRN 2006), however, the reasons behind this disorder are still not known.

2.2.14 Memory Changes

What we frequently reflect of as "memory" in everyday habit is in point of fact "long-term memory," but there also exist short-term and sensory memory processes.

Different types of memory of each individual depend on their exacting approach of action, but there is a kind of coordination in the method of memorization, and hence these can be defined as three essential stepladders in lasting memory formation. It may exist as sensory, short-term, or long-term memory, rather than the unitary process as a sequence of three stages. It is referred to as the modal or multi-store or Atkinson-Shiffrin model, post Richard Atkinson and Richard Shiffrin. This model was developed in the year of 1968 and remains one of the most influential models for memory study. It is often referred to as the method of memory [www.lukemastin.com/humanmemory].

2.3 Features of Alzheimer's Disease

There is an emergent belief that Alzheimer's disease like other chronic disorders develops because of multiple factors rather than a solitary factor; with the exception of certain cases, as a result of genetic mutations (Fig. 2.1). But it is not clearly understood exactly what initiates the AD process or why several normal changes, which are usually linked with aging, become much more prominent and disparaging in individuals and with aging become much more extreme and destructive in people and manifest as a disease. The dangerous menacing factor for Alzheimer's disease is aging. Majority of the people diagnosed with this disorder lie in the age group of 65 or higher. Individuals below this age group rarely suffer from this disease. Although, this disease is prevalent in higher age groups but aging alone is not the solitary factor behind the development of this disease and is not sufficient to cause it. If the person is older than 80 and AD takes grip, the physical and mental

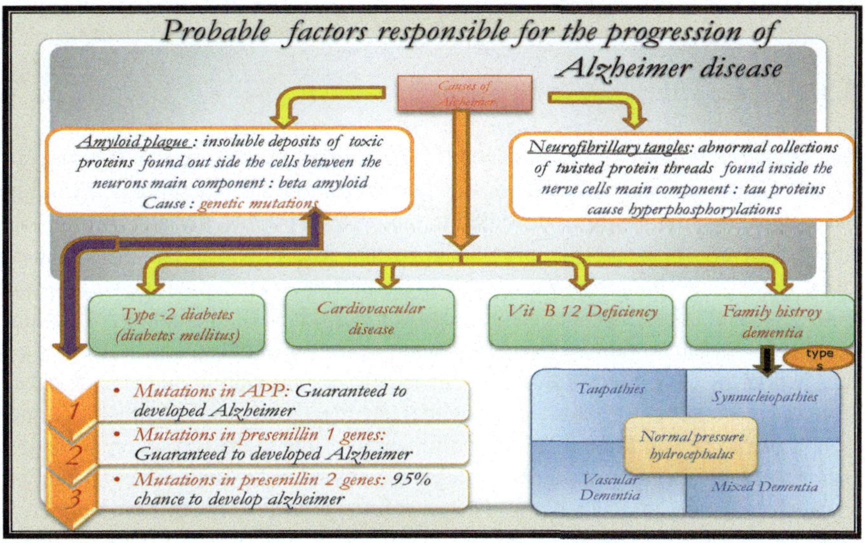

Fig. 2.1 Potential factors responsible for Alzheimer's progression

transformations that transpire from the point of diagnosis to death are faster and complete within a short period of as little as 3 or 4 years (ALZHEIMER'S DISEASE 2008).

The metabolic processes that are essential to keep nerve cells in a healthy state are disrupted during Alzheimer's disease, thereby, ceasing these cells to function and hence finally causing death of these cells. When the metabolic pathways are disrupted, these cells stop functioning in a proper manner, thus, ultimately causing the failure of memory, changes in personality along with disruption of some other daily routine activities. There is a profusion of two unusual structures, viz., amyloid plaques and neurofibrillary tangles, comprising of misfolded proteins. Certain specific regions of the brain that play a key role in the memory process are the predominant ones that get affected because of these unusual structures. The third main trait of AD leading to diminishing cell function and cell death is due to the loss of connections between cells (ALZHEIMER'S DISEASE 2008).

2.3.1 Amyloid Plaques

Amyloid plaques are found in the spaces between the brain's nerve cells. The main component of plaque is a toxic protein peptide or fragment known as beta-amyloid. With aging, these plaques may be present in some of the older individuals; however, the frequency of plaques in the people suffering from AD is significantly higher as compared to those found in aging individuals. However, if these plaques themselves cause AD or are some by-products of these plaques involved in AD is still unknown and is given quite a lot of emphasis in recent times. Some genetic mutations can also cause accelerated production of beta-amyloid, thereby causing rare inherited forms of AD.

2.3.2 Neurofibrillary Tangles

The second trait of AD, also described by Dr. Alzheimer, is neurofibrillary tangles. These refer to anomalous collections of twisted protein threads present within the nerve cells, this deformity is attributed to tau protein. For healthy neurons to function effectively, microtubules play an important role providing support to these cells. These microtubules play a key role in the nutrient transport, along with certain other essential chemicals such as neurotransmitters. In AD, uncharacteristic phosphorylation of tau protein is observed as a large number of phosphate molecules get attached to the tau protein. Owing to this hyperphosphorylation, tau gets detached from microtubules and comes closer to other tau threads, thereby causing the formation of paired helical filaments which ultimately results in the formation of tangles inside the cell. The microtubules disintegrate, and hence, the entire internal support network collapses, and hence, these nerve cells lose their communication potential (ALZHEIMER'S DISEASE 2008).

2.4 Other Causes of Alzheimer's Disease

2.4.1 Genetic Mutations and Family History Are Also a Cause of Alzheimer's Disease

There are rare cases of Alzheimer (less than unitary percent) buildup owing to mutations in any three of the specific genes that play a key role in AD (Bekris et al. 2010). Any change in the normal sequence of the chemical pairs of the bases that make up the genes is known as genetic mutation. These mutations engage the genes for the amyloid precursor protein (APP) and the genes for the presenilin 1 and presenilin 2 proteins. Individuals carrying mutations in these genes are definite to develop AD, particularly individuals with mutation in presenilin 2 genes having as high as 95% chances of developing it (Goldman et al. 2011). Thus, people having mutations in either of the three genes are predisposed to develop AD usually before the age of 65 (Alzheimer's Association 2015), in rarest cases the disorder occurring at an early age, say 30.

2.4.2 Family History

There is an important role of family history in the development of this disease though it is not mandatory for an individual to develop it. However, people, who have their parents, brothers, or sisters suffering from AD, are more liable to suffer from this disorder as compared to the ones who do not have any connection with this disease (Green et al. 2002; Loy et al. 2014). The people having more than one first-degree relative with Alzheimer have a further high risk of this disorder (Lautenschlager et al. 1996). When diseases are inherited in families, either hereditary (genetic) or acquired by shared environment and lifestyle factors or all of these may have an important role in the incidence of disease.

2.4.3 Cardiovascular Disease Risk Factors

The heart in its proper functional state ensures that adequate amount of blood is pumped through the blood vessels and it is the duty of healthy blood vessels to make sure that there is an adequate supply of oxygen and other nutrients to the brain. Many factors that amplify the threat of cardiovascular diseases are also somehow connected with an increased risk of dementia. These factors include smoking (Anstey et al. 2007; Pendlebury and Rothwell 2009; Rusanen et al. 2011), midlife obesity (Anstey et al. 2011; Rönnemaa et al. 2011; Loef and Walach 2013), and diabetes (Pendlebury and Rothwell 2009; Wu et al. 2008; Reitz et al. 2011; Gudala et al. 2013; Vagelatos and Eslick 2013). There are certain evidences supporting the fact that impaired glucose processing also somehow leads to an increased dementia risk (Launer et al. 2000; Sajeev et al. 2014; Ninomiya et al. 2011). There is also growing evidence implicating midlife hypertension (Rönnemaa et al. 2011;

Ninomiya et al. 2011; Debette et al. 2011) and midlife high cholesterol levels as risk factors (Solomon et al. 2009; Meng et al. 2014).

Alternatively, the factors which shield the heart may also somehow protect the brain, thereby reducing the accountability of Alzheimer and progression of other dementias; physical activity is one of such factors (Reitz et al. 2011; Willis et al. 2012; Harrington et al. 2014).

2.4.4 Vitamin B12 Deficiency or Is It Alzheimer's?

Owing to the essentiality of vitamins in a variety of biochemical processes, they play a key role in maintenance of good health. Thus, there is a possibility that these vitamins can play a major role in the possible prevention and treatment of these dementias. Hence, there is a growing research underlying the relation between vitamins and dementias (Koseoglu 2011).

Several studies have shown that B vitamins deficiency, including thiamine (B1), riboflavin (B2), niacin (B3), pyridoxine (B6), folate (B9), and cobalamin (B12), is somehow linked with cognitive dysfunction. Recently, vitamins D and K are also found to be associated with cognitive functioning. Vitamin D has also been found to be essential for development of healthy brain and proper working; sufficient amounts of vitamin D protect brain cells, thereby reducing inflammation. Brain also requires vitamin K for normal development and proper functioning; many reports have suggested a role of this vitamin in Alzheimer's pathogenesis.

Lack of some of B vitamins, including thiamine (B1), riboflavin (B2), niacin (B3), pyridoxine (B6), folate (B9), and cobalamin (B12), is linked with cognitive dysfunction in many observational studies (Riedel et al. 1998); especially deficiency of thiamine is linked with lactic acid accumulation, decline in oxygen uptake, impairment in cholinergic activity, and reduction in transketolase activity, thereby leading to memory loss and other cognitive function (Willis et al. 2012; Micheau et al. 1985). Cobalamin is mandatory for generation of neurons and its deficiency may lead to nervous system degeneration (Herrmann and Obeid 2007). Folate and vitamin B12 are mandatory cofactors for the methionine/Hcy cycle in the brain.

2.4.5 Diabetes Mellitus And Alzheimer's Disease

Diabetes mellitus (DM) is also someway related with cognitive decline (Elias et al. 1997; Gregg et al. 2000; Knopman et al. 2001; Fontbonne et al. 2001; Arvanitakis et al. 2004; Yaffe et al. 2004; Logroscino et al. 2004; Elias et al. 2005) in anatomical measures of brain aging which manifest as a whole-brain atrophy and hippocampal atrophy (Araki et al. 1994; den Heijer et al. 2003; Schmidt et al. 2004) and with an increased risk of stroke and vascular dementia (VaD) (Luchsinger et al. 2001; Hassing et al. 2002; Haan et al. 2003; Schnaider Beeri et al. 2004; Xu et al. 2004). The connection sandwiched between diabetes and Alzheimer is less obvious as there were certain studies that find a connection (Schnaider Beeri et al. 2004;

Leibson et al. 1997; Brayne et al. 1998; Ott et al. 1999; Peila et al. 2002) while others did not (Yoshitake et al. 1995; MacKnight et al. 2002). These variations in studies can be owed to the differences in age groups, sexes, ethnicity, and also the risk factor pattern of varying population profile. These are some criteria that can be used to define DM, dementia, VaD, and AD. It was lately found that there were deposits of amylin in the brain tissues of patients suffering from type-2 diabetes (T2D) and cognitive impairment (Jabir et al. 2014). Amylin is a hormone that is synthesized and cosecreted along with insulin by pancreatic β-cells. The individuals with obesity or prediabetic insulin resistance have an excessive secretion of amylin, and this excess amylin oligomerizes, thereby causing inflammation in pancreatic islets and contributing to T2D development (Kamal et al. 2014).

Individuals with type-2 diabetes are also susceptible to cerebrovascular injury and cognitive decline (Ott et al. 1999; Srodulski et al. 2014; Craft 2009; Nelson et al. 2009; Luchsinger 2012). Human amylin hormone can be toxic if it accumulates in tissues which may drop back in its function through oligomerization and amyloid formation. For example, the majority of patients with T2D have copious deposition of amylin amyloid in the pancreas (Bell 1959; Westermark 1972; Höppener et al. 2000). Pancreatic islets, if showing elevated β-cell apoptosis encouraging for amylin amyloid and decreased β-cell region, suggest a role of amylin amyloid in the formation of T2D (Jurgens et al. 2011). Amylin is synthesized and cosecreted along with insulin by pancreatic β-cells (Kahn et al. 1990) and plays a composite role in regulating the peripheral energy balance. However, some of the metabolic effects of insulin are opposite to those of amylin (Leighton and Cooper 1988; Molina et al. 1990; Zierath et al. 1992; Westermark et al. 2011).

2.4.5.1 Other Alzheimer's Symptoms

In Alzheimer's disease, some degenerative changes occur that influence areas of the brain playing a key role in controlling thought process and memory along with language, thus leading to signs and symptoms in the behavior of individuals. In AD, very often, physical activities such as control of bowel and bladder are also influenced. However, the problem with AD is that there is far high variability in the symptoms on the individualistic level and also the speed of development varies among individuals to a great extent. The symptoms have different durations for different individuals and also the changes in behavior vary to some extent. Typically there is a slow development of the symptoms of Alzheimer's disease; the period between the onset of the disease and death spanning 5–20 years (Alzheimer Society of Canada 2012).

Alzheimer's disease is the most universal cause of dementia among old age individuals. Dementia is the loss of cognitive functions, viz., thinking, recalling, reasoning, and behavioral abilities, to such a degree that it hinders a person's normal everyday life and routine activities. It can range from moderate stage, where only functioning of an individual is affected, to a most severe stage, when the individual becomes completely dependent on others for the normal routine activities.

2.4.5.2 Stages of Alzheimer's

There are four phases that illustrate this devastating disease. It is not mandatory that all the individuals suffering from AD have a definite pattern of these stages and symptoms, but certainly, there will be a progressive decline of the patient's cognitive function. AD generally lasts from 3 to 20 years, cognitive function steadily decreasing during this span of time. In many cases of Alzheimer, other complications such as pneumonia, heart failure, or infections are the reasons behind death.

During early stages of Alzheimer's, patients do not have memory impairment nor do they have any clear signs of other cognitive decline. But the patients during the middle stages sense as if they have memory lapses, viz., they are unable to recall common words or names or the place of bike or car keys, sunglasses, or other everyday objects. The problem that lies is that all these lapses in normal activities are not prevalent during medical examination rather only apparent to friends and family.

2.4.5.3 Common Inabilities Include

1. Word or name-finding difficulty in family and close friends.
2. Reduced ability to recall names of unfamiliar people.
3. Decreased knowledge of recent events.
4. Reduced potential of managing complicated tasks such as marketing activities, finances, etc.
5. Short memory of personal history.
6. The affected person may seem acquiescent and quiet, particularly in socially or mentally demanding scenarios.
7. Unsure about the dates and weekdays coupled with confusion to recall at what place they are.
8. Usually require support with eating or using the toilet.

2.5 Alzheimer's Disease Results in Severe Cognitive Decline During Its Advanced Stage

Memory problems keep on worsening in combination with striking changes in personality which force affected people for widespread aid in normal day-to-day activities. At this stage, individuals may:

1. Face difficulty in recalling even the most recent activities as well as the environment.
2. Reminisce their personal history faultily, although they usually remember their own name.
3. Seldom forget spouse names but normally can differentiate familiar and unfamiliar faces.
4. Unable to get dressed properly on their own; say putting shoes on wrong feet without supervision.
5. Have increasing episodes of urinary or fecal incontinence.

6. Faces striking changes in personality coupled with changes in behavioral aspects, viz., suspiciousness and delusions, hallucinations (seeing or hearing things that are not really there); or neurotic, repetitive behaviors such as hand-wringing or tissue shredding.
7. Loss of walking ability without support, sitting ability unless assisted, and smiling ability coupled with the ability to hold head in an upright position. Reflexes become feeble and muscle becomes rigid. Swallowing ability is also affected and impairment is observed (Alzheimer's Association 2004, 2007).

2.6 Prevention and Treatment of Alzheimer's Disease

Case-control and longitudinal studies suggest certain approaches that can be undertaken to reduce risk factors that are associated with AD and other dementias (Islam et al. 2017a). There is some evidence that vitamin E supplements are beneficial (and possibly also vitamin C). It is thought that AD can be prevented by a well-balanced diet including sufficient folate and vitamin B12, physical and mental alertness, evading injuries to head, alcohol intake in moderate amounts, bypassing surplus exposure to aluminum, avoid smoking, and keeping elevated blood pressure, diabetes, and other vascular risk factors under control (Fig. 2.2).

Currently, data is not sufficient to sustain the usage of agents that lower the cholesterol level, anti-inflammatory agents, proteins (Islam et al. 2017b), estrogen, caffeine, or ginkgo biloba to prevent AD and dementia, but from future perspective, some of these agents are under clinical trials, and hence these recommendations can be altered with time. Anti-amyloid vaccination and gene therapy are some of the

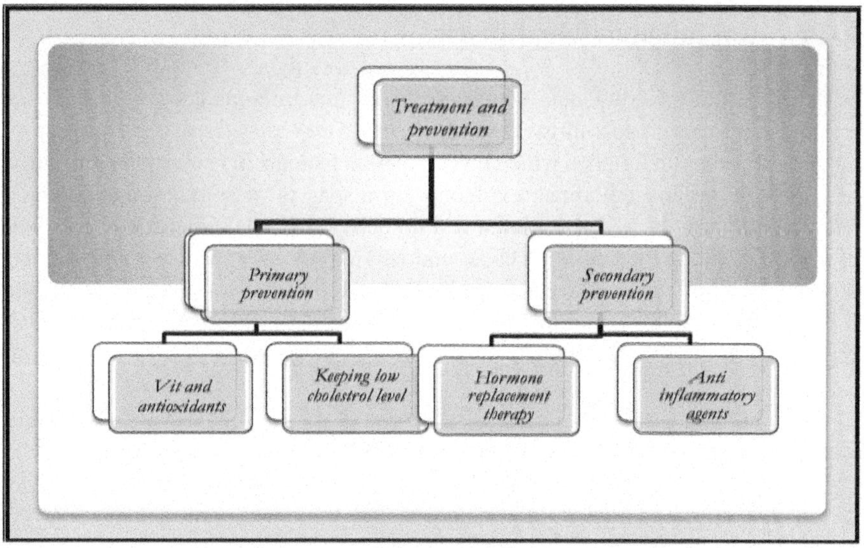

Fig. 2.2 Available prevention options for the management of Alzheimer's disease

approaches that appear to be very promising from future perspective. Addressing groups that lie in high-risk zones, i.e., individuals with mild cognitive impairment, may amplify the advantage of preventative approaches. This neurodegenerative diseases affects many people, and currently no cure therapy is available. It could be of great help to the society if AD and related conditions could be prevented or/and treated. Enzymatic approach has also been used in the prevention of AD (Islam and Tabrez 2017), specific inhibitors of cholinesterase are potential tool for AD management (Jabir et al. 2018). Primary prevention refers to preventing a disease from occurring so people currently without AD remain healthy. This can be achieved by delaying the onset of AD. In secondary prevention, the disorder is hindered by means of early detection, diagnosis, and treatment by using polyphenols and nanotechnological therapeutic approaches (Ashraf et al. 2015; Greig et al. 2013).

Tertiary prevention mitigates the aftereffects of disease by containing disability and dependency and maintaining an acceptable quality of life. Such studies are currently going on and some risk factors are strongly associated with AD and are amendable, so more proven preventative approaches are probably not far away.

2.6.1 Cholesterol Lowering

Cholesterol is required for proper functioning of brain, however, when excess cholesterol was fed to brain of rabbits pathological changes similar to AD were observed. The exact mechanism by which cholesterol-lowering drugs prevents AD is not known but some studies suggest that these drugs affect the processing of amyloid precursor protein (Prof. Colin 2002).

2.6.2 Anti-inflammatory Agents

In several studies, individuals who were given anti-inflammatory agents for the treatment of rheumatoid arthritis have shown to have a reduced risk of AD. Recently, a large study was undertaken which involved 8000 patients. It showed that the intake of non-steroidal anti-inflammatory agents for a span of more than 2 years diminished the chances of AD by 80% but had no effect on vascular dementia (Int' Veld et al. 2001). Inflammatory changes around amyloid plaques (McGeer and McGeer 1995) are observed which can be seen as the potent site of action for anti-inflammatory drugs. Owing to vast side effects of these anti-inflammatory drugs, prevention trials are a prerequisite.

2.6.3 Hormone Replacement Therapy (HRT)

Estrogen is quite ineffective in established AD treatment (Mulnard et al. 2000; Henderson et al. 2000). There are viable methods through which estrogen may wield a shielding effect; as the brain cells have estrogen receptors and estrogen enhances brain synapses in animal models, thereby increasing the blood flow to the brain. It has also been reported that if estrogen is used for a long time, there is a general trend of more cardiovascular events in women (Cholerton et al. 2002); therefore, any preventive role for dementia must be in balance with all the other potential risks.

2.6.4 Vitamins/Antioxidants

It has been observed that there is a link between cancer and AD via nitric oxide induced oxidative stress (Aliev et al. 2013). For a preventive measure of AD, it has been shown that a greater antioxidant intake, including vitamin E, is somehow connected with a reduced risk of AD. The Chicago Health and Aging Project during over a 4-year follow-up showed that people taking higher amount of vitamin E from food had lower risk of developing AD up to 70% (Morris et al. 2002). A Portuguese study, based on 54 cases and 54 controls, has suggested that regular intake of coffee like two to three cups a day has a protective effect against AD (Maia and de Mendonca 2002).

According to some studies, there is a high possibility that individuals with head injuries (loss of consciousness) are more liable to AD. For instance, head injury was found to increase the risk of AD tenfold with loss of perception during the Multi-Institutional Research in Alzheimer's Genetic Epidemiology (MIRAGE) study of 2233 AD patients and 14 668 family controls (Guo et al. 2000).

This is one of the present highlighted areas of dementia research and many studies have revealed that elevated homocysteine levels, low folate, or low B12 are risk factors for dementia. Vitamin B coupled with folate food can diminish the levels of homocysteine; however, there is no proof of the exact advantage obtained from these approaches, and a similar scenario applies for vitamin B6 or thiamine supplements. Currently, the intake of (nontoxic) oral B12 supplemented with foliate appears to be defensive but it may be ineffective; though a good diet may serve the purpose in a better way. It is still unknown if the homocysteine level is a risk factor on its own, though deficiencies of B12 and folate lead to higher levels of homocysteine which is an independent risk factor for vascular diseases (Bots et al. 1997). Aluminum is neurotoxic (Rifat et al. 1990), but there is no concrete evidence regarding it causing AD. Drinking water is the major source of our exposure as it varies in respect of the aluminum concentrations. There are certain overseas studies which show enough evidence for the connection between AD and concentration of aluminum in drinking water. Most cardiovascular risk factors, including hypertension and diabetes, amplify the risk of vascular dementia cognitive impairment and AD in general (Skoog 1994; Skoog et al. 1996). Also, vascular disease intensifies the severity of dementia; particularly systolic hypertension at the age of 40 seems to be a potential risk factor if compared with later part of life (Storey and Masters 1995).

At this stage, it seems reasonable to treat cardiovascular risk factors to not only avoid dementia but also other ailments such as heart attacks and strokes. When homocysteine levels are reduced, it somehow helps in reducing the cardiovascular risk. Several studies have found an involvement between a history of depression and an amplified risk of developing AD (Kokmen et al. 1991; Speck et al. 1995; Devanand et al. 1996; Steffens et al. 1997). In a pooled analysis of case-control studies of AD, Jorm and colleagues found depression was a risk factor for AD (Jorm et al. 1991).

2.6.5 Targeting Prevention

The aforementioned approaches can be undertaken for population in general or in high-risk groups especially for those having high blood pressure or earlier suffering from strokes. Another category of prevention, called indicated prevention, is suitable for individuals having early or minimal symptoms of cognitive decline (Jorm 2002). In this respect, mild cognitive impairment (MCI) is characterized by mild cognitive loss and is a major risk factor for AD, with about 15% converting to AD each year (Ritchie and Touchon 2000). If the above approaches are undertaken in high-risk groups, these could certainly be more effective. In addition, medications that are undertaken to treat established AD may inhibit or diminish development from MCI (mild cognitive impairment) to AD and in this way itself will act as preventive.

2.7 Conclusion

With the existing knowledge of preventive measures, if not all but certain cases of AD can definitely be prevented. There is little harm and significant impending benefits in taking vitamin E supplements and possibly also vitamin C along with balanced diet containing adequate folate and B12. AD can also be prevented to some extent by physical and mental alteration, avoiding head injury, continuing moderate alcohol intake, keeping off excess aluminum exposure, stopping smoking, and keeping blood pressure and other vascular risk factors controlled. However, much more data are required from longer and larger prospective trials that involve use of new agents and approaches to efficiently treat AD and the other dementias in upcoming years.

Summary Key Points

1. Currently, there is no interference that is complete on its own to prevent Alzheimer's disease.
2. Administration of anti-inflammatory drugs, vitamin E, and vitamin C and containment of homocysteine levels by giving folate supplements, preventing injuries to head, maintenance of physical and leisure activities, and control on cardiovascular risk factors may help in reducing the probability of Alzheimer's disease.
3. From future perspective, anti-amyloid approaches, viz., vaccination, hold an important role in Alzheimer treatment.

Acknowledgments We greatly acknowledge facility provided by Aligarh Muslim University, Aligarh, India; University of Quebec, Trois-Rivieres, Canada; King Fahd Medical Research Center (KFMRC), Jeddah; King Abdulaziz University, Jeddah; and King Saud University, Riyadh, Saudi Arabia for carrying out this work.

References

Aliev G, Obrenovich ME, Tabrez S, Jabir NR, Reddy P, Li Y, Burnstock G, Cacabelos R, Kamal MA (2013) Link between cancer and Alzheimer disease via oxidative stress induced by nitric oxide-dependent mitochondrial DNA over-proliferation and deletion. Oxidative Med Cell Longev 2013:962984

Alzheimer Society of Canada (2012) Stages of Alzheimer's disease (what happens). Toronto. www.alzheimer.ca/en/About-dementia/Alzheimer-s-disease/Stages-of-Alzheimer-s-disease

Alzheimer's Association (2004) All rights reserved. Stages of Alzheimer's Disease Fact Sheet

Alzheimer's Association © (2007) All rights reserved. 1.800.272.3900|www.alz.org

Alzheimer's Association (2015) Alzheimer's disease facts and figures. Alzheimers Dement 11(3):332

ALZHEIMER'S DISEASE Unraveling the Mystery, U.S. Department of Health and Human Services NIH Publication Number: 08–3782 September 2008

Alzheimer's Disease Education & Referral (ADEAR) Center (2012) A Service of the National Institute on aging, National Institutes of Health, U.S. Department of Health and Human Services. NIH Publication No. 11-6423

Anstey KJ, von Sanden C, Salim A, O'Kearney R (2007) Smoking as a risk factor for dementia and cognitive decline: a meta-analysis of prospective studies. Am J Epidemiol 166(4):367–378

Anstey KJ, Cherbuin N, Budge M, Young J (2011) Body mass index in midlife and late-life as a risk factor for dementia: a meta-analysis of prospective studies. Obes Rev 12(5):e426–e437. Review

Araki Y, Nomura M, Tanaka H et al (1994) MRI of the brain in diabetes mellitus. Neuroradiology 36:101–103

Arvanitakis Z, Wilson RS, Bienias JL, Evans DA, Bennett DA (2004) Diabetes mellitus and risk of Alzheimer disease and decline in cognitive function. Arch Neurol 61:661–666

Ashraf GM, Tabrez S, Jabir NR, Firoz CK, Ahmad S, Hassan I, Alexiou A, Kamal MA (2015) An overview on global trends in nano-technological approaches for Alzheimer therapy. Curr Drug Metab 16(8):719–727

Ashraf GM, Chibber S, Mohammad ZSK, Tabrez S, Ahmad A, Shakil S, Mustaq G, Baeesa S, Kamal MA (2016) Recent updates on the association between Alzheimer disease and vascular dementia. Med Chem 12(3):226–237

Bekris LM, Yu CE, Bird TD, Tsuang DW (2010) Genetics of Alzheimer disease. J Geriatr Psychiatry Neurol 23(4):213–227

Bell ET (1959) Hyalinization of the islets of Langerhans in nondiabetic individuals. Am J Pathol 35(4):801–805

Bots ML, Launer LJ, Lindemans J, Hofman A, Grobbee DE (1997) Homocysteine, atherosclerosis and prevalence cardiovascular disease in the elderly: the Rotterdam study. J Intern Med 242:339–347

Brayne C, Gill C, Huppert FA et al (1998) Vascular risks and incident dementia: results from a cohort study of the very old. Dement Geriatr Cogn Disord 9

Byrd C, RN, MSN, ANP, ACNP, CNRN (2006) Normal Pressure Hydrocephalus DEMENTIA'S CAUSE. Nurse Pract 31(7):28–29

Cholerton B, Gleason CE, Baker LD, Asthana S (2002) Estrogen and Alzheimer's disease: the story so far. Drugs Aging 19:405–427

Craft S (2009) The role of metabolic disorders in Alzheimer disease and vascular dementia: two roads converged. Arch Neurol 66(3):300–305

Debette S, Seshadri S, Beiser A, Au R, Himali JJ, Palumbo C et al (2011) Midlife vascular risk factor exposure accelerates structural brain aging and cognitive decline. Neurology 77:461–468
den Heijer T, Vermeer SE, van Dijk EJ et al (2003) Type 2 diabetes and atrophy of medial temporal lobe structures on brain MRI. Diabetologia 46:1604–1610
Devanand DP, Sano M, Tang MX, Taylor S, Gurland BJ, Wilder D et al (1996) Depressed mood and the incidence of Alzheimer's disease in the elderly living in the community. Arch Gen Psychiatry 53:175–182
Elias PK, Elias MF, D'Agostino RB et al (1997) NIDDM and blood pressure as risk factors for poor cognitive performance: the Framingham study. Diabetes Care 20:1388–1395
Elias MF, Elias PK, Sullivan LM, Wolf PA, D'Agostino RB (2005) Obesity, diabetes and cognitive deficit: the Framingham heart study. Neurobiol Aging 26(Suppl 1):11–16
Fontbonne A, Berr C, Ducimetiere P, Alperovitch A (2001) Changes in cognitive abilities over a 4-year period are unfavorably affected in elderly diabetic subjects: results of the epidemiology of vascular aging study. Diabetes Care 24:366–370
Goldman JS, Hahn SE, Bird T (2011) Genetic counseling and testing for Alzheimer disease: joint practice guidelines of the American College of Medical Genetics and the National Society of Genetic Counselors. Genet Med 13:597–605
Green RC, Cupples LA, Go R, Benke KS, Edeki T, Griffith PA, Williams M, Hipps Y, Graff-Radford N, Bachman D, Farrer LA (2002) Risk of dementia among white and African American relatives of patients with Alzheimer disease. JAMA 287:329–336
Gregg EW, Yaffe K, Cauley JA et al (2000) Is diabetes associated with cognitive impairment and cognitive decline among older women? Study of osteoporotic fractures research group. Arch Intern Med 160:174–180
Greig NH, Kamal MA, Jabir NR, Tabrez S, Nasim FH, Abuzenadah AM, Aliev G (2013) Specific cholinesterase inhibitors: A potential tool to assist in management of Alzheimer disease. Front Drug Des Discovery 6:366–386
Gudala K, Bansal D, Schifano F, Bhansali A (2013) Diabetes mellitus and risk of dementia: a meta-analysis of prospective observational studies. Diabetes Investig 4(6):640–650
Guo Z, Cupples LA, Kurz A, Auerbach SH, Volicer L, Chui H et al (2000) Head injury and the risk of AD in the MIRAGE study. Neurology 54:1316–1323
Haan MN, Mungas DM, Gonzalez HM, Ortiz TA, Acharya A, Jagust WJ (2003) Prevalence of dementia in older latinos: the influence of type 2 diabetes mellitus, stroke and genetic factors. J Am Geriatr Soc 51:169–177
Hakim S, Adams RD (1965) The special clinical problem of symptomatic hydrocephalus with normal cerebrospinal fluid pressure. Observations on cerebrospinal fluid hydrodynamics. J Neurol Sci 2:307–327
Harrington M, Weuve J, Jackson JW, Blacker D (2014) Physical activity. The AlzRisk Database. Alzheimer Research Forum. Available at: http://www.alzrisk.org
Hassing LB, Johansson B, Nilsson SE et al (2002) Diabetes mellitus is a risk factor for vascular dementia, but not for Alzheimer's disease: a population-based study of the oldest old. Int Psychogeriatr 14:239–248
Henderson VW, Paganini-Hill A, Miller BL, Elble RJ, Reyes PF, Shoupe D et al (2000) Estrogen for Alzheimer's disease in women: randomized, double-blind, placebo-controlled trial. Neurology 54:295–301
Herrmann W, Obeid R (2007) Biomarkers of folate and vitamin B(12) status in cerebrospinal fluid. Clin Chem Lab Med 45(12):1614–1620, ISSN 1434-6621
Hickey J (2003) The Clinical Practice of Neurological and Neurosurgical Nursing, 5th edn. Williams & Wilkins, Philadelphia
Höppener JWM, Ahren B, Lips CJM (2000) Islet amyloid and type 2 diabetes mellitus. N Engl J Med 343(6):411–419
Igor OK (2014) Review article, Alzheimer's disease: a clinical and basic science review. MSRJ 4:24

Int' Veld BA, Ruitenberg A, Hofman A, Launer LJ, van Duijn CM, Stijnen T et al (2001) Nonsteroidal anti-inflammatory drugs and the risk of Alzheimer's disease. N Engl J Med 345:1515–1521

Islam B, Tabrez S (2017) Management of Alzheimer's disease- an insight of the enzymatic and other novel potential targets. Int J Biol Macromol 97:700–709

Islam B, Khan MS, Jabir NR, Kamal MA, Tabrez S (2017a) Elucidating treatment of Alzheimer's disease *via* different receptors. Curr Top Med Chem 17(12):1400–1407

Islam B, Zaidi SK, Kamal MA, Tabrez S (2017b) Exploration of various proteins for the treatment of Alzheimer's disease. Curr Drug Metab 18(9):808–813

Jabir NR, Kamal MA, Abuzenadah AM, Gan SH, Alama MN, Baeesa SS, Tabrez S (2014) Alzheimer's and type 2 diabetes treatment *via* common enzyme targeting. CNS Neurol Disord Drug Targets 13:299–304

Jabir NR, Firoz CK, Baeesa SS, Ashraf GM, Kamal MA, Tabrez S (2015) Synopsis on the linkage of Alzheimer's and Parkinson's disease with Chronic diseases. CNS Neurosci Ther 21(1):1–7

Jabir NR, Khan FR, Tabrez S*. (2018) Cholinesterase targeting by polyphenols: a therapeutic approach for the treatment of Alzheimer's disease. CNS Neurosci Ther 24:753–762

Jorm AF (2002) Prospects for the prevention of dementia. Aust J Ageing 21:9–13

Jorm AF, van Duijn CM, Chandra V, Fratiglioni L, Graves AB, Heyman A et al (1991) Psychiatric history and related exposures as risk factors for Alzheimer's disease: a collaborative re-analysis of case control studies. Int J Epidemiol 20(suppl. 2):S43–S47

Jurgens CA, Toukatly MN, Fligner CL, Udayasankar J, Subramanian SL, Zraika S, Aston-Mourney K, Carr DB, Westermark P, Westermark GT, Kahn SE, Hull RL (2011) β-cell loss and β-cell apoptosis in human type 2 diabetes are related to islet amyloid deposition. Am J Pathol 178(6):2632–2640

Kahn SE, D'Alessio DA, Schwartz MW, Fujimoto WY, Ensinck JW, Taborsky GJ Jr, Porte D Jr (1990) Evidence of cosecretion of islet amyloid polypeptide and insulin by beta-cells. Diabetes 39(5):634–638

Kamal MA, Priyamvada S, Arivarasu NA, Jabir NR, Tabrez S, Greig NH (2014) Linking Alzheimer's disease and type 2 diabetes mellitus *via* aberrant insulin signalling and inflammation. CNS Neurol Disord Drug Targets 13:338–346

Knopman D, Boland LL, Mosley T et al (2001) Cardiovascular risk factors and cognitive decline in middle-aged adults. Neurology 56:42–48

Kokmen E, Beard CM, Chandra V, Offord KP, Schoenberg BS, Ballard DJ (1991) Clinical risk factors for Alzheimer's disease: a population-based case-control study. Neurology 41:1393–1397

Emel Koseoglu (2011). The relations between the vitamins and Alzheimer Dementia, Alzheimer's disease pathogenesis-core concepts, shifting paradigms and therapeutic targets, Dr. Suzanne De La Monte (Ed.),

Launer LJ, Ross GW, Petrovitch H, Masaki K, Foley D, White LR et al (2000) Midlife blood pressure and dementia: the Honolulu- Asia aging study. Neurobiol Aging 21(1):49–55

Lautenschlager NT, Cupples LA, Rao VS, Auerbach SA, Becker R, Burke J et al (1996) Risk of dementia among relatives of Alzheimer's disease patients in the MIRAGE study: what is in store for the oldest old? Neurology 46(3):641–650

Leibson CL, Rocca WA, Hanson VA et al (1997) Risk of dementia among persons with diabetes mellitus: a population-based cohort study. Am J Epidemiol 145:301–308

Leighton B, Cooper GJS (1988) Pancreatic amylin and calcitonin gene-related peptide causes resistance to insulin in skeletal muscle in vitro. Nature 355(6191):632–635

Loef M, Walach H (2013) Midlife obesity and dementia: meta-analysis and adjusted forecast of dementia prevalence in the United States and China. Obesity (Silver Spring) 21(1):E51–E55

Logroscino G, Kang JH, Grodstein F (2004) Prospective study of type 2 diabetes and cognitive decline in women aged 70-81 years. BMJ 328:548

Loy CT, Schofield PR, Turner AM, Kwok JBJ (2014) Genetics of dementia. Lancet 383:828–840

Luchsinger JA (2012) Type 2 diabetes and cognitive impairment: linking mechanisms. J Alzheimers Dis 30(2):S185–S198

Luchsinger JA, Tang MX, Stern Y, Shea S, Mayeux R (2001) Diabetes mellitus and risk of Alzheimer's disease and dementia with stroke in a multiethnic cohort. Am J Epidemiol 154:635–641

MacKnight C, Rockwood K, Awalt E, McDowell I (2002) Diabetes mellitus and the risk of dementia, Alzheimer's disease and vascular cognitive impairment in the Canadian study of health and aging. Dement Geriatr Cogn Disord 14:77–83

Maia L, de Mendonca A (2002) Does caffeine intake protect from Alzheimer's disease? Eur J Neurol 9:377–382

McGeer PL, McGeer EG (1995) The inflammatory response system of brain: implications for therapy of Alzheimer's and other neurodegenerative diseases. Brain Res Rev 21:195–218

Meng XF, Yu JT, Wang HF, Tan MS, Wang C, Tan CC, Tan L (2014) Midlife vascular risk factors and the risk of Alzheimer's disease: a systematic review and meta-analysis. J Alzheimers Dis 42(4):1295–1310

Micheau J, Durkin TP, Destrade C, Rolland Y, Jaffard R (1985) Chronic administration of sulbutiamine improves long term memory formation in mice: possible cholinergic mediation. Pharmacol Biochem Behav 23(2):195–198, ISSN 0091-3057

Molina JM, Cooper GJS, Leighton B, Olefsky JM (1990) Induction of insulin resistance in vivo by amylin and calcitonin gene-related peptide. Diabetes 39(2):260–265

Morris MC, Evans DA, Bienias JL, Tangney CC, Bennett DA, Aggarwal N et al (2002) Dietary intake of antioxidant nutrients and the risk of incident Alzheimer's disease in a bi-racial community study. JAMA 287:3230–3237

Mulnard RA, Cotman CW, Kawas C, van Dyck CH, Sano M, Doody R et al (2000) Estrogen replacement therapy for treatment of mild to moderate Alzheimer's disease: a randomized controlled trial. Alzheimer's Disease Cooperative Study. JAMA 283:1007–1015

Nelson PT, Smith CD, Abner EA, Schmitt FA, Scheff SW, Davis GJ, Keller JN, Jicha GA, Davis D, Wang-Xia W, Hartman A, Katz DG, Markesbery WR (2009) Human cerebral neuropathology of Type 2 diabetes mellitus. Biochim Biophys Acta 1792(5):454–469

Ninomiya T, Ohara T, Hirakawa Y, Yoshida D, Doi Y, Hata J et al (2011) Midlife and late-life blood pressure and dementia in Japanese elderly: the Hisayama study. Hypertension 58(1):22–28

Ott A, Stolk RP, Van Harskamp F, Pols HA, Hofman A, Breteler MM (1999) Diabetes mellitus and the risk of dementia: the Rotterdam study. Neurology 53(9):1937–1942

Peila R, Rodriguez BL, Launer LJ (2002) Type 2 diabetes, APOE gene, and the risk for dementia and related pathologies: the Honolulu-Asia aging study. Diabetes 51:1256–1262

Pendlebury ST, Rothwell PM (2009) Prevalence, incidence, and factors associated with pre-stroke and post-stroke dementia: a systematic review and meta-analysis. Lancet Neurol 8(11):1006–1018

Nowotny Petra, Washington University School of Medicine, Saint Louis, Missouri, USA, Jennifer M Kwon, Washington University School of Medicine, Saint Louis, Missouri, USA Alison M Goate, ENCYCLOPEDIA OF LIFE SCIENCES/ & 2001 Nature Publishing Group/www.els.net

Prof. Colin Masters, Professor of Pathology, University of Melbourne, personal communication, April 2002

Reitz C, Brayne C, Mayeux R (2011) Epidemiology of Alzheimer disease. Nat Rev Neurol 7(3):137–152

Riedel WJ, Jorissen B, Nutrients L (1998) Age and cognitive function. Curr Opin Clin Nutr Metab Care 1(6):579–585, ISSN 163-1950

Rifat SL, Eastwood MR, McLachlan DRC, Corey PN (1990) Effect of exposure of miners to aluminium powder. Lancet 336:1162–1165

Ritchie K, Touchon J (2000) Mild cognitive impairment: conceptual basis and current nosological status. Lancet 355:225–228

Rönnemaa E, Zethelius B, Lannfelt L, Kilander L (2011) Vascular risk factors and dementia: 40-year follow-up of a population-based cohort. Dement Geriatr Cogn Disord 31(6):460–466

Rusanen M, Kivipelto M, Quesenberry CP, Zhou J, Whitmer RA (2011) Heavy smoking in midlife and long-term risk of Alzheimer disease and vascular dementia. Arch Intern Med 171(4):333–339

Sajeev G, Weuve J, McQueen MB, Blacker D (2014) Diabetes. The AlzRisk Database Alzheimer Research Forum Available at: http://www.alzrisk.org

Schmidt R, Launer LJ, Nilsson LG et al (2004) Magnetic resonance imaging of the brain in diabetes: the cardiovascular determinants of dementia (CASCADE) study. Diabetes 53:687–692

Schnaider Beeri M, Goldbourt U, Silverman JM et al (2004) Diabetes mellitus in midlife and the risk of dementia three decades later. Neurology 63:1902–1907

Skoog I (1994) Risk factors for vascular dementia: a review. Dementia 5:137–144

Skoog I, Lernfelt B, Landahl S, Palmertz B, Andreasson LA, Nilsson L et al (1996) 15-year longitudinal study of blood pressure and dementia. Lancet 347:1141–1145

Solomon A, Kivipelto M, Wolozin B, Zhou J, Whitmer RA (2009) Midlife serum cholesterol and increased risk of Alzheimer's and vascular dementia three decades later. Dement and Geriatr Disord 28:75–80

Speck CE, Kukull WA, Brenner DE, Bowen JD, McCormick WC, Teri L et al (1995) History of depression as a risk factor for Alzheimer's disease. Epidemiology 6:366–369

Srodulski S, Sharma S, Bachstetter AB, Brelsfoard JM, Pascual C, Xie XS, Saatman KE, Van Eldik LJ, Despa F (2014) Neuroinflammation and neurologic deficits in diabetes linked to brain accumulation of amylin. Mol Neurodegener 9:30. https://doi.org/10.1186/1750-1326-9-30. PubMed PMID: 25149184; PubMed Central

Steffens DC, Plassman BL, Helms MJ, Welsh-Bohmer KA, Saunders AM, Breitner JC (1997) A twin study of late-onset depression and apolipoprotein E e4 as risk factors for Alzheimer's disease. Biol Psychiatry 41:851–856

Storey E, Masters C (1995) Amyloid, aluminium and the aetiology of Alzheimer's disease. MJA 163:256–259

THE DEMENTIAS, Hope through research, National Institute of Neurological Disorders and Stroke National Institute on Aging, NIH publication no: 132252, 2013

U.S. Department of Health and Human Services, National Institutes of Health, National Institute on Aging, NIH Publication No. 11 5441 June 2011

Vagelatos NT, Eslick GD (2013) Type 2 diabetes as a risk factor for Alzheimer's disease: the confounders, interactions, and neuropathology associated with this relationship. Epidemiol Rev 35(1):152–160

Verrees M, Selman W (2004) Management of normal pressure hydrocephalus. Am Fam Phyikian 70:1071–1078

Westermark P (1972) Quantitative studies on amyloid in the islets of Langerhans. Ups J Med Sci 77(2):91–94

Westermark P, Andersson A, Westermark GT (2011) Islet amyloid polypeptide, islet amyloid, and diabetes mellitus. Physiol Rev 91(3):795–826

Willis BL, Gao A, Leonard D, DeFina LF, Berry JD (2012) Midlife fitness and the development of chronic conditions in later life. Arch Intern Med 172(17):1333–1340

Wu W, Brickman AM, Luchsinger J, Ferrazzano P, Pichiule P, Yoshita M et al (2008) The brain in the age of old: the hippocampal formation is targeted differentially by diseases of late life. Ann Neurol 64:698–706

Xu WL, Qiu CX, Wahlin A, Winblad B, Fratiglioni L (2004) Diabetes mellitus and risk of dementia in the Kungsholmen project: a 6-year follow-up study. Neurology 63:1181–1186

Yaffe K, Blackwell T, Kanaya AM, Davidowitz N, Barrett-Connor E, Krueger K (2004) Diabetes, impaired fasting glucose, and development of cognitive impairment in older women. Neurology 63:658–663

Yoshitake T, Kiyohara Y, Kato I et al (1995) Incidence and risk factors of vascular dementia and Alzheimer's disease in a defined elderly Japanese population: the Hisayama study. Neurology 45:1161–1168

Zierath JR, Galuska D, Engstrom A, Johnson KH, Betsholtz C, Westermark P, Wallberg-Henriksson H (1992) Human islet amyloid polypeptide at pharmacological levels inhibits insulin and phorbol ester-stimulated glucose transport in vitro incubated human muscle strips. Diabetologia 35(1):26–31

Diet and Nutrition in Alzheimer's Disease and Healthy Aging

3

Muhammed Bule, Muhammad Ajmal Shah, Ahmed Abdulahi Abdurahman, Malik Saad Ullah, Shahid Shah, Adnan Amin, and Kamal Niaz

Abstract

According to UN report of the world aging, older people are those 60 years of age and over, though others defined those over 85 as "oldest old." Aging is a complex and gradual process which involves degeneration of cells mainly due to cellular redox reactions that result in detrimental biochemical and genetic alterations. Studies have related the risk of age-related diseases with increased level of oxidants. Dietary polyphenols like flavonoids are strong antioxidant that acts through interacting with reactive oxygen species producing reactive metals.

M. Bule
Department of Pharmacy, College of Medicine and Health Sciences, Ambo University, Ambo, Ethiopia

M. A. Shah (✉)
Department of Pharmacognosy, Faculty of Pharmaceutical Sciences, Government College University, Faisalabad, Pakistan
e-mail: ajmalshah@gcuf.edu.pk

A. A. Abdurahman
Food and Nutrition Society of Ethiopia (FoNSE), Addis Ababa, Ethiopia

M. S. Ullah
Department of Pharmaceutical Chemistry, Faculty of Pharmaceutical Sciences, Government College University, Faisalabad, Pakistan

S. Shah
Department of Pharmacy Practice, Faculty of Pharmaceutical Sciences, Government College University, Faisalabad, Pakistan

A. Amin
Department of Pharmacognosy, Faculty of Pharmacy, Gomal University, Dera Ismail Khan, Pakistan

K. Niaz
Department of Pharmacology and Toxicology, Faculty of Bio-Sciences, Cholistan University of Veterinary and Animal Sciences (CUVAS), Bahawalpur, Pakistan

G. M. Ashraf, A. Alexiou (eds.), *Biological, Diagnostic and Therapeutic Advances in Alzheimer's Disease*, https://doi.org/10.1007/978-981-13-9636-6_3

All organisms need organic and inorganic nutrients such as proteins, carbohydrates, lipids, vitamins, minerals, and water. These nutrients have a significant role in regular biological activities like metabolism, growth, and repair. Protein is the most essential among the three macronutrients (i.e. carbohydrates, protein, and fat) responsible for aging. Moreover, bioactive chemicals from plants are important, though not essential, and they have been referred to as "life span essential," since they have beneficial effect in healthy aging. Therefore, dietary interventions along with avoiding risk factors will reduce the risk of age-related degenerative diseases and increase healthy life span among the elderly.

Keywords
Alzheimer disease · Aging · Nutrition · Protein · Minerals · Vitamins · Omega fatty acids

3.1 Introduction

Basically, along with the chronological aging process, aging is a result of both biological and environmental factors, and there is no clear definition of aged and elderly. The United Nations report of the world aging mentioned that older people are those 60 years of age and over, while some studies defined those over 85 as "oldest old." This classification varies widely in different countries around the world in relation to the varying life expectancies (Clements and Carding 2018). Yet, in the year 2050, the estimated population of older adults in world is 2 billion, which is almost 21% of the total population and more than double of the 841 million aged populations in 2013. The number of those aged over 80 will also increase three fold between 2013 and 2050 becoming 392 million (Shlisky et al. 2017). In elderly people, a reduced muscle and body mass, increasing volume of extracellular fluid, as well as decline in body cell mass are widely observed. The cognitive and physical functional status, nutritional and endocrine status, quality of life, and comorbidity in elderly people must be well defined. These changes shall not be perceived as simple imbalance between energy intake-expenditure process, since they are results of complex metabolic processes that also include hormones like growth and sex hormones (Baumgartner 2000; Balagopal et al. 1997).

Alzheimer's disease (AD) is a very prevalent form of dementia that has affected more than 5.4 million people only in the USA. Production and accumulation of the β amyloid (βA) senile plaque is the pathological hallmark of AD. Although various therapeutic approaches have been tested against the production and deposition of βA in the past 20+ years, their clinical efficacy is not promising (Solfrizzi et al. 2017). To date, none of the therapeutic interventions succeeded in averting the onset or progression of AD. On the contrary, optimal use of essential nutrients that aid proper functioning of the brain is considered vital for neuronal health and its normal functioning (Mohajeri et al. 2015). Hence, nutritional approaches are currently the promising options toward managing AD risk factors. In this regard, nutritional pattern like the Mediterranean diet that comprise good proportion of vegetables, fruits, whole grain, legumes, nuts, or n-3 PUFA is the most recommended (Alles et al. 2012). Diet is an important factor among the flexible environmental factors; however, the experimental data from literatures on the impact of individual food item or nutrient on AD is not consistent. This is due to the complex nature of human dietary approach where combination of nutrients and food items are consumed at once (Gu et al. 2010). Alternatively, mechanistic studies, epidemiologic analyses, and randomized controlled trials have laid foundation underlining the neuronal health-promoting effect of docosahexaenoic acid (DHA) and micronutrients such as the vitamin B family and vitamins E, C, and D, during aging. In addition to being essential, these nutrients are cheap, are easy to obtain, and have lesser side effects. Besides, they have established mode of actions and are widely accepted by the public (Mohajeri et al. 2015).

3.2 Overview of Health and Aging

Aging is a complex and gradual process which involves degeneration of cells mainly due to cellular redox reactions that result in detrimental biochemical and genetic alterations. The cellular metabolic reactions produce reactive oxygen species (ROS) and reactive nitrogen species (RNS) that act through oxidative stress (OS) and are known to destroy cellular biomolecules such as proteins, lipids, carbohydrates, and nucleic acids. Therefore, cellular OS is responsible for aging and chronic diseases including diabetes, cardiovascular disease (CVD), cancer, AD, Parkinson's diseases (PD), and other age-related diseases (Shlisky et al. 2017; Thapa and Carroll 2017). The increasing aged population of the world is becoming a major public health burden. One of the age-related public health challenge due to increased life span is a sharp increase of age-related chronic diseases worldwide (Rajaram et al. 2017). A decreased appetite in relation to aging leads to energy imbalance between the physiological energy demands and energy expenditure in older adults. As compared to younger adults of similar level of activity, height, and weight, older adults consume lesser food items (energy), although the body's demand for micronutrients remained high (Mohajeri et al. 2015). The resulting malnutrition is associated with impaired muscle function, immune dysfunction, poor wound healing, anemia, delayed recovering from surgery, impaired muscle function, decreased bone mass, reduced cognitive function, higher hospital and readmission rate, and mortality. Currently, malnutrition is rising among the older population being 16% in those aged 65 and over and 2% among the 85 years and older (Ahmed and Haboubi 2010).

The energy consumed and body weight have impact on the quality and span of old age. Despite the controversies on humans, caloric restriction positively affects life span in different animal species. On the contrary, however, some researchers have suggested mild and excess obesity is associated with lower risk of mortality in older adults. Hence, this debatable topic needs further research to clearly state the impact of energy intake and weight on the age/mortality/among the olds (Shlisky et al. 2017). Mitochondrial functions have significant role in aging and age-related chronic ailments such as AD, and thus it is an important therapeutic target to focus on during intervention. As a result of the process of aging, OS and the level of oxidized molecules rise, and the energy balance in the brain becomes impaired. In this process of aging, mitochondria plays a major role as they are the center of energy production and hence springs ROS in the cells and promotes age-related diseases. Accordingly, various animal model studies on aging and aging studies on human skeletal muscles have reported mitochondrial dysfunction in the brain (Gu et al. 2010). In addition, certain actions that control cellular metabolism are integrated with the molecular mechanisms regulating synaptic functions. For instance, mitochondrial metabolic activities can influence cognitive functions because of the fact that they are involved in some features of synaptic plasticity. Similarly, synaptic plasticity declines whenever there is excess calorie due to increase in ROS formation beyond the buffering capacity of the cell. Thus, reasonable reduction of caloric intake can reduce the risk of oxidative damage in the brain (Gómez-Pinilla 2008).

3.3 Role of Nutrition in Healthy Aging

The estimates of the year 2100 shows that nearly one third of the world population will be 60 years and older (Hardman et al. 2015). However, as the size of population reaching advanced age expands, the public health problems rise due to age-related diseases like coronary heart disease (CHD), stroke, AD, and PD. Majority of these diseases resulted from sustained OS and low-grade inflammation (López-Miranda et al. 2010). According to the WHO, avoiding the main risk factors for the chronic disorders (smoking, lack of exercise, and poor diet) can reduce the risk of stroke, cardiovascular diseases (CVD), and type II diabetes by 80% (Shlisky et al. 2017). Although the effect of caloric restriction is not well defined in human, it has been found robust and reproducible in mitigating age-related conditions as well as extending life span in animal models. The impact of caloric restriction on human aging is the most important area to be studied but it is still a gap in modern biogerontology. Yet, various epidemiological studies and short-term human studies underlined the vital role of caloric restriction in human health (Willcox et al. 2007). Moreover, newer study findings demonstrate that caloric restrictions have an impact on life span that differs based on genotype, sex, and the type of diet. Despite all the controversies, caloric restriction is generally deemed to have beneficial influence on age-related health and health span, and it positively acts across a range of taxa from yeast to humans (Simpson et al. 2017).

Observational studies have related the risk of age-related diseases with increased level of oxidants, and conversely they showed that the association of increase in consumption of antioxidant-containing diet was with reduced chronic disease incidence. Hence, consuming dietary monounsaturated fatty acids (MUFA) and n-3 polyunsaturated fatty acids (n-3 PUFA), in addition to fish, fruits, vegetables, and adequate coffee, can benefit the body (Solfrizzi et al. 2017). Increase in consumption of fruits and vegetables or antioxidant-rich diet comprising vitamins, fiber, carotenoids, and magnesium has been associated with raised serum level of nutrients, decline in OS, and lower C-reactive protein (CRP) levels, which is a cognition-related inflammatory marker. Besides, the vegetable- and fruit-derived flavonoids have a good neuroprotective, cardioprotective, and anti-inflammatory activities (Hardman et al. 2015). Thus, individuals with a risk of CVD shall consume diet low in saturated and trans-fats and high in fruits, vegetables, and whole grains. Functional foods and supplements such as red yeast rice, soy protein, marine-derived omega-3 fatty acids, plant sterols, green tea, and probiotic yogurt are helpful in statin-intolerant patients who failed to attain target Low density lipoprotien (LDL) level (Hunter and Hegele 2017; Bule et al. 2018). Pertaining to the type of diet, people who comply with the Mediterranean diet style benefit in the long term through reducing the chance of cognitive decline from normal to mild cognitive impairment (MCI) and then to dementia. The Mediterranean diet mainly consists of fruits, vegetables, nuts, cereals, legumes, fish, and olive oil, with a moderate intake of alcohol and lower consumption of red meat and poultry (Hardman et al. 2015). Furthermore, some animal studies have also recommended that membranes with higher MUFA content are less affected by ROS and protect aged cells. In this regard, olive oil has beneficial effect against OS due to its high oleic acid, which is enriched in lipoproteins and cellular membranes (López-Miranda et al. 2010). A reduced incidence of MCI, lesser incidence of vascular events, as well as lower risk of mortality have been associated to adherence to Mediterranean diet style (Hardman et al. 2015; Prinelli et al. 2015).

3.4 Role of Vitamins and Minerals in Health

In order to sustain life, all organisms need organic and inorganic nutrients such as proteins, carbohydrates, lipids, vitamins, minerals, and water. These nutrients have significant role in regular biological activities like metabolism, growth, and repair. Unlike other nutrients vitamins and minerals can't serve direct energy production but they act as cofactors in a variety of biological functions including hormonal signaling and mitochondrial energy metabolism. We obtain the minerals and vitamins via diet since human body can't synthesize them. In case of deficiency of vitamins and minerals, a range of cellular activities will be halted resulting in a number of deficiency diseases (Lee et al. 2015). For instance, the deficiency of vitamin B6, B12, and folate causes rise in homocysteine, which is a precursor of methionine and cysteine, concentration via various pathways. Homocysteine is in charge of AD pathways through vascular mechanisms because it is active in brain tissue. Therefore, through reducing homocysteine level in the brain, increased intake of folate is believed to be associated with minimal risk of stroke (Luchsinger and Mayeux 2004). On the other hand, strong antioxidants like carotenoids, vitamin C, and vitamin E are potent electron donors that quench radical reactions in blood and plasma. Vitamin C is a strong inhibitor of lipid peroxidation; likewise, vitamin E is also reported to decrease isoprostanes (biomarkers for lipid peroxidation) levels in animal models. Vitamin C has been shown to speed up the synthesis of vitamin E in membranes and lipoproteins. Moreover, vitamin C regenerates vitamin E (α-tocopherol) and reduces α-tocopherolxyl free radical reactions in membranes. In addition, vitamin C and vitamin E act in conjunction to prevent lipid peroxidation (Thapa and Carroll 2017). Other antioxidant molecules like β-carotene, dietary polyphenols (flavonoids), and Se also reduce OS in neurons. The antioxidant, anti-inflammatory, and neuroprotective effect of vitamin D has also been widely reported (Alles et al. 2012).

The other most important nutrients are the mineral elements such as Fe, Ca, Mg, Mn, Cu, Zn, Se, P, K, Na, S, Cl, I, Mb, Cr, and F. These trace mineral elements serve as enzyme cofactors and they are named "antioxidant micronutrients" because of their antioxidant functions. For instance, Se is an integral part of selenoproteins in humans, which plays a key role as body's own antioxidant defense (e.g., glutathione peroxidase) and protects the body against CVD or cancer (Kozarski et al. 2015; Gil and Gil 2015). Zn is a central cofactor to over 300 enzymes that have various roles including DNA and RNA metabolism and serves a major function in the stabilization protein structures. Cu is also a catalytic cofactor involved in various important enzymatic reactions due to its tendency to change its oxidation state (Gil and Gil 2015). Furthermore, the antioxidant action of various other micronutrients in mitochondria is significant. For example, the coenzyme α-lipoic acid has essential function in maintaining energy homeostasis in mitochondria. It is mainly available in animal meat including in the liver, heart, kidney, and vegetables like potato, broccoli, and spinach (Gómez-Pinilla 2008).

3.5 Dietary Supplements, Herbs, and Functional Foods in Health

A broader range of items are included under the term "nutraceutical." Nutraceuticals are dietary products such as functional foods, dietary supplements, and medicinal foods (formulated foods that can be consumed upon recommendation by physician or dietician in relation to certain disease condition) (LaRocca et al. 2017). According to the US Department of Agriculture, functional foods are defined as "natural or processed foods that contain known or unknown biologically active compounds, which, in defined, effective nontoxic amounts, provide a clinically proven and documented health benefit for the prevention, management or treatment of chronic disease." The functional food comprises items like probiotic yogurt and fortified grain products. On the other hand, dietary supplements as defined by the FDA are "a product intended for ingestion that contains a 'dietary ingredient' intended to add further nutritional value to (supplement) the diet." For instance, vitamins, minerals, herbs, extracts, metabolites, and amino acids are dietary supplements (Hunter and Hegele 2017). Generally, the underlying concepts behind nutraceutical are as follows: i. some dietary patterns have beneficial effects because of the abundant particular nutrient composition they have and ii. the strong activity of these molecules against OS and inflammation is important to prevent physiological/arterial dysfunction (LaRocca et al. 2017).

Plants produce enormous amount of health-promoting compounds including polyphenols, dietary fibers, antioxidant molecules, vitamins, and minerals (Filannino et al. 2018). Basically, food items of plant origin provides a higher nutritive value, low caloric density, and low-energy constituents (e.g., minerals, vitamins, dietary fibers) as well as being rich sources of bioactive phytochemicals such as carotenoids, sterols, polyphenols, and glucosinolates (Table 3.1) (Manach et al. 2017). Dietary polyphenols like flavonoids are strong antioxidant that acts through interacting with ROS-producing reactive metals. For example, curcumin is a dietary polyphenol with strong antioxidant activity that have been used against a number of age-related diseases like AD for more than a century. Curcumin acts via chelating reactive metal ions like Fe^{2+}, thus diminishing their oxidative power and the resulting OS due to free radicals. Ginkgo biloba extract has also been on use to treat age-related disease. Ginkgo biloba extracts various polyphenols such as flavonoids, and terpenes interact with superoxide anion, hydroxyl, and peroxyl free radicals to quench the radical chain reactions in mitochondrial respiratory chain function. Besides, it has been reported that ginkgo biloba extract prevents neuronal apoptosis due to OS via increasing the rate of neurotransmitter uptake. In AD, the improvement in cognitive function along with neuroprotective activity in animal model of Ginkgo biloba extract is related to downregulation of toxic Aβ aggregates (Thapa and Carroll 2017; Tiwari et al. 2018). The important effect of tea polyphenols on cognitive function of aged people is attributed to its flavonoid (catechin) content. There is a profound inverse association between tea drinking and the risk of dementia, AD, and PD. Green tea catechins administered for seven months to old Wister rats protected the rats from memory and spatial learning decline, probably as a consequence of its strong antioxidant potential (Mandel et al. 2012). In general, bioactive chemicals from plants are important though not essential, and they have been referred to as "life span essential," since they have beneficial effect in healthy aging. Most plant secondary metabolites are commonly available in vegetables and fruits; however, their concentration varies widely with the type of the plant, for example, isoflavones are abundant in soya, lycopene in tomatoes, glucosinolates in cruciferous vegetables, and anthocyanins in berries (Manach et al. 2017).

3.6 Protein and Energy in Healthy Aging

Protein is possibly the most essential macronutrient responsible for aging among the three essential macronutrients (i.e., carbohydrates, protein, and fat). A further breakdown studies revealed that particular amino acids like tryptophan and methionine, which are common in dietary protein, fed to laboratory animals improved the animal's health and delayed aging (Simpson et al. 2017). In order to avoid age-related loss of lean body mass, the daily-recommended dietary allowance (RDA) of protein for adults is 0.8g/kg of body weight. Studies have demonstrated that in aged people, taking protein above the RDA helps to increase muscle strength, muscle mass, and muscle function, boosts immune status, and improves wound healing (Ahmed and Haboubi 2010). Despite the

Table 3.1 Natural Products, nutrients, and foods with direct and indirect action on AD

Foods	Biological activity	Natural products	Biological activity	Nutrients	Biological activity
Fish and seafood Solfrizzi et al. (2017), Gómez-Pinilla (2008), Hennebelle et al. (2013), and Cederholm et al. (2013)	↑AD pathology	Rice bran extract (Hagl et al. (2015)	Improves mitochondrial function	B12 (Eastley et al. (2000)	Improves cognitive function
Salad dressing Gu et al. (2010)	↓Risk of AD	Curcumin Thapa and Carroll (2017)	↓OS	Lower SFA Gu et al. (2010)	↓Risk of AD
Olive oil Abbatecola et al. (2018)	↓Risk of MCI	Ginkgo Thapa and Carroll (2017)	Neuroprotective	Niacin Morris et al. (2004)	↓Incidence of AD and cognitive decline
Eggs Solfrizzi et al. (2017)	↓Risk of MCI	*Red yeast rice extract* Hunter and Hegele (2017)	↓LDL	Ascorbic acid Monacelli et al. (2017)	Delay in AD onset
Meat Xu et al. (2015)	↑Memory function	Catechin Mandel et al. (2012)	↓Risk of AD and dementia	Docosahexaenoic acid (DHA) Solfrizzi et al. (2017), Gómez-Pinilla (2008), Hennebelle et al. (2013), and Cederholm et al. (2013)	Improves cognitive function and AD
Nuts Dong et al. (2016) and Rajaram et al. (2016)	↓Risk of MCI, better overall cognition	Caffeine Beydoun et al. (2014) and Travassos et al. (2015)	↓Risk of MCI	Vitamin E and tocopherol forms Gu et al. (2010), Luchsinger and Mayeux (2004), and Morris et al. (2005)	↓Risk of AD and cognitive decline
Tomatoes Gu et al. (2010)	↓Risk of AD	*Policosanol* Hunter and Hegele (2017)	↓LDL	Folate Alles et al. (2012) and Gu et al. (2010)	↓Risk of AD
Cruciferous vegetables Gu et al. (2010)	↓Risk of AD, ↓ROS	*Galanthus nivalis* extract (galantamine) Brown et al. (2016)	Treating AD	β-Carotene Gu et al. (2010), Dai et al. (2006), and Kesse-Guyot et al. (2014)	↓Risk of AD, ↓ROS
Poultry Shakersain et al. (2016)	↓Risk of AD	Flavonoids Alles et al. (2012)	↓OS	PUFA Gu et al. (2010) and Dai et al. (2006)	↓Risk of AD

Low fat dairy Shakersain et al. (2016) and Wengreen et al. (2013)	↓Risk of MCI	Jujube extract Chen et al. (2017)	Improves memory and learning	MUFA Gu et al. (2010) and Dai et al. (2006)	↓Risk of AD
Whole grain Abbatecola et al. (2018)	↓Hypertension, ↓Risk of MCI	Phytoestrogens (genistin, daidzein) Zhao et al. (2002)	Neuroprotective	Vitamin D Alles et al. (2012)	Anti-inflammatory and neuroprotective
Fruits Dong et al. (2016)	↓Risk of MCI	Anthocyanins Shih et al. (2010)	Improves cognitive function	Vitamin C Gu et al. (2010) and Dai et al. (2006)	↓Risk of AD, ↓ROS
Red fruits Abbatecola et al. (2018)	↓Hypertension, ↓Risk of MCI	Choline Jones et al. (2017)	Improves cognition	Vitamin B1 Rodríguez et al. (2001)	↓AD progression
Dark and green leafyvegetables Gu et al. (2010)	↓Risk of AD, ↓ROS	Ferulic acid Jones et al. (2017)	Prevents AD	Low Zn level Jones et al. (2017)	OS
Tea Mandel et al. (2012) and Dai et al. (2006)	Neuroprotection			Al Barnard et al. (2014)	↑ Risk of AD
Coffee Solfrizzi et al. (2017) and Araújo et al. (2015)	↓Risk of MCI			Excess Fe and Cu Barnard et al. (2014)	Causes cognitive problems

evidences of 0.8 g protein/kg/day, others have recommended that slightly higher level of protein intake (i.e., 1.0–1.3g/kg/day) is helpful in old age. This increase in protein consumption among elderly offers ability to maintain nitrogen balance, improve insulin action, offset a potentially lower energy intake, and improve protein synthesis. However, these recommendations could pose a potential risk of nephrotoxicity or renal malfunction. In this regard, sulfur-containing proteins have been associated with adverse events when taken at higher amount (45% energy), yet moderate amount of protein-containing diet (20–35% energy) appears to be safe (Paddon-Jones et al. 2008). The common manifestation of early aging is a change in body composition, such as decrease in lean mass and increase in fat mass. The most significant age-related risk that could lead to functional impairment and mortality is a fast reduction in skeletal muscle mass (sarcopenia). A number of age-related factors can contribute to a fast drop in skeletal muscle mass; however, insufficient protein in the daily diet is the most significant factor that accelerates this process. On top of this, most elderly people do not consume enough amount of protein in their diet as per the RDA (Houston et al. 2008). Thus, energy dense supplements can be used as meal replacement therapy in elderly people so as to maintain the daily protein balance along with controlled energy and essential nutrient intake. In cases where there is accelerated protein catabolism (e.g., cachexia, trauma, and sarcopenia), the required amino acid should be given as a supplement; however, for most of the elderly individuals, the problem of skeletal muscle protein anabolism can be alleviated by serving high protein meal in their daily diet (Paddon-Jones et al. 2008).

The current sedentary life trends in the world have resulted in increased risk of metabolic disorders like type II diabetes and obesity. The situation is even worse because of the wide use of ready-made foods of high caloric content in addition to the sedentary life style; consequently an imbalance between the energy intake and expenditure will contribute to health risks such as CVD, diabetes, and other age-related disorders (vel Szic et al. 2015). Caloric restriction (CR) sometimes referred to as "energy restriction in the absence of malnutrition" has beneficial impact on various physiological parameters during aging. CR decreases arterial oxidative stress and inflammation, and increases nitric oxide bioavailability. As a result, the arterial functions will be improved leading to a modulation in the body's key energy sensing networks, which is mostly affected with aging. In addition, CR boosts the endogenous antioxidant and anti-inflammatory defense system. Therefore, CR represents a novel approach toward modulating age-related disorders along with consuming nutritional supplements and nutraceuticals (LaRocca et al. 2017).

3.7 Nutrition in the Prognosis of Alzheimer Disease (AD)

3.7.1 Glucose and Oxidizing Agents

The complex nature of diet make it difficult to whether CR can be achieved through decreased caloric intake or reducing a particular macronutrient or both (Simpson et al. 2017). Current research findings showed that controlling the caloric composition of the diet has a potential to affect the cognitive functions due to the fact that the cellular metabolic activities in the mitochondria can regulate certain aspects of synaptic plasticity (Gómez-Pinilla 2008). In this regard, studies have observed a severe decline in glucose metabolism in AD-affected brain regions. In vivo imaging studies of patients with dementia also depicted a progressive decline of glucose metabolism and lowering of blood flow in the affected brain regions. However, the cerebral energy pool is slightly affected in the process of normal aging. Impaired glucose metabolism in the brain leads to limited synthesis of aspartate, glutamate, gamma-aminobutyric acid (GABA), glycine, acetylcholine, and ATP production (Münch et al. 1998). Furthermore, a diet containing low fats and highly processed carbohydrate causes a swift increase in blood glucose level upon consumption. Thus, the resulting glycation will impair serum proteins by forming modified proteins called advanced glycation end products (AGEs). For instance, glycated hemoglobin (HbA1c) is an AGE protein, the level of which in serum serves as a standard in diabetes test (Suzuki et al. 2010). Furthermore, other mechanisms of glucose in the pathogenesis of AD have been suggested such as the process that induces OS via nonenzymatic reaction of glucose to form AGEs on long-lived protein deposits and consequently disrupting glucose metabolism. Accordingly, more than any other protein in the body, apolipoproteins are susceptible to glycation. Although high level of glucose is mostly suggested as the main cause of AGEs, fructose is by far the worst reducing agent as compared to glucose (Seneff et al. 2011). Naturally, glucose is the least reactive sugar among those usable by the body, and probably that is why it remained an evolutionary biological energy carrier. One of the major challenges of AGE formation is that once it is formed, it is irreversible, and it leads to protein deposition and amyloidosis since it is a protease-resistant cross-linking of proteins and peptides (Münch et al. 1998). The major factor contributing to Aβ plaque deposition in AD is the high level of AGEs in the patients' brain. Since apolipoproteins are susceptible to glycation, glycated ApoE is excessively available in the cerebrospinal fluid (CSF) of AD patients (Shuvaev et al. 2001). This indicates that ApoE could be the initial stage in Alzheimer's cascade. Moreover, various proteins in neurons and astrocytes are vulnerable to glycation upon exposure to excess glucose (Li and Dickson 1997). The consequent loss of function, vulnerability to oxidative stress, and lesser degradation and disposal of AGEs and their presence in microscopic slices where Aβ is absent show that glycation is an early hallmark of AD (Seneff et al. 2011).

3.7.2 Impaired Glutamate and AD

Glutamate-induced excitotoxic cascade is one of the pathologic mechanisms associated with AD. Several studies have also suggested that impaired glutamatergic transmission is the hallmark of Alzheimer's disease. Initially, Aβ and tau activate the N-methyl-D-aspartate (NMDA) receptors. Then the activated NMDA receptors boost the making of Aβ and tau (Lesné et al. 2005; Amadoro et al. 2006). Increase in excitatory amino acids in the synapse that results in activation of glutamatergic receptors and increased ROS level are cascades of inter-related pathways that lead to AD, PD, multiple sclerosis, and ischemia. The ionotropic and metabotropic glutamatergic receptors are both shown to play a role in etiology of Alzheimer's disease. Moreover, abnormal glutamate transport function has been identified in mutant amyloid-β protein precursor (AβPP) transgenic (tg) mice models of AD (Jacob et al. 2007). The impairment of synapses in AD encompasses Aβ-induced OS in the synapse that damages the glutamate receptors and the function of membrane ion along with a decline in mitochondrial efficiency because of the OS (Mattson 2004). In postmortem brain of APP tg mouse models of AD, the expression of glutamate transporters has been found low. Recent studies have also stipulated that Aβ-mediated synaptic suppression is partially due to inhibition of the GABAA receptor. These findings indicated that Aβ inhibits glutamate uptake, increases extracellular glutamate, and thus activates extrasynaptic GluN2B receptors and it also diminishes GABAA receptor-mediated inhibition, finally resulting in neuronal hyperexcitability (Lei et al. 2016). Moreover, a number of studies have reported diminished glutamate uptake in the frontal and temporal cortices of AD brains. These changes were suggested as the main reason for the rise in extracellular glutamate level (Cassano et al. 2012). Results of in vivo imaging studies demonstrated that AD patients brain show a significant shrinkage in the temporal, parietal, and frontal lobes of the cortex in addition to sulcal widening and ventricular enlargement. Such atrophies results from loss of pyramidal neurons and their synapses as well as the surrounding neuropil. Biochemical evidence identified glutamate as the neurotransmitter of these pathways, and hence it is obviously understandable that the glutamatergic neurons degenerate in AD (Francis 2003).

Medicinal plants and alternative medicines have been used to treat AD since the pharmacological interventions does not have satisfactory option for AD. *Hericium erinaceus* is an edible and medicinal mushroom which has been used to treat a number of diseases because of its potential antitumor, antimutagenic, antioxidant, hypolipidemic, immunomodulatory, and neuroprotective activities. According to Zhang J. et.al. (2016), *Hericium erinaceus* prevents neurodegenerative diseases. *Hericium erinaceus* prevented DPC12 cells against neurotoxicity induced by L-Glu. Additional observations on AlC13- and D-gal-induced AD mice have proved that *Hericium erinaceus* has neuroprotective activity that involves neurotransmitter modulation (Zhang et al. 2016). Furthermore, it has been reported that ascorbic acid release mediated by neurons is related to glutamate metabolism and kinetics in the brain. Especially, the extracellular release of ascorbic acid is directly linked to astrocyte swelling mediated by glutamate receptors' increased sodium uptake. Henceforth, ascorbic acid in the brain and the CSF acts against glutamate excitotoxicity via antioxidant and neuroprotective mechanism (Nualart et al. 2014).

3.7.3 Role of Dietary Fats in AD

The brain is rich in lipids, although the content of fat differs among tissues. The content and composition of fatty acids (FAs) play a key role in brain functions like cognition. Thus, knowing the lipid composition of the brain is an important step toward understanding the role of consuming altered fatty acid (FA) on its function. The dry weight of the gray matter, white matter, and myelin contains about 40%, 66%, and 81% lipid, respectively (Cederholm et al. 2013). Of all the nutrients essential to brain, some of the ω-3 FAs such as docosahexaenoic acid (DHA) have a special importance. Seafood and fatty fish are rich sources of both DHA and eicosapentaenoic acid (EPA) (Hennebelle et al. 2013). The DHA content in the brain has inverse relation with age, although it normally increases over the first two decades (Cederholm et al. 2013). Regularly consuming vegetables and fish fats of high PUFA, moderate unsaturated fats, and low saturated fat can lower the incidence of CVD and AD. Conversely, increased hydrogenated and saturated fats consumption is associated with insulin resistance, which is indirectly related to higher incidence of AD (Luchsinger and Mayeux 2004). Additionally, a high fat and low carbohydrate diet (Ketogenic diet) has been suggested to be effective against a range of neurologic diseases. In clinical settings, ketogenic diet has been utilized to treat AD, PD, epilepsy, and autism (Ma et al. 2018). Moreover, there are a number of evidences from preclinical studies showing the therapeutic activities of ketogenic diet for various disease conditions including AD, ischemia, and traumatic brain injury (Van der Auwera et al. 2005; Prins et al. 2005; Puchowicz et al. 2008).

3.7.4 Role of Omega-3 Fatty Acids in AD

The impact of dietary factors on various brain processes is through modulating neurotransmitter pathways, membrane fluidity, synaptic transmission, and signal transduction pathways. On the other hand, the direct effect of lipids on the brain function is attracting attention in addition to their indirect action through their impact on the CVS physiology. ω-3- PUFAs are cell membrane constituents, and they are essential nutrients of the brain. The dietary deficiency of ω-3 FAs has been linked with various severe cognitive disorders such as dementia, schizophrenia, depression, bipolar disorder, dyslexia, and attention-deficit disorder. Among the ω-3 FAs, DHA is an integral part of neuronal membranes, but the human body cannot synthesize DHA, and it has to obtain from dietary sources (Gómez-Pinilla 2008). ω-3 PUFA have various therapeutic benefits including anti-inflammatory, antioxidant at lower doses, and anti-amyloidogenesis (Alles et al. 2012). On the contrary, low consumption of ω-3- FAs has been widely associated to increased risk of suboptimal brain development and cardiovascular diseases (Hennebelle et al. 2014). As a major component in neuronal membranes, DHA has a number of physiological activities including sustaining the normal cell membrane fluidity and structure and also plays a role in cellular functions and response (Calder 2012). Furthermore, ω-3 FAs activate transcription factors like PPAR-g so as to modulate gene expressions that in turn controls the inflammatory mechanisms (Alles et al. 2012). In addition, animal studies have reported that the level of hippocampal brain-derived neurotrophic factor (BDNF) is increased and cognitive function is enhanced in brain trauma rodent models treated with dietary DHA supplements. Mechanistically, DHA might enhance cognitive functions via aiding synaptic plasticity and/or enhancing synaptic membrane fluidity; its actions on metabolism might also contribute through enhancing glucose utilization and decreasing OS. However, the exact mechanisms of action through which the ω-3- FAs (particularly that of DHA) act on brain plasticity and cognition are not fully understood (Gómez-Pinilla 2008).

3.7.5 Role of Dietary Antioxidants in AD

The brain is prone to OS due to its high metabolic burden and its rich content of oxidizable materials including PUFA of the neuronal membranes. ROS are related to neuronal impairment in AD. Rise in oxidative stress and excess production and deposition of Aβ are early events in AD. Aβ deposition mainly results in reduced iron and copper level in the brain tissues, which in turn causes cascade of processes like production of hydrogen peroxide, increased OS, and neuronal damage (Luchsinger and Mayeux 2004). Quite a number of antioxidant foods have been known for their beneficial effects on neuronal functions (Fig. 3.1). For instance, barriers have been used for their strong antioxidant potential, yet the individual compounds evaluated so far are few, namely, the tannins procyanidin, and prodelphinidin, some anthocyanins and phenolic compounds. Moreover, curcumin is another dietary antioxidant well known as preservative and herbal medicine mainly in Indian subcontinent. Curcumin has a strong antioxidant property and reduces nitric-oxide-based radicals as well as preventing lipid peroxidation in the brain. In addition, animal model studies have been demonstrated to reduce memory deficit in AD and brain trauma. Curcumin is safe even at doses higher than the dose used in animal

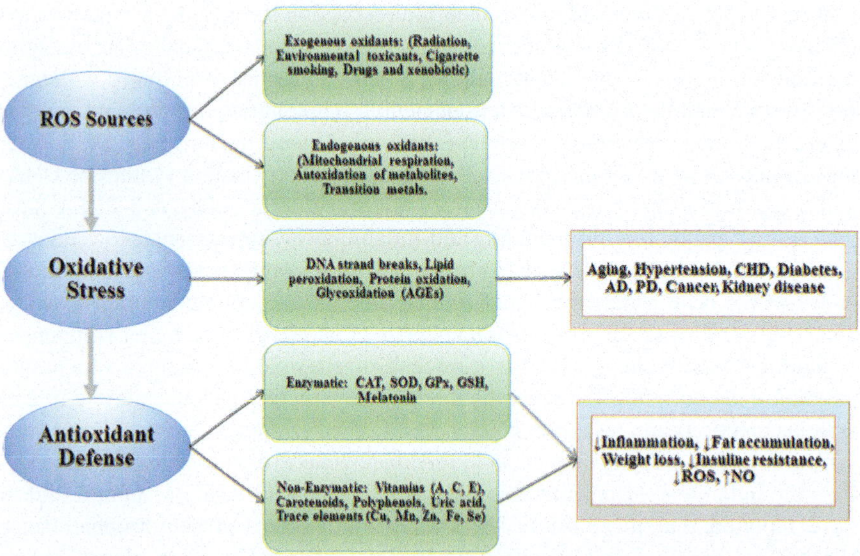

Fig. 3.1 The possible effects of oxidative stress and antioxidant nutrients. CAT: catalase, SOD: superoxide dismutase, GPx: Glutathione peroxidase, GSH: glutathione

models (Gómez-Pinilla 2008). Various studies have confirmed that vegetable and fruit juices available commercially also have high level of antioxidant polyphenolic compounds. This might be due to the high-pressure mechanical extraction during processing of vegetable and fruit juices that drives out all the antioxidant compounds from the peels and pulp in addition to the main fruit and vegetable fluids. According to various in vitro, in vivo, and clinical studies, a stronger neuroprotective activity than antioxidant vitamins has been reported in polyphenols from grape, apple, and citrus fruit juices (Dai et al. 2006). Dietary antioxidants have been related to reduce risk of stroke and thus AD, since stroke is mostly associated to higher incidence of AD through cerebrovascular events that links antioxidants, vitamins, and AD (Luchsinger and Mayeux 2004).

3.8 Conclusion

In today's world, the population size of aged people is growing faster. In the coming few decades, their numbers will double leaving the age-related disease to become a public health burden. Cognitive dysfunctions, in particular AD, are one of the most prominent the public health threat related to aging. Various pharmacological therapies have been used to counter the risk of AD and to prevent its progression. However, none is efficient in mitigating the problem of AD in relation to aging. Yet, nutritional interventions have attracted researcher's attention since most of the antioxidant-containing vegetables and fruits are important in reducing the incidence of AD. The basic pathology studies on AD have claimed that the major contributing factor in the onset and progression of AD is oxidative stress. Henceforth, controlling the environmental risk factors responsible for OS and consuming dietary antioxidants including dietary supplements and nutraceuticals has been proved effective in treating AD. Especially, the importance of caloric restriction and consuming PUFAs, proteins, vegetables, and fruits are reported in many observational studies. Therefore, dietary interventions along with avoiding risk factors will reduce the risk of age-related degenerative diseases and increase healthy life span among elderly.

Acknowledgments This article is the outcome of an in-house financially nonsupported study.

Author Contributions All authors have directly participated in the planning or drafting of the manuscript and read and approved the final version.

Conflict of Interest The authors declare no conflict of interest.

References

Abbatecola AM, Russo M, Barbieri M (2018) Dietary patterns and cognition in older persons. Curr Opin Clin Nutr Metab Care 21(1):10–13

Ahmed T, Haboubi N (2010) Assessment and management of nutrition in older people and its importance to health. Clin Interv Aging 5:207

Alles B et al (2012) Dietary patterns: a novel approach to examine the link between nutrition and cognitive function in older individuals. Nutr Res Rev 25(2):207–222

Amadoro G et al (2006) NMDA receptor mediates tau-induced neurotoxicity by calpain and ERK/MAPK activation. Proc Natl Acad Sci 103(8):2892–2897

Araújo LF et al (2015) Inconsistency of association between coffee consumption and cognitive function in adults and elderly in a cross-sectional study (ELSA-Brasil). Nutrients 7(11):9590–9601

Balagopal P, Proctor D, Nair KS (1997) Sarcopenia and hormonal changes. Endocrine 7(1):57–60

Barnard ND et al (2014) Dietary and lifestyle guidelines for the prevention of Alzheimer's disease. Neurobiol Aging 35:S74–S78

Baumgartner RN (2000) Body composition in healthy aging. Ann N Y Acad Sci 904(1):437–448

Beydoun MA et al (2014) Caffeine and alcohol intakes and overall nutrient adequacy are associated with longitudinal cognitive performance among US adults–3. J Nutr 144(6):890–901

Brown JV et al (2016) Multi-nutrient fortification of human milk for preterm infants. Cochrane Database Syst Rev (5):Cd000343

Bule MH et al (2018) Microalgae as a source of high-value bioactive compounds. Front Biosci(Sch Ed) 10:197–216

Calder PC (2012) Mechanisms of action of (n-3) fatty acids, 2. J Nutr 142(3):592S–599S

Cassano T et al (2012) Glutamatergic alterations and mitochondrial impairment in a murine model of Alzheimer disease. Neurobiol Aging 33(6):1121. e1–1121. e12

Cederholm T, Salem N Jr, Palmblad J (2013) Omega-3 fatty acids in the prevention of cognitive decline in humans. Adv Nutr 4(6):672–676

Chen J et al (2017) A review of dietary Ziziphus jujuba fruit (Jujube): developing health food supplements for brain protection. Evid Based Complement Alternat Med 2017

Clements SJ, Carding SR (2018) Diet, the intestinal microbiota, and immune health in aging. Crit Rev Food Sci Nutr 58(4):651–661

Dai Q et al (2006) Fruit and vegetable juices and Alzheimer's disease: the kame project. Am J Med 119(9):751–759

Dong L et al (2016) Diet, lifestyle and cognitive function in old Chinese adults. Arch Gerontol Geriatr 63:36–42

Eastley R, Wilcock GK, Bucks RS (2000) Vitamin B12 deficiency in dementia and cognitive impairment: the effects of treatment on neuropsychological function. Int J Geriatr Psychiatry 15(3):226–233

Filannino P, Di Cagno R, Gobbetti M (2018) Metabolic and functional paths of lactic acid bacteria in plant foods: get out of the labyrinth. Curr Opin Biotechnol 49:64–72

Francis PT (2003) Glutamatergic systems in Alzheimer's disease. Int J Geriatr Psychiatry 18(S1):S15–S21

Gil A, Gil F (2015) Fish, a Mediterranean source of n-3 PUFA: benefits do not justify limiting consumption. Br J Nutr 113(Suppl 2):S58–S67

Gómez-Pinilla F (2008) Brain foods: the effects of nutrients on brain function. Nat Rev Neurosci 9(7):568

Gu Y et al (2010) Food combination and Alzheimer disease risk: a protective diet. Arch Neurol 67(6):699–706

Hagl S et al (2015) Beneficial effects of ethanolic and hexanic rice bran extract on mitochondrial function in PC12 cells and the search for bioactive components. Molecules 20(9):16524–16539

Hardman RJ et al (2015) A randomised controlled trial investigating the effects of Mediterranean diet and aerobic exercise on cognition in cognitively healthy older people living independently within aged care facilities: the lifestyle intervention in independent living aged care (LIILAC) study protocol [ACTRN12614001133628]. Nutr J 14:53

Hennebelle M et al (2013) Ageing and apoE change DHA homeostasis: relevance to age-related cognitive decline. Proc Nutr Soc 73(1):80–86

Hennebelle M et al (2014) Ageing and apoE change DHA homeostasis: relevance to age-related cognitive decline. Proc Nutr Soc 73(1):80–86

Houston DK et al (2008) Dietary protein intake is associated with lean mass change in older, community-dwelling adults: the health, aging, and body composition (health ABC) study. Am J Clin Nutr 87(1):150–155

Hunter PM, Hegele RA (2017) Functional foods and dietary supplements for the management of dyslipidaemia. Nat Rev Endocrinol 13(5):278

Jacob C et al (2007) Alterations in expression of glutamatergic transporters and receptors in sporadic Alzheimer's disease. J Alzheimers Dis 11(1):97–116

Jones JM et al (2017) Impact of minerals, phytochemicals, specific grain-based foods, and dietary patterns on mild cognitive impairment, Alzheimer's disease, and Parkinson's disease. Cereal Foods World 62(3):104–114

Kesse-Guyot E et al (2014) Carotenoid-rich dietary patterns during midlife and subsequent cognitive function. Br J Nutr 111(5):915–923

Kozarski M et al (2015) Antioxidants of edible mushrooms. Molecules 20(10):19489–19525

LaRocca TJ, Martens CR, Seals DR (2017) Nutrition and other lifestyle influences on arterial aging. Ageing Res Rev 39:106–119

Lee D et al (2015) Effects of nutritional components on aging. Aging Cell 14(1):8–16

Lei M et al (2016) Soluble Aβ oligomers impair hippocampal LTP by disrupting glutamatergic/GABAergic balance. Neurobiol Dis 85:111–121

Lesné S et al (2005) NMDA receptor activation inhibits α-secretase and promotes neuronal amyloid-β production. J Neurosci 25(41):9367–9377

Li YM, Dickson DW (1997) Enhanced binding of advanced glycation endproducts (AGE) by the ApoE4 isoform links the mechanism of plaque deposition in Alzheimer's disease. Neurosci Lett 226(3):155–158

López-Miranda J et al (2010) Olive oil and health: summary of the II international conference on olive oil and health consensus report, Jaén and Córdoba (Spain) 2008. Nutr Metab Cardiovasc Dis 20(4):284–294

Luchsinger JA, Mayeux R (2004) Dietary factors and Alzheimer's disease. The Lancet Neurology 3(10):579–587

Ma D et al (2018) Ketogenic diet enhances neurovascular function with altered gut microbiome in young healthy mice. 8(1):6670

Manach C et al (2017) Addressing the inter-individual variation in response to consumption of plant food bioactives: towards a better understanding of their role in healthy aging and cardiometabolic risk reduction. Mol Nutr Food Res 61(6):1600557

Mandel SA et al (2012) Molecular mechanisms of the neuroprotective/neurorescue action of multi-target green tea polyphenols. Front Biosci (Schol Ed) 4:581–598

Mattson MP (2004) Pathways towards and away from Alzheimer's disease. Nature 430(7000):631

Mohajeri MH, Troesch B, Weber P (2015) Inadequate supply of vitamins and DHA in the elderly: implications for brain aging and Alzheimer-type dementia. Nutrition 31(2):261–275

Monacelli F et al (2017) Vitamin C, aging and Alzheimer's disease. Nutrients 9(7):670

Morris MC et al (2004) Dietary niacin and the risk of incident Alzheimer's disease and of cognitive decline. J Neurol Neurosurg Psychiatry 75(8):1093–1099

Morris MC et al (2005) Relation of the tocopherol forms to incident Alzheimer disease and to cognitive change. Am J Clin Nutr 81(2):508–514

Münch G et al (1998) Alzheimer's disease–synergistic effects of glucose deficit, oxidative stress and advanced glycation endproducts. J Neural Transm 105(4–5):439–461

Nualart F et al (2014) Vitamin C transporters, recycling and the bystander effect in the nervous system: SVCT2 versus gluts. J Stem Cell Res Ther 4(5):209

Paddon-Jones D et al (2008) Role of dietary protein in the sarcopenia of aging. Am J Clin Nutr 87(5):1562S–1566S

Prinelli F et al (2015) Mediterranean diet and other lifestyle factors in relation to 20-year all-cause mortality: a cohort study in an Italian population. Br J Nutr 113(6):1003–1011

Prins M, Fujima L, Hovda D (2005) Age-dependent reduction of cortical contusion volume by ketones after traumatic brain injury. J Neurosci Res 82(3):413–420

Puchowicz MA et al (2008) Neuroprotection in diet-induced ketotic rat brain after focal ischemia. J Cereb Blood Flow Metab 28(12):1907–1916
Rajaram S et al (2016) The walnuts and healthy aging study (WAHA): protocol for a nutritional intervention trial with walnuts on brain aging. Front Aging Neurosci 8:333
Rajaram S et al (2017) The Walnuts and Healthy Aging Study (WAHA): protocol for a nutritional intervention trial with walnuts on brain aging. Front Aging Neurosci 8:333
Rodríguez JL, Qizilbash N, López-Arrieta J (2001) Thiamine for Alzheimer's disease. Cochrane Database Syst Rev 2
Seneff S, Wainwright G, Mascitelli L (2011) Nutrition and Alzheimer's disease: the detrimental role of a high carbohydrate diet. Eur J Intern Med 22(2):134–140
Shakersain B et al (2016) Prudent diet may attenuate the adverse effects of Western diet on cognitive decline. Alzheimers Dement 12(2):100–109
Shih P-H et al (2010) Antioxidant and cognitive promotion effects of anthocyanin-rich mulberry (Morus atropurpurea L.) on senescence-accelerated mice and prevention of Alzheimer's disease. J Nutr Biochem 21(7):598–605
Shlisky J et al (2017) Nutritional considerations for healthy aging and reduction in age-related chronic disease. Adv Nutr (Bethesda, Md) 8(1):17–26
Shuvaev VV et al (2001) Increased protein glycation in cerebrospinal fluid of Alzheimer's disease2. Neurobiol Aging 22(3):397–402
Simpson SJ et al (2017) Dietary protein, aging and nutritional geometry. Ageing Res Rev 39:78–86
Solfrizzi V et al (2017) Relationships of dietary patterns, foods, and micro- and macronutrients with Alzheimer's disease and late-life cognitive disorders: a systematic review. J Alzheimers Dis 59(3):815–849
Suzuki R et al (2010) Diabetes and insulin in regulation of brain cholesterol metabolism. Cell Metab 12(6):567–579
Thapa A, Carroll NJ (2017) Dietary modulation of oxidative stress in Alzheimer's disease. Int J Mol Sci 18:7
Tiwari R et al (2018) Herbal Immunomodulators-a remedial panacea for designing and developing effective drugs and medicines: current scenario and future prospects. Curr Drug Metab 19(3):264–301
Travassos M et al (2015) Does caffeine consumption modify cerebrospinal fluid amyloid-β levels in patients with Alzheimer's disease? J Alzheimers Dis 47(4):1069–1078
Van der Auwera I et al (2005) A ketogenic diet reduces amyloid beta 40 and 42 in a mouse model of Alzheimer's disease. Nutr Metab 2(1):28
vel Szic KS et al (2015) From inflammaging to healthy aging by dietary lifestyle choices: is epigenetics the key to personalized nutrition. Clin Epigenetics 7(1):33
Wengreen H et al (2013) Prospective study of dietary approaches to stop hypertension- and mediterranean-style dietary patterns and age-related cognitive change: the Cache County study on memory, health and aging. Am J Clin Nutr 98(5):1263–1271
Willcox BJ et al (2007) Caloric restriction, the traditional Okinawan diet, and healthy aging. Ann N Y Acad Sci 1114(1):434–455
Xu X et al (2015) Tofu intake is associated with poor cognitive performance among community-dwelling elderly in China. J Alzheimers Dis 43(2):669–675
Zhang J et al (2016) The neuroprotective properties of Hericium erinaceus in glutamate-damaged differentiated PC12 cells and an Alzheimer's disease mouse model. Int J Mol Sci 17(11):1810
Zhao L, Chen Q, Brinton RD (2002) Neuroprotective and neurotrophic efficacy of phytoestrogens in cultured hippocampal neurons. Exp Biol Med 227(7):509–519

Carbon Nanostructure-Based Materials: A Novel Tool for Detection of Alzheimer's Disease

4

Mohammad Ashfaq, Neetu Talreja, Divya Chuahan, and Werayut Sriturayanich

Abstract

Alzheimer's disease is one of the most common forms of dementia and is an overwhelming neurodegenerative disease. Usually, this neurological condition is categorized by a loss of cognitive functions and recognized pathophysiological hallmarks in the brain such as accumulation of extracellular (amyloid-β (Aβ) peptide) and intracellular (neurofibrillar tangles of hyperphosphorylated τ protein) protein. Presently, a huge number (approximately 35 million) of people are affected by this form of dementia and neurodegenerative disease. The situation becomes more dangerous and will lead to social burden in the near future. Therefore, there is a need to develop effective therapy and early diagnostic approaches. In this con-

M. Ashfaq (✉)
School of Life Science, BS Abdur Rahaman Crescent Institute of Science and Technology, Chennai, India

Department of Mechanical Engineering, Faculty of Engineering, Chulalongkorn University, Bangkok, Thailand

Biomedical Engineering Research Center, Faculty of Engineering, Chulalongkorn University, Bangkok, Thailand
e-mail: ashfaqm.sls@crescent.education

N. Talreja
Department of Bio-nanotechnology, Gachon University, Seongnam, South Korea

D. Chuahan
Department of Chemistry, Punjab University, Chandigarh, India

W. Sriturayanich
Department of Mechanical Engineering, Faculty of Engineering, Chulalongkorn University, Bangkok, Thailand

Biomedical Engineering Research Center, Faculty of Engineering, Chulalongkorn University, Bangkok, Thailand
e-mail: werayut.s@chula.ac.th

G. M. Ashraf, A. Alexiou (eds.), *Biological, Diagnostic and Therapeutic Advances in Alzheimer's Disease*, https://doi.org/10.1007/978-981-13-9636-6_4

text, biosensor has become an indispensable tool in biomarker sensing, in which sensor electrode materials play critical roles in accurate and selective sensing. Carbon nanostructure-based materials have the potential ability for biosensor platform because of their excellent mechanical, electronic, optical, and easy functionalization ability. Herein, we discuss the various carbon nanostructure-based materials, biomarkers of Alzheimer's disease, carbon nanostructured material-based biosensor, and future prospects in the detection of Alzheimer's disease.

Keywords

Alzheimer's disease · Nanomaterials · Biosensor · Biomarkers · Neurological disease

4.1 Introduction

Alzheimer's disease (AD) is one of the leading forms of dementia (~80%) cases and neurodegenerative disease in old-age people. The aging of the world's population and a number of dementia is approximately expected from 35 to 65 million by the year 2030. According to the Centers for Disease Control and Prevention, cases of AD will be expected to develop in every 33 seconds by the year 2050 and estimated prevalence of around 14 million. The early symptoms of Alzheimer's disease such as disorientation, dysfunction, and memory impairment lead to worst cognitive impairment subsequently severe dementia and death (Alves et al. 2012; Cummings et al. 2018; Hajipour et al. 2017; Huang and Mucke 2012; Brookmeyer et al. 2007). Figure 4.1 shows the schematic representation of healthy and Alzheimer's disease brain. The figure shows the severe cortical shrinkage, enlarged ventricles, shrinkage of the hippocampus, deposition of amyloid plaques, and disintegrating microtubules that are observed in Alzheimer's disease brain compared with a healthy brain. Alzheimer's disease mainly occurs because of the accumulation of extracellular (amyloid-β (Aβ) peptide) and intracellular (neurofibrillar tangles of hyperphosphorylated τ protein) protein. Usually, the progression of Alzheimer's disease occurs in three stages: (1) early stage (mild), (2) middle stage (moderate), and (3) late stage (severe) (Frisoni et al. 2017; Hansen 2014; Dubois et al. 2009; Sangubotla and Kim 2018). Figure 4.2 shows the schematic representation of the different stages of Alzheimer's disease. The figure shows the progression of the disease from medial temporal lobe that spread to lateral temporal and partial lobes (early stage) after that disease spread to the frontal lobe (moderate stage), and finally, disease spread at occipital lobe (severe stage) (Sangubotla and Kim 2018; Iqbal et al. 2015). Alzheimer's disease affects people in different ways, as every patient has different symptoms in every stage of the disease; attributed Alzheimer's disease is a continuous process that leads to cognitive impairment. The histopathological analysis is one of the main characteristics of Alzheimer's disease (Kumar et al. 2015; Kurz and Perneczky 2011; Yiannopoulou and Papageorgiou 2013; Hardy 2009).

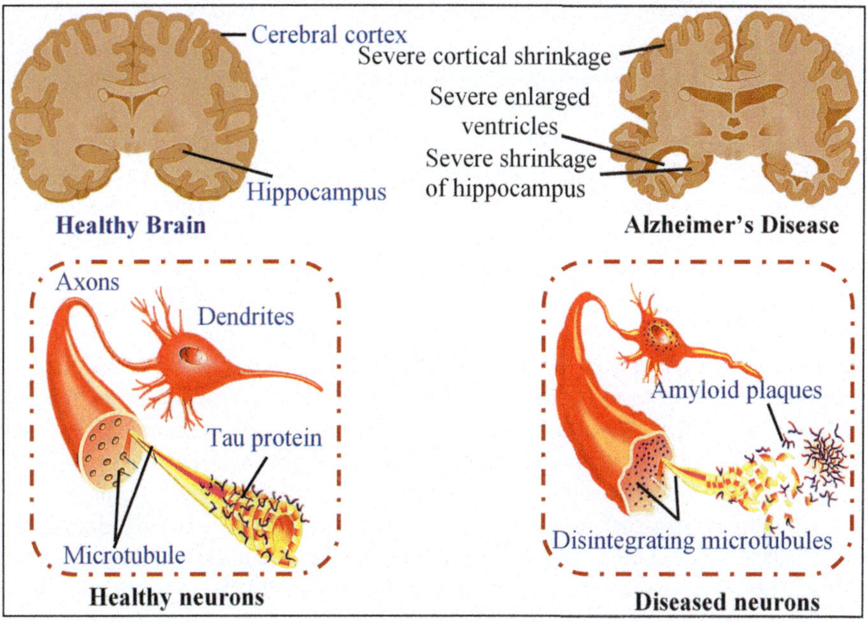

Fig. 4.1 Schematic illustration of healthy and Alzheimer's disease brain

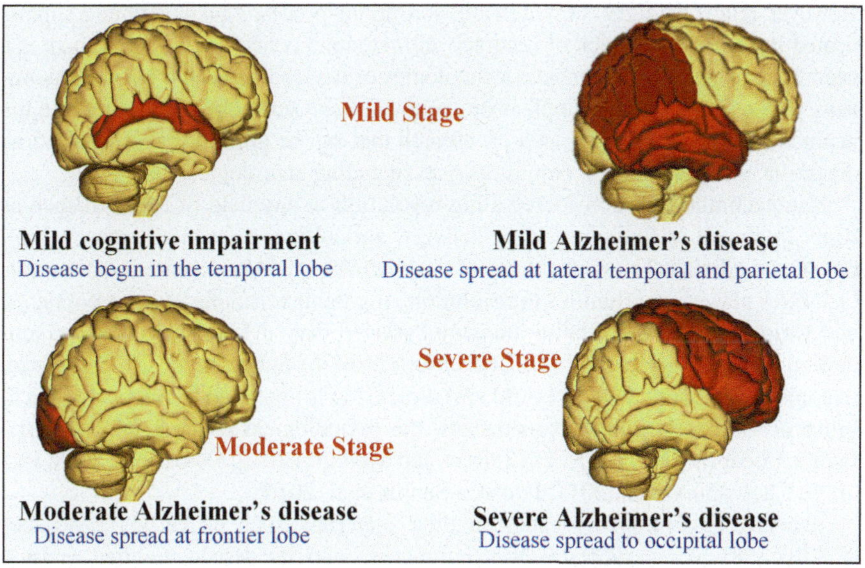

Fig. 4.2 Schematic representation of different stages of Alzheimer's disease

However, accurate diagnosis is not possible until autopsy. On the other hand, accurate diagnosis can be done by an opinion of a multidisciplinary specialist such as geriatrician, psychiatrist, and neurologist. Usually, an opinion of the specialist depends on the clinical blood assay, neuroimaging, and neuropsychological data (Barber 2010; Jellinger et al. 2008). However, it is time-consuming, expensive, and depends on various special needs of research in this field. In this context, diagnostic on the basis of various biomarkers is one of the important areas of research in Alzheimer's disease.

The biomarkers are any detectable biological analyte that might be used to diagnose or predict the pathological condition of the particular disease. Various biomarkers such as imaging of β-amyloid protein deposition, magnetic resonance imaging (MRI) of brain volume, genotyping, cerebrospinal fluid (CSF), and blood (plasma/serum) are related with risk of disease. The identification of biomarkers (neurological indicators) is one of the important aspects for the detection of every stage of Alzheimer's disease. These biomarkers are useful to diagnose the early stage of the disease, thereby offering effective treatments (Sharma and Singh 2016; Isaac et al. 2017; Mantzavinos and Alexiou 2017; Bjerke and Engelborghs 2018).

Several processes such as surface plasmon resonance (SPR), capillary electrophoresis, spectroscopic ellipsometry, nanoparticle-based dot-blot assay, field-effect transistor (FET), enzyme-linked immunosorbent assay (ELISA), resonance light scattering, and electrophoresis have been developed to detect Alzheimer's disease, especially Aβ peptide. Several research focuses on the development of an electrochemical sensor for the detection of Aβ peptide using expensive electrode materials (Albert et al. 2011a, b; Ringman et al. 2015; Choi et al. 2005; Kumar et al. 2015). However, relatively higher cost, the time-consuming process which required a sophisticated instrument, and lack of accuracy still remain a concern. Therefore, there is a need to develop newer diagnostic technologies or device that should be an economically viable, fast, required simple instrument with high sensitivity. In this context, the nanomaterial is an emerging area of research that can be utilized for the prevention, diagnosis, and treatment of various diseases including neurological disorder.

Nanotechnology is considered a big revolution in any field of research such as water treatment technologies, drug delivery, agriculture, electronics, energy, biology, and medicine. The applications of nanotechnology to healthcare or nanomedicine offer newer opportunities to revolutionizing the medical diagnosis, treatments, and various therapies including imaging, targeted drug delivery, and tissue regeneration (Nobile and Nobile 2017; Zäch et al. 2006; Sahoo et al. 2007). Indeed, nanomaterials-based device (1–100 nm) dimension is approved for clinical applicability of various products. Nonetheless, the toxicological concern might restrict their applicability (Kim et al. 2018; Rees and Moghimi 2012; Ovais et al. 2018; Jain 2005; Chowdhury et al. 2017; Fonseca-Santos et al. 2015).

Recently, carbon-based nanostructured materials such as carbon nanotubes (CNTs), carbon nanofibers (CNFs), graphene, and its derivative offer exciting opportunities for enhancing biosensor performance ability due to their unique characteristics like higher surface area, good electrical, mechanical and optical properties, easy functionalization and biocompatibility (Sankararamakrishnan et al. 2016; Ashfaq et al. 2013, 2014, 2016; Talreja et al. 2014; Shi and Fang 2018). Therefore, incorporation of carbon-based nanostructured materials provides newer biosensor platform for various diseases including Alzheimer's disease. Herein, we discuss various biomarkers of Alzheimer's disease, carbon-based nanostructured materials, and its biosensor device for early detection of disease. This book chapter also focuses on the limitation of carbon nanostructured material-based biosensor and future prospects in the detection of Alzheimer's disease.

4.2 Biomarker for Alzheimer's Disease

The biomarker is considered as an indicator of biological and pathogenesis process (normal or abnormal). The biomarker also served as an indicator of pharmacological responses on health or disease. The diagnosis utility of biomarkers depends on their sensitivity, specificity, and ease of use (Lashley et al. 2018; Lista et al. 2013; Thambisetty et al. 2011). Moreover, various biomarkers are used as risk factors of the disease rather than a true disease. The ideal biomarkers should contain following criteria that reflect (1) pathophysiological process of the brain, (2) aging processes, (3) high sensitivity, and (4) high specificity, (5) react with pharmacological intervention, (6) are easily operational, and (7) are noninvasive or minimally invasive. For Alzheimer's disease several biomarkers including imaging of β-amyloid protein deposition, MRI of brain volume, genotyping, CSF, and blood (plasma/serum) related with risk of disease (Thambisetty et al. 2011; Thambisetty and Lovestone 2010; Humpel 2011). These neurological indicators are essential for understanding each and every stage of Alzheimer's disease for early diagnosis and treatment of disease.

4.2.1 CSF Biomarkers

The CSF biomarkers reflect the molecular changes in the extracellular and interstitial environments of a brain. The CSF biomarkers for Alzheimer's disease indicate the pathogenic processes within the brain. The CSF biomarkers contain τ protein and Aβ 42 peptide that reflects neuron degeneration and Aβ metabolism as well as plaque formation, respectively. Usually, CSF biomarkers are obtained through lumbar puncture. However, post-lumbar puncture causes headache and other complications associated with elderly people. In this context, various studies focus on small gauge needles and atraumatic techniques used for lumbar puncture that reduced the post-lumbar complications (Blasko et al. 2006; Anoop et al. 2010). The τ protein, P-tau, and Aβ 42 are mainly CSF-based biomarkers that have been widely assessed for Alzheimer's disease. Interestingly, all studies suggested increasing the level of T-tau and P-tau in CSF, whereas decreasing the level of Aβ 42 in Alzheimer's disease with dementia. The ELISA process was used for the detection of τ protein because it can detect all τ protein isomers. The τ protein in CSF reflects degeneration of neurons. The CSF biomarkers showed the high predictive tools for neurodegenerative diseases, with the sensitivity of 96% for Alzheimer's disease (Humpel 2011; Olson and Humpel 2010; Hock et al. 2000; Anoop et al. 2010).

4.2.2 Blood Plasma Biomarkers

Blood plasma can be easily isolated by centrifugation process, and easier sampling makes it an ideal biomarker for several investigations. The plasma biomarker is reliable biomarkers for Alzheimer's disease. Several blood biomarkers have proposed, and changes in the expression level might be correlated with Alzheimer's disease compared with a healthy control sample. For example, complement factor H (CFH), Aβ, α2-macroglobulin (α2M), α1-antichymotrypsin, and α1-antitrypsin showed high expression level in Alzheimer's disease patients compared with a healthy person. Both CFH and α2M protein present in senile plaque (Bauer et al. 1991; Kuo et al. 2000; Hye et al. 2006; Fukumoto et al. 2003; Dekker et al. 2017). On the other hand, some protein levels like apolipoprotein A1 decrease expression level in Alzheimer's disease patients compared with a healthy person. However, irreproducibility might occur because of the analytical process used in a different laboratory. The Aβ protein is one of the most popular plasma biomarkers as it is the fundamental element of senile plaque. The study suggested that a higher ratio of Aβ42/Aβ40 attributed risk of Alzheimer's disease. The Aβ proteins might be measured in plasma and relate with cerebral β-amyloidosis. The Aβ proteins basically are influenced by its secretion from platelets and cerebral tissue (Hye et al. 2006; Fukumoto et al. 2003). Moreover, the hydrophobic nature of Aβ protein easily binds with plasma protein that might create interfaces.

4.2.3 Urine Biomarkers

The urine or blood-based biomarkers might be advantageous because there is no need for invasive, inexpensive, and required less time for analysis. Usually, changes in the level of neural thread protein (NTP) have been reliable for recognition as a biomarker for Alzheimer's disease. The concentration of NTP protein increases with various stages of Alzheimer's disease. Moreover, AD7c-NTP-associated NTP in CSF also showed accurate data. However, more study is required for effects on AD7c-NTP levels during therapeutic intervention. On the other hand, urine sample collection is easy and noninvasive, compared to CSF and plasma (Munzar et al. 2002; Tuppo et al. 2001; Levy et al. 2007; Ghanbari et al. 1998; Dekker et al. 2017). However, the relatively lower concentration of protein and high level of salt restrict their use as a biomarker.

4.3 Carbon Nanostructure-Based Materials

The combination of nanotechnology, biotechnology, and medicine is an emerging area of research that has been utilized in various aspects mainly prevention, treatment, and detection of various diseases including Alzheimer's disease. Significant research has been done so far in the development of nanomedicine, thereby requires better knowledge or understanding the cellular interaction with nanomaterials and various cell lines or targeted cell lines.

Recently, carbon nanostructure-based materials such as carbon nanotubes (CNTs), carbon nanofibers (CNFs), graphene, and its derivative were synthesized by using chemical vapor deposition (CVD) and exfoliation process. These carbon-based nanomaterials were extensively used in various applications such as water remediation (Talreja et al. 2014; Kumar et al. 2011; Saraswat et al. 2012; Khare et al. 2013; Singh et al. 2013; Talreja et al. 2016), drug delivery, agriculture (Ashfaq et al. 2017a; Kumar et al. 2018), wound dressing (Ashfaq et al. 2017b; Bhadauriya et al. 2018), sensor (Kumar and Talreja 2018; Ashfaq et al. 2018), and antibiotic materials (Ashfaq et al. 2016), etc., due to its unique characteristic and structure (composed of sp2-bonded carbon atom). Usually, carbon nanostructure-based materials characterized mainly of four types on the basis of its dimensionality: (1) zero-dimensional (0-D) like atomic cluster and fullerene; (2) one-dimensional (1-D) like carbon nanorods, nanofibers, nanotubes, and multilayer; (3) two-dimensional (2-D) like graphene and its derivative; and (4) three-dimensional (3-D) like carbon sponge (Afreen et al. 2018). Figure 4.3 shows different carbon-based nanostructured materials. Carbon nanostructure-based materials might provide new insights into biological applications mainly targeted drug delivery, controlled delivery of the drug, and sensor. However, large-scale synthesis, dispersion of nanomaterials in aqueous solution or polymeric composite, and biocompatibility still remain a concern.

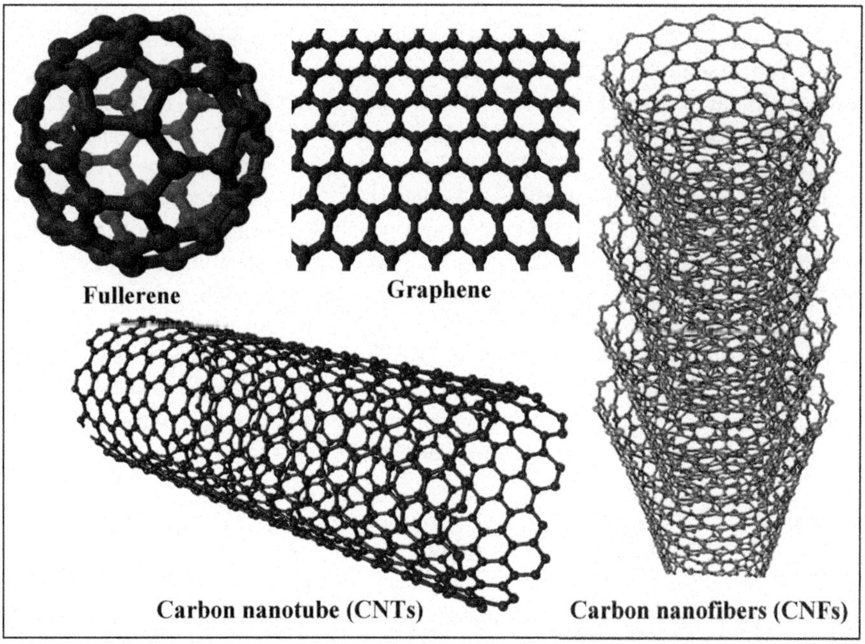

Fig. 4.3 Schematic representation of carbon-based nanostructured materials

4.3.1 Carbon Nanotubes (CNTs)

The CNTs hold promising applications and a lot of attention toward researchers since their discovery by Iijima due to their unique characteristics like high surface area, excellent mechanical ability, tubular structure, and electrical, thermal, and optical properties. Moreover, high biocompatibility of the CNTs is suitable for various end applications including nanomedicine, drug delivery, tissue remodeling, agriculture, and water treatment technologies. The high tensile strength, hollow porous structure, large surface area, and high electronic conductivity of CNTs make them a suitable candidate for electrode materials. Moreover, easy functionalization with functional group or protein considered as smart electrode materials for fabrication of sensing device.

4.3.2 Carbon Nanofibers (CNFs)

Recently, metal nanoparticle-dispersed CNFs are relatively newer materials in the carbon family. The transitional metals like copper (Cu), Zinc (Zn), nickel (Ni), iron (Fe), and bimetallic nanoparticle-dispersed CNFs are grown using the CVD process. The different metal nanoparticles dispersed CNFs extensively used in various end applications such as environmental remediation (chromium, arsenic, vitamin B_{12}, fluoride, and salicylic acid) (Talreja et al. 2014; Kumar et al. 2011; Saraswat et al. 2012; Talreja et al. 2016), biological applications such as drug delivery (Ashfaq et al. 2014), micronutrient delivery (Ashfaq et al. 2017a), agriculture (Kumar et al. 2018), wound dressing (Ashfaq et al. 2017b), controlling of bacterial infections and diabetes (Bhadauriya et al. 2018; Ashfaq et al. 2016), lithium-ion battery (Sharma et al. 2013), hydrogen storage (Yadav et al. 2017), microbial fuel cells (Singh et al. 2016), chemical and biosensors (Kumar and Talreja 2018; Ashfaq et al. 2018). The CNFs also have a unique characteristic like CNTs. The CNFs hold the metal nanoparticles thereby it can control the release of nanoparticles (Ashfaq et al. 2014). Moreover, dispersion of metal nanoparticles in CNFs enhances the applicability toward end applications like the electrical conductivity and antibacterial, antifungal, and catalytic ability compared with CNTs. The slow release of metal nanoparticles from CNFs also aided advantages in various applications like drug delivery and agriculture system. Furthermore, CNFs also have insignificant toxicity among all carbon- (activated carbon, activated fibers, and CNTs) or non-carbon (silica, zeolite, and alumina)-based materials (Ashfaq et al. 2013). Therefore, CNFs might be potential tools for various end applications mainly in the diagnosis and treatment of various diseases.

4.3.3 Graphene and Its Derivative

Graphene and its derivatives like graphene oxide (GO) and reduced graphene oxide (rGO) are one of the most extensively used two-dimensional materials with sp^2-bonded carbon atom. Graphene derivative is similar to graphene analog, but it contains oxygen-containing functional groups like carboxyl, carbonyl, and a hydroxyl. Moreover, graphene and its derivative have excellent electrical, mechanical, thermal, and optical ability, thereby attracting researchers in end applications such as energy, solar cells, lithium-ion battery, supercapacitors, fuel cells, drug delivery, medicine, bioimaging, chemical sensor, and biosensor. The synthesis of graphene mainly occurs by both top-down and bottom-up approaches. Usually, graphite rod milled by the ball-milling process after that graphene synthesized by micromechanical cleavage, scotch tape exfoliation, and liquid-phase exfoliation process (Omar et al. 2019; Afreen et al. 2018; Nag et al. 2018; Mohan et al. 2018; Ren et al. 2018). The liquid-phase exfoliation process is one of the most reliable processes to produce high-quality graphene. However, flake size and yield still remain a concern. In this context, CVD process was used for growing graphene on the substrate (Cu/Ni/Si) to enhance the yield of graphene. Nonetheless, controlling the thickness and detachment from the substrate still remain a concern. The extraordinary optical properties like photoluminescence and fluorescence quenching and Raman properties make them suitable candidates for biosensing or imaging applications (Nag et al. 2018; Mohan et al. 2018; Ren et al. 2018). Moreover, higher surface area achieved higher absorption or binding of functional groups, thereby providing new insight for highly sensitive and accurate biosensor technology platform. In this context, these carbon nanostructure-based materials might have the potential ability to develop novel diagnostic probes.

4.4 Carbon Nanostructured Material-Based Biosensor for Alzheimer's Disease

Alzheimer's disease is the most common neurodegenerative disease categorized by protein aggregation like Aβ plaques and tau protein, thereby resulting in atrophy of brain and loss of cognitive function. Usually, the development of the pathophysiological process of Alzheimer's disease was analyzed by using structural and molecular imaging process. Currently, three types of amyloid-PET tracers have been approved by the FDA and European Medical Agency (EMA) to eliminate Alzheimer's disease from other dementia.

The early-stage diagnosis is one of the most important and beneficial in neurological or Alzheimer's disease, thereby enhancing therapeutic efficacy. In this context, various biosensors such as electrochemical, optical, fluorescent, surface plasmon resonance (SPR), localized surface plasmon resonance (LSPR), and chemiluminescence biosensor have the potential ability to detect different biomarkers of Alzheimer's disease. Electrochemical biosensor was extensively used to detect Alzheimer's disease due to their ease of use, inexpensiveness, sensitivity, and accurateness. Moreover, higher surface area, high transfer of an electron, high chemical stability and biocompatibility, and easy functionalization with a functional group make it a suitable candidate for biosensor (Houmani et al. 2018; Dubois et al. 2015; Sabbagh et al. 2017; Weller and Budson 2018). Herein, our focus is on the carbon nanostructured materials such as CNTs, CNFs, and graphene and its derivative-based biosensors for the detection of different biomarkers of Alzheimer's disease.

4.4.1 Carbon-Based Biosensor for the Detection of Amyloid-β (Aβ) Peptide

The Aβ peptide is one of the potential biomarkers for the detection of Alzheimer's disease (early stage or progression of the disease). In this context, various carbon nanomaterials-based biosensors have been developed for the detection of Aβ peptide as a biomarker of Alzheimer's disease. For example, Vestergaard et al. (2005) fabricate first electrochemical detection of Aβ1–40 and Aβ1–42 protein using glassy carbon electrode. Another study also focuses on the fabrication of reusable biosensor by using magnetic nitrogen-doped graphene modified with Au electrode (MNG) for the detection of an Aβ42 biomarker of Alzheimer's disease. The detection of biomarkers was based on the specific recognition of elements using Aβ1–28 (Aβab) antibodies. The produce MNG-based biosensor has high sensitivity with the detection limit of 5 pg/mL. Moreover, the produce MNG-based biosensor enhanced the detection limit as well as decreased the cost and response time (Li et al. 2016). Figure 4.4 shows the schematic representation of MNG-based electrochemical immunosensor for the detection of Alzheimer's disease. Another study focuses on the fabrication of carbon fiber microelectrodes for the detection of Aβ protein in the CSF sample. The Aβ1–40 and Aβ1–42 antibodies are used for selectivity and accuracy of the carbon fiber microelectrode-based immunosensor. The produce immunosensor (Aβ1–40 and Aβ1–42) has high sensitivity with the detection limit of 20–140 nM, respectively (Prabhulkar et al. 2012).

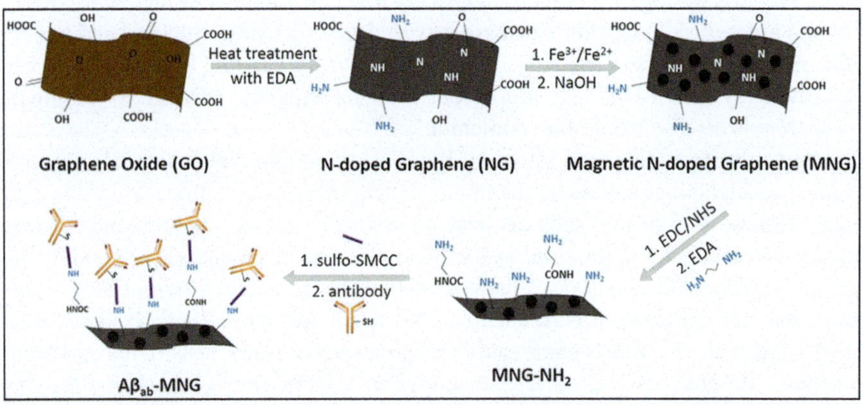

Fig. 4.4 Schematic illustration of MNG-based biosensor for the detection of Alzheimer's disease. (Image taken with permission (Li et al. 2016). This work is licensed under a Creative Commons Attribution 4.0 International License)

4.4.2 Carbon-Based Biosensor for the Detection of MicroRNAs (miRNA)

The miRNAs have the important role in the development and specification of neurons. Several studies suggested that miRNA might be used as a biomarker for the detection of Alzheimer's disease (Zhao et al. 2015; Kumar et al. 2013). In this context, various carbon-based biosensors focus on the detection of miRNA as a biomarker for Alzheimer's disease. For example, Azimzadeh et al. (2017) fabricated electrochemical biosensor by using electrochemically reduced graphene oxide with gold wires and modified with doxorubicin (DOX)-loaded screen-printed carbon electrode (SPCE) for the detection of serum miR-137 as a biomarker of Alzheimer's disease. The produce electrochemical biosensor characterized by using cyclic voltammetry (CV) and electrochemical impedance spectroscopy (EIS) suggested high sensitivity with the detection limit of 1.7 fM. Isin et al. (2017) fabricated graphene oxide (GO)-based biosensor for the detection of miRNA-34a as a biomarker of Alzheimer's disease. For this, activated pencil graphite electrodes were modified with GO for the production of the biosensor. The produce GO-based biosensor was characterized by differential pulse voltammetry (DPV), and CV attributed high sensitivity against miRNA-34a with the detection limit of 7.52 μg/mL. Congur et al. (2015) fabricated GO modified with pencil graphite electrodes for the detection of miRNA-34a. The produce GO-based biosensor was characterized by CV and EIS techniques. The data suggested that produce GO-based biosensor was highly sensitive and selective against miRNA-34a, and the detection limit was found to be 1.9 mg/mL. Erdem et al. (2015) fabricated carbon nanofiber-enriched disposable screen-printed electrodes (CNF-SPEs) for the detection of miRNA-34a. The produce CNF-SPE-based biosensor was characterized by DPV and EIS process. The data suggested the single use and highly sensitive disposable sensor with the detection limit of 10.98 μg/mL. Seo et al. (2017) fabricated graphene-based biosensor for the detection of miRNA-1306. For this, graphene was grown by using ambient air thermal CVD process and then transfer onto the poly (methyl methacrylate) (PMMA) to produce PMMA/graphene film. The produce PMMA/graphene film binds with NH_2-conjugated miRNAs for the detection of miRNA-1306 from samples. The produce PMMA/graphene film-based biosensor was characterized by EIS measurement. The data suggested that produce PMMA/graphene film-based biosensor is highly sensitive and with the lowest detection limit (0.8 fM).

Moreover, Vilela et al. (2017) developed GO-based optical biosensor for the detection of miRNA as a biomarker of cancer and Alzheimer's disease. For this, whole cell lysate was prepared and then mixed with up-conversion nanoparticles and GO. The prepared solution binds with mRNAs associated with Alzheimer's disease (BACE-1) or prostate cancer (PCA3). The produce GO-based optical biosensor has the potential ability to detect miRNA with changes in fluorescence intensity.

4.4.3 Carbon-Based Biosensor for the Detection of Chloride Ions (Cl^-)

The Cl^- ions play important roles in the neuron physiology that closely relates with Alzheimer's disease. The increased chloride intracellular channel-1 (CLIC1) protein with Cl^- ion activity attributed closed relation between the Cl^- ions and neurological diseases. In this context, few studies have been the focus for the detection of Cl^- ions as a biomarker of Alzheimer's disease (Cárdenas-Aguayo et al. 2014; Thal et al. 2008; Toivari et al. 2011; Jalonen et al. 1997). For example, Dong et al. (2017) developed CNT-based biosensor for the detection of Cl^- ions from the live mouse brain. For this, silver (Ag) nanoparticles, methylene blue, and DNA were used as a recognition element of the biosensor. The three-dimensional structure was composed of single-walled carbon nanotubes (SWCNTs) and Au leaves deposited onto the surface of carbon fiber microelectrode to produce CNT-based biosensor. The

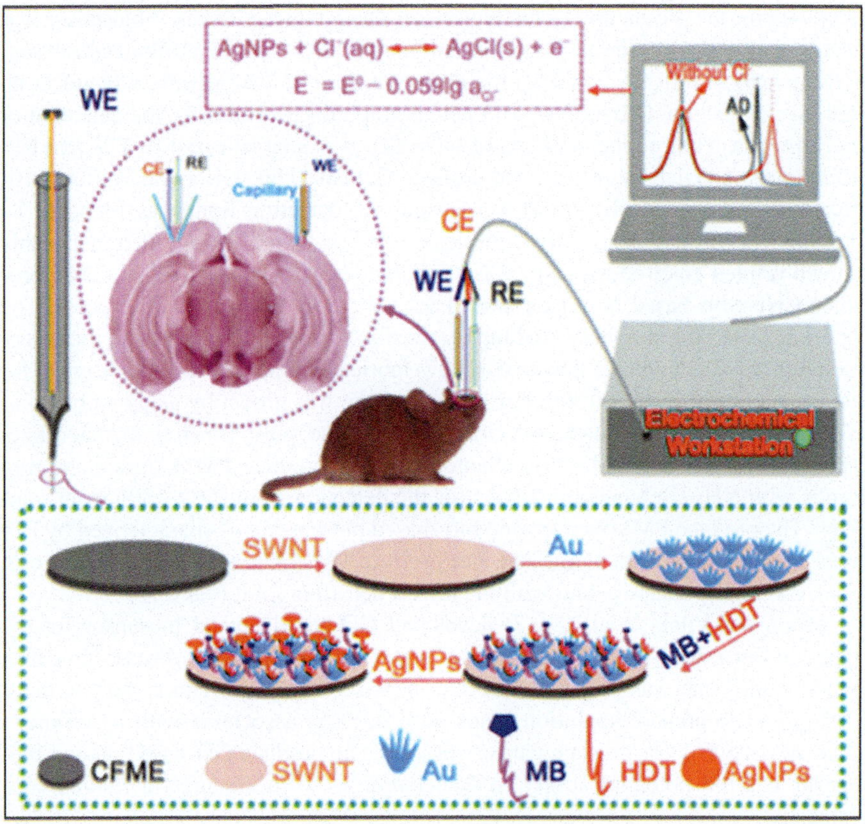

Fig. 4.5 Development of CNTs-based label-free biosensor for the detection of Cl^- level in live mouse brain as the biomarker of Alzheimer's disease. (Image was taken with permission (Dong et al. 2017) Copyright © 2016 American Chemical Society)

produce biosensor has the potential ability to detect Cl^- ions from the brain. The data suggested that the level of Cl^- ions decreases in the striatum and hippocampus by approximately 19.8 ± 0.5 and 27.2 ± 0.3%, respectively. In general, accuracy and reliability of the electrochemical carbon-based biosensor have the potential ability with early detection of Alzheimer's disease that might be useful in future clinical applications (Fig. 4.5).

4.5 Conclusion and Future Prospects

The fabrication of biosensor devices for early detection of Alzheimer's disease and treatment monitoring is one of the most important aspects of diagnosis and therapy. Usually, biomarkers offer valuable information about diseases. However, the sensitivity, accuracy, low cost, and reliable point of care detection are the most important challenge nowadays. Several sensing techniques are used for the detection of various biomarkers. The carbon-based nanomaterials have the potential ability to become a suitable candidate for sensing electrode materials due to its extraordinary mechanical, electrical, optical properties, high biocompatibility, and easy functionalization with several functional groups. The carbon nanomaterials-based biosensor platform might be beneficial for the detection of various diseases especially cancer and Alzheimer's disease. However, a lot of research focuses on the development of newer biosensing platform for early detection of Alzheimer's disease. The carbon nanomaterials-based biosensor are sensitive, specific, accurate, low cost, and adaptable for the detection of Aβ peptide, miRNA, and Cl^- ions as a biomarker for Alzheimer's disease. Future progress in this area is expected to develop newer biomarkers or newer carbon-based hybrid materials-based biosensor to improve sensitivity, accuracy, and specificity, thereby enhancing clinical applicability. Therefore, carbon nanostructured materials offer newer biosensing platform for clinical diagnosis and therapy of Alzheimer's disease.

Acknowledgment The authors acknowledge support from Chulalongkorn University through Chulalongkorn Academic Advancement into its Second Century Project (Small Medical Device) and support from NPDF, SERB, Department of Science and Technology, New Delhi, India, in the form of a research grant (PDF/2016/003602).

References

Afreen S, Omar RA, Talreja N, Chauhan D, Ashfaq M (2018) Carbon-based nanostructured materials for energy and environmental remediation applications. In: Prasad R, Aranda E (eds) Approaches in bioremediation: the new era of environmental microbiology and Nanobiotechnology. Springer, Cham, pp 369–392. https://doi.org/10.1007/978-3-030-02369-0_17

Albert MS, DeKosky ST, Dickson D, Dubois B, Feldman HH, Fox NC, Gamst A, Holtzman DM, Jagust WJ, Petersen RC, Snyder PJ, Carrillo MC, Thies B, Phelps CH (2011a) The diagnosis of mild cognitive impairment due to Alzheimer's disease: recommendations from the National Institute on Aging-Alzheimer's association workgroups on diagnostic guidelines for Alzheimer's disease. Alzheimers Dement 7(3):270–279. https://doi.org/10.1016/j.jalz.2011.03.008

Albert MS, DeKosky ST, Dickson D, Dubois B, Feldman HH, Fox NC, Gamst A, Holtzman DM, Jagust WJ, Petersen RC, Snyder PJ, Carrillo MC, Thies B, Phelps CH (2011b) The diagnosis of mild cognitive impairment due to Alzheimer's disease: recommendations from the National Institute on Aging-Alzheimer's association workgroups on diagnostic guidelines for Alzheimer's disease. Alzheimers Dement 7(3):270–279. https://doi.org/10.1016/j.jalz.2011.03.008

Alves L, Correia ASA, Miguel R, Alegria P, Bugalho P (2012) Alzheimer's disease: a clinical practice-oriented review. Front Neurol 3:63–63. https://doi.org/10.3389/fneur.2012.00063

Anoop A, Singh PK, Jacob RS, Maji SK (2010) CSF biomarkers for Alzheimer's disease diagnosis. Int J Alzheimers Dis 2010:606802. https://doi.org/10.4061/2010/606802

Ashfaq M, Singh S, Sharma A, Verma N (2013) Cytotoxic evaluation of the hierarchical web of carbon micronanofibers. Ind Eng Chem Res 52(12):4672–4682. https://doi.org/10.1021/ie303273s

Ashfaq M, Khan S, Verma N (2014) Synthesis of PVA-CAP-based biomaterial in situ dispersed with Cu nanoparticles and carbon micro-nanofibers for antibiotic drug delivery applications. Biochem Eng J 90:79–89. https://doi.org/10.1016/j.bej.2014.05.016

Ashfaq M, Verma N, Khan S (2016) Copper/zinc bimetal nanoparticles-dispersed carbon nanofibers: a novel potential antibiotic material. Mater Sci Eng C 59:938–947. https://doi.org/10.1016/j.msec.2015.10.079

Ashfaq M, Verma N, Khan S (2017a) Carbon nanofibers as a micronutrient carrier in plants: efficient translocation and controlled release of Cu nanoparticles. Environ Sci Nano 4(1):138–148. https://doi.org/10.1039/C6EN00385K

Ashfaq M, Verma N, Khan S (2017b) Highly effective Cu/Zn-carbon micro/nanofiber-polymer nanocomposite-based wound dressing biomaterial against the P. aeruginosa multi- and extensively drug-resistant strains. Mater Sci Eng C 77:630–641. https://doi.org/10.1016/j.msec.2017.03.187

Ashfaq M, Verma N, Khan S (2018) Novel polymeric composite grafted with metal nanoparticle-dispersed CNFs as a chemiresistive non-destructive fruit sensor material. Mater Chem Phys 217:216–227. https://doi.org/10.1016/j.matchemphys.2018.06.056

Azimzadeh M, Nasirizadeh N, Rahaie M, Naderi-Manesh H (2017) Early detection of Alzheimer's disease using a biosensor based on electrochemically-reduced graphene oxide and gold nanowires for the quantification of serum microRNA-137. RSC Adv 7(88):55709–55719. https://doi.org/10.1039/C7RA09767K

Barber RC (2010) Biomarkers for early detection of Alzheimer disease. J Am Osteopath Assoc 110(9_suppl_8):S10–S15

Bauer J, Strauss S, Schreiter-Gasser U, Ganter U, Schlegel P, Witt I, Yolk B, Berger M (1991) Interleukin-6 and α-2-macroglobulin indicate an acute-phase state in Alzheimer's disease cortices. FEBS Lett 285(1):111–114. https://doi.org/10.1016/0014-5793(91)80737-N

Bhadauriya P, Mamtani H, Ashfaq M, Raghav A, Teotia AK, Kumar A, Verma N (2018) Synthesis of yeast-immobilized and copper nanoparticle-dispersed carbon Nanofiber-based diabetic wound dressing material: simultaneous control of glucose and bacterial infections. ACS Appl Bio Mater 1(2):246–258. https://doi.org/10.1021/acsabm.8b00018

Bjerke M, Engelborghs S (2018) Cerebrospinal fluid biomarkers for early and differential Alzheimer's disease diagnosis. J Alzheimers Dis 62(3):1199–1209. https://doi.org/10.3233/JAD-170680

Blasko I, Lederer W, Oberbauer H, Walch T, Kemmler G, Hinterhuber H, Marksteiner J, Humpel C (2006) Measurement of thirteen biological markers in CSF of patients with Alzheimer's disease and other dementias. Dement Geriatr Cogn Disord 21(1):9–15. https://doi.org/10.1159/000089137

Brookmeyer R, Johnson E, Ziegler-Graham K, Arrighi HM (2007) Forecasting the global burden of Alzheimer's disease. Alzheimers Dement 3(3):186–191. https://doi.org/10.1016/j.jalz.2007.04.381

Cárdenas-Aguayo MC, Gómez-Virgilio L, DeRosa S, Meraz-Ríos MA (2014) The role of tau oligomers in the onset of Alzheimer's disease neuropathology. ACS Chem Neurosci 5(12):1178–1191. https://doi.org/10.1021/cn500148z

Choi EJ, Bang H, Im JH, Chung SJ, Lee J-H (2005) A case of biopsy-proven early-onset Alzheimer's disease with Hemiparkinsonism. J Clin Neurol (Seoul, Korea) 1(1):97–100. https://doi.org/10.3988/jcn.2005.1.1.97

Chowdhury A, Kunjiappan S, Panneerselvam T, Somasundaram B, Bhattacharjee C (2017) Nanotechnology and nanocarrier-based approaches on treatment of degenerative diseases. Int Nano Lett 7(2):91–122. https://doi.org/10.1007/s40089-017-0208-0

Congur G, Eksin E, Erdem A (2015) Impedimetric detection of microRNA at graphene oxide modified sensors. Electrochimica Acta 172:20–27. https://doi.org/10.1016/j.electacta.2015.03.210

Cummings J, Lee G, Ritter A, Zhong K (2018) Alzheimer's disease drug development pipeline: 2018. Alzheimers Dement (New York, N Y) 4:195–214. https://doi.org/10.1016/j.trci.2018.03.009

Dekker AD, Fortea J, Blesa R, De Deyn PP (2017) Cerebrospinal fluid biomarkers for Alzheimer's disease in down syndrome. Alzheimers Dement (Amsterdam, Netherlands) 8:1–10. https://doi.org/10.1016/j.dadm.2017.02.006

Dong H, Zhang L, Liu W, Tian Y (2017) Label-free electrochemical biosensor for monitoring of chloride ion in an animal model of Alzhemier's disease. ACS Chem Neurosci 8(2):339–346. https://doi.org/10.1021/acschemneuro.6b00296

Dubois B, Picard G, Sarazin M (2009) Early detection of Alzheimer's disease: new diagnostic criteria. Dialogues Clin Neurosci 11(2):135–139

Dubois B, Padovani A, Scheltens P, Rossi A, Dell'Agnello G (2015) Timely diagnosis for Alzheimer's disease: a literature review on benefits and challenges. J Alzheimers Dis 49(3):617–631. https://doi.org/10.3233/JAD-150692

Erdem A, Eksin E, Congur G (2015) Indicator-free electrochemical biosensor for microRNA detection based on carbon nanofibers modified screen printed electrodes. J Electroanal Chem 755:167–173. https://doi.org/10.1016/j.jelechem.2015.07.031

Fonseca-Santos B, Gremião MPD, Chorilli M (2015) Nanotechnology-based drug delivery systems for the treatment of Alzheimer's disease. Int J Nanomedicine 10:4981–5003. https://doi.org/10.2147/IJN.S87148

Frisoni GB, Boccardi M, Barkhof F, Blennow K, Cappa S, Chiotis K, Démonet J-F, Garibotto V, Giannakopoulos P, Gietl A, Hansson O, Herholz K, Jack CR Jr, Nobili F, Nordberg A, Snyder HM, Ten Kate M, Varrone A, Albanese E, Becker S, Bossuyt P, Carrillo MC, Cerami C, Dubois B, Gallo V, Giacobini E, Gold G, Hurst S, Lönneborg A, Lovblad K-O, Mattsson N, Molinuevo J-L, Monsch AU, Mosimann U, Padovani A, Picco A, Porteri C, Ratib O, Saint-Aubert L, Scerri C, Scheltens P, Schott JM, Sonni I, Teipel S, Vineis P, Visser PJ, Yasui Y, Winblad B (2017) Strategic roadmap for an early diagnosis of Alzheimer's disease based on biomarkers. Lancet Neurol 16(8):661–676. https://doi.org/10.1016/S1474-4422(17)30159-X

Fukumoto H, Tennis M, Locascio JJ, Hyman BT, Growdon JH, Irizarry MC (2003) Age but not diagnosis is the main predictor of plasma amyloid β-protein levels. Arch Neurol 60(7):958–964. https://doi.org/10.1001/archneur.60.7.958

Ghanbari H, Ghanbari K, Beheshti I, Munzar M, Vasauskas A, Averback P (1998) Biochemical assay for AD7C-NTP in urine as an Alzheimer's disease marker. J Clin Lab Anal 12(5):285–288. https://doi.org/10.1002/(SICI)1098-2825(1998)12:5<285::AID-JCLA6>3.0.CO;2-5

Hajipour MJ, Santoso MR, Rezaee F, Aghaverdi H, Mahmoudi M, Perry G (2017) Advances in Alzheimer's diagnosis and therapy: the implications of nanotechnology. Trends Biotechnol 35(10):937–953. https://doi.org/10.1016/j.tibtech.2017.06.002

Hansen N (2014) Brain stimulation for combating Alzheimer's disease. Front Neurol 5:80–80. https://doi.org/10.3389/fneur.2014.00080

Hardy J (2009) The amyloid hypothesis for Alzheimer's disease: a critical reappraisal. J Neurochem 110(4):1129–1134. https://doi.org/10.1111/j.1471-4159.2009.06181.x

Hock C, Heese K, Müller-Spahn F, Huber P, Riesen W, Nitsch RM, Otten U (2000) Increased CSF levels of nerve growth factor in patients with Alzheimer's disease. Neurology 54(10):2009. https://doi.org/10.1212/WNL.54.10.2009

Houmani N, Vialatte F, Gallego-Jutglà E, Dreyfus G, Nguyen-Michel V-H, Mariani J, Kinugawa K (2018) Diagnosis of Alzheimer's disease with electroencephalography in a differential framework. PLoS One 13(3):e0193607. https://doi.org/10.1371/journal.pone.0193607

Huang Y, Mucke L (2012) Alzheimer mechanisms and therapeutic strategies. Cell 148(6):1204–1222. https://doi.org/10.1016/j.cell.2012.02.040

Humpel C (2011) Identifying and validating biomarkers for Alzheimer's disease. Trends Biotechnol 29(1):26–32. https://doi.org/10.1016/j.tibtech.2010.09.007

Hye A, Lynham S, Thambisetty M, Causevic M, Campbell J, Byers HL, Hooper C, Rijsdijk F, Tabrizi SJ, Banner S, Shaw CE, Foy C, Poppe M, Archer N, Hamilton G, Powell J, Brown RG, Sham P, Ward M, Lovestone S (2006) Proteome-based plasma biomarkers for Alzheimer's disease. Brain 129(11):3042–3050. https://doi.org/10.1093/brain/awl279

Iqbal K, Liu F, Gong C-X (2015) Tau and neurodegenerative disease: the story so far. Nat Rev Neurol 12:15. https://doi.org/10.1038/nrneurol.2015.225. https://www.nature.com/articles/nrneurol.2015.225#supplementary-information

Isaac MT, Vamvakas S, Isaac MB (2017) Diagnostic biomarkers for Alzheimer's disease: a regulatory view. Lancet Neurol 16(8):580–581. https://doi.org/10.1016/S1474-4422(17)30223-5

Isin D, Eksin E, Erdem A (2017) Graphene oxide modified single-use electrodes and their application for voltammetric miRNA analysis. Mater Sci Eng C 75:1242–1249. https://doi.org/10.1016/j.msec.2017.02.166

Jain KK (2005) The role of nanobiotechnology in drug discovery. Drug Discov Today 10(21):1435–1442. https://doi.org/10.1016/S1359-6446(05)03573-7

Jalonen TO, Charniga CJ, Wielt DB (1997) β-Amyloid peptide-induced morphological changes coincide with increased K+ and Cl− channel activity in rat cortical astrocytes. Brain Res 746(1):85–97. https://doi.org/10.1016/S0006-8993(96)01189-4

Jellinger KA, Janetzky B, Attems J, Kienzl E (2008) Biomarkers for early diagnosis of Alzheimer disease: 'ALZheimer ASsociated gene'–a new blood biomarker? J Cell Mol Med 12(4):1094–1117. https://doi.org/10.1111/j.1582-4934.2008.00313.x

Khare P, Talreja N, Deva D, Sharma A, Verma N (2013) Carbon nanofibers containing metal-doped porous carbon beads for environmental remediation applications. Chem Eng J 229:72–81. https://doi.org/10.1016/j.cej.2013.04.113

Kim H, Lee JU, Song S, Kim S, Sim SJ (2018) A shape-code nanoplasmonic biosensor for multiplex detection of Alzheimer's disease biomarkers. Biosen Bioelectron 101:96–102. https://doi.org/10.1016/j.bios.2017.10.018

Kumar D, Talreja N (2018) Nickel nanoparticles-doped rhodamine grafted carbon nanofibers as colorimetric probe: naked eye detection and highly sensitive measurement of aqueous Cr3+ and Pb2+. Korean J Chem Eng. https://doi.org/10.1007/s11814-018-0139-0

Kumar V, Talreja N, Deva D, Sankararamakrishnan N, Sharma A, Verma N (2011) Development of bi-metal doped micro- and nano multi-functional polymeric adsorbents for the removal of fluoride and arsenic(V) from wastewater. Desalination 282:27–38. https://doi.org/10.1016/j.desal.2011.05.013

Kumar P, Dezso Z, MacKenzie C, Oestreicher J, Agoulnik S, Byrne M, Bernier F, Yanagimachi M, Aoshima K, Oda Y (2013) Circulating miRNA biomarkers for Alzheimer's disease. PLoS One 8(7):e69807. https://doi.org/10.1371/journal.pone.0069807

Kumar A, Singh A, Ekavali SA (2015) A review on Alzheimer's disease pathophysiology and its management: an update. Pharmacol Rep 67(2):195–203. https://doi.org/10.1016/j.pharep.2014.09.004

Kumar R, Ashfaq M, Verma N (2018) Synthesis of novel PVA–starch formulation-supported Cu–Zn nanoparticle carrying carbon nanofibers as a nanofertilizer: controlled release of micronutrients. J Mater Sci 53(10):7150–7164. https://doi.org/10.1007/s10853-018-2107-9

Kuo Y-M, Kokjohn TA, Kalback W, Luehrs D, Galasko DR, Chevallier N, Koo EH, Emmerling MR, Roher AE (2000) Amyloid-β peptides interact with plasma proteins and erythrocytes:

implications for their quantitation in plasma. Biochem Biophys Res Commun 268(3):750–756. https://doi.org/10.1006/bbrc.2000.2222

Kurz A, Perneczky R (2011) Novel insights for the treatment of Alzheimer's disease. Prog Neuro-Psychopharmacol Biol Psychiatry 35(2):373–379. https://doi.org/10.1016/j.pnpbp.2010.07.018

Lashley T, Schott JM, Weston P, Murray CE, Wellington H, Keshavan A, Foti SC, Foiani M, Toombs J, Rohrer JD, Heslegrave A, Zetterberg H (2018) Molecular biomarkers of Alzheimer's disease: progress and prospects. Disease Models & Mechanisms 11(5):dmm031781. https://doi.org/10.1242/dmm.031781

Levy S, McConville M, Lazaro GA, Averback P (2007) Competitive ELISA studies of neural thread protein in urine in Alzheimer's disease. J Clin Lab Anal 21(1):24–33. https://doi.org/10.1002/jcla.20159

Li S-S, Lin C-W, Wei K-C, Huang C-Y, Hsu P-H, Liu H-L, Lu Y-J, Lin S-C, Yang H-W, Ma C-CM (2016) Non-invasive screening for early Alzheimer's disease diagnosis by a sensitively immunomagnetic biosensor. Sci Rep 6:25155. https://doi.org/10.1038/srep25155. https://www.nature.com/articles/srep25155#supplementary-information

Lista S, Faltraco F, Prvulovic D, Hampel H (2013) Blood and plasma-based proteomic biomarker research in Alzheimer's disease. Prog Neurobiol 101–102:1–17. https://doi.org/10.1016/j.pneurobio.2012.06.007

Mantzavinos V, Alexiou A (2017) Biomarkers for Alzheimer's disease diagnosis. Curr Alzheimer Res 14(11):1149–1154. https://doi.org/10.2174/1567205014666170203125942

Mohan VB, Lau K-T, Hui D, Bhattacharyya D (2018) Graphene-based materials and their composites: a review on production, applications and product limitations. Compos Part B Eng 142:200–220. https://doi.org/10.1016/j.compositesb.2018.01.013

Munzar M, Levy S, Rush R, Averback P (2002) Clinical study of a urinary competitve ELISA for neural thread protein in Alzheimer disease. Neurol Clin Neurophysiol 2002(1):2–8. https://doi.org/10.1162/15268740252769709

Nag A, Mitra A, Mukhopadhyay SC (2018) Graphene and its sensor-based applications: a review. Sens Actuators A Phys 270:177–194. https://doi.org/10.1016/j.sna.2017.12.028

Nobile S, Nobile L (2017) Nanotechnology for biomedical applications: recent advances in neurosciences and bone tissue engineering. Polym Eng Sci 57(7):644–650. https://doi.org/10.1002/pen.24595

Olson L, Humpel C (2010) Growth factors and cytokines/chemokines as surrogate biomarkers in cerebrospinal fluid and blood for diagnosing Alzheimer's disease and mild cognitive impairment. Exp Gerontol 45(1):41–46. https://doi.org/10.1016/j.exger.2009.10.011

Omar RA, Afreen S, Talreja N, Chauhan D, Ashfaq M (2019) Impact of nanomaterials in plant systems, in plant nanobionics: volume 1. In Prasad R (ed) Advances in the understanding of nanomaterials research and applications. Springer International Publishing, Cham, pp 117–140

Ovais M, Zia N, Ahmad I, Khalil AT, Raza A, Ayaz M, Sadiq A, Ullah F, Shinwari ZK (2018) Phytotherapeutic and Nanomedicinal approaches to cure Alzheimer's disease: present status and future opportunities. Front Aging Neurosci 10(284). https://doi.org/10.3389/fnagi.2018.00284

Prabhulkar S, Piatyszek R, Cirrito JR, Wu Z-Z, Li C-Z (2012) Microbiosensor for Alzheimer's disease diagnostics: detection of amyloid beta biomarkers. J Neurochem 122(2):374–381. https://doi.org/10.1111/j.1471-4159.2012.07709.x

Rees M, Moghimi SM (2012) Nanotechnology: from fundamental concepts to clinical applications for healthy aging. Maturitas 73(1):1–4. https://doi.org/10.1016/j.maturitas.2012.05.010

Ren S, Rong P, Yu Q (2018) Preparations, properties and applications of graphene in functional devices: a concise review. Ceram Int 44(11):11940–11955. https://doi.org/10.1016/j.ceramint.2018.04.089

Ringman JM, Liang L-J, Zhou Y, Vangala S, Teng E, Kremen S, Wharton D, Goate A, Marcus DS, Farlow M, Ghetti B, McDade E, Masters CL, Mayeux RP, Rossor M, Salloway S, Schofield PR, Cummings JL, Buckles V, Bateman R, Morris JC, Dominantly Inherited Alzheimer N (2015) Early behavioural changes in familial Alzheimer's disease in the dominantly inher-

ited Alzheimer network. Brain J Neurol 138(Pt 4):1036–1045. https://doi.org/10.1093/brain/awv004
Sabbagh MN, Lue L-F, Fayard D, Shi J (2017) Increasing precision of clinical diagnosis of Alzheimer's disease using a combined algorithm incorporating clinical and novel biomarker data. Neurol Ther 6(Suppl 1):83–95. https://doi.org/10.1007/s40120-017-0069-5
Sahoo SK, Parveen S, Panda JJ (2007) The present and future of nanotechnology in human health care. Nanomed Nanotechnol Biol Med 3(1):20–31. https://doi.org/10.1016/j.nano.2006.11.008
Sangubotla R, Kim J (2018) Recent trends in analytical approaches for detecting neurotransmitters in Alzheimer's disease. TrAC Trends in Analytical Chemistry 105:240–250. https://doi.org/10.1016/j.trac.2018.05.014
Sankararamakrishnan N, Chauhan D, Dwivedi J (2016) Synthesis of functionalized carbon nanotubes by floating catalytic chemical vapor deposition method and their sorption behavior toward arsenic. Chem Eng J 284:599–608. https://doi.org/10.1016/j.cej.2015.08.145
Saraswat R, Talreja N, Deva D, Sankararamakrishnan N, Sharma A, Verma N (2012) Development of novel in situ nickel-doped, phenolic resin-based micro–nano-activated carbon adsorbents for the removal of vitamin B-12. Chem Eng J 197:250–260. https://doi.org/10.1016/j.cej.2012.05.046
Seo DH, Pineda S, Fang J, Gozukara Y, Yick S, Bendavid A, Lam SKH, Murdock AT, Murphy AB, Han ZJ, Ostrikov K (2017) Single-step ambient-air synthesis of graphene from renewable precursors as electrochemical genosensor. Nat Commun 8:14217. https://doi.org/10.1038/ncomms14217. https://www.nature.com/articles/ncomms14217#supplementary-information
Sharma N, Singh AN (2016) Exploring biomarkers for Alzheimer's disease. J Clin Diagn Res 10(7):KE01–KE06. https://doi.org/10.7860/JCDR/2016/18828.8166
Sharma AK, Khare P, Singh JK, Verma N (2013) Preparation of novel carbon microfiber/carbon nanofiber-dispersed polyvinyl alcohol-based nanocomposite material for lithium-ion electrolyte battery separator. Mater Sci Eng C 33(3):1702–1709. https://doi.org/10.1016/j.msec.2012.12.083
Shi J, Fang Y (2018) 9 – biomedical applications of graphene. In: Zhu H, Xu Z, Xie D, Fang Y (eds) Graphene. Academic, pp 215–232. https://doi.org/10.1016/B978-0-12-812651-6.00009-4
Singh S, Ashfaq M, Singh RK, Joshi HC, Srivastava A, Sharma A, Verma N (2013) Preparation of surfactant-mediated silver and copper nanoparticles dispersed in hierarchical carbon micro-nanofibers for antibacterial applications. New Biotechnol 30(6):656–665. https://doi.org/10.1016/j.nbt.2013.05.002
Singh S, Modi A, Verma N (2016) Enhanced power generation using a novel polymer-coated nanoparticles dispersed-carbon micro-nanofibers-based air-cathode in a membrane-less single chamber microbial fuel cell. Int J Hydrogen Energy 41(2):1237–1247. https://doi.org/10.1016/j.ijhydene.2015.10.099
Talreja N, Kumar D, Verma N (2014) Removal of hexavalent chromium from water using Fe-grown carbon nanofibers containing porous carbon microbeads. J Water Process Eng 3:34–45. https://doi.org/10.1016/j.jwpe.2014.08.001
Talreja N, Verma N, Kumar D (2016) Carbon bead-supported ethylene Diamine-functionalized carbon nanofibers: an efficient adsorbent for salicylic acid. Clean (Weinh) 44(11):1461–1470. https://doi.org/10.1002/clen.201500722
Thal DR, Griffin WST, Braak H (2008) Parenchymal and vascular Abeta-deposition and its effects on the degeneration of neurons and cognition in Alzheimer's disease. J Cell Mol Med 12(5B):1848–1862. https://doi.org/10.1111/j.1582-4934.2008.00411.x
Thambisetty M, Lovestone S (2010) Blood-based biomarkers of Alzheimer's disease: challenging but feasible. Biomark Med 4(1):65–79. https://doi.org/10.2217/bmm.09.84
Thambisetty M, Simmons A, Hye A, Campbell J, Westman E, Zhang Y, Wahlund L-O, Kinsey A, Causevic M, Killick R, Kloszewska I, Mecocci P, Soininen H, Tsolaki M, Vellas B, Spenger C, Lovestone S, for the AddNeuroMed c (2011) Plasma biomarkers of brain atrophy in Alzheimer's disease. PLoS One 6(12):e28527. https://doi.org/10.1371/journal.pone.0028527
Toivari E, Manninen T, Nahata AK, Jalonen TO, Linne M-L (2011) Effects of transmitters and amyloid-beta peptide on calcium signals in rat cortical astrocytes: Fura-2AM measurements

and stochastic model simulations. PLoS One 6(3):e17914–e17914. https://doi.org/10.1371/journal.pone.0017914

Tuppo EE, Forman LJ, Spur BW, Chan-Ting RE, Chopra A, Cavalieri TA (2001) Sign of lipid peroxidation as measured in the urine of patients with probable Alzheimer's disease. Brain Res Bull 54(5):565–568. https://doi.org/10.1016/S0361-9230(01)00450-6

Vestergaard M, Kerman K, Saito M, Nagatani N, Takamura Y, Tamiya E (2005) A rapid label-free electrochemical detection and kinetic study of Alzheimer's amyloid Beta aggregation. J Am Chem Soc 127(34):11892–11893. https://doi.org/10.1021/ja052522q

Vilela P, El-Sagheer A, Millar TM, Brown T, Muskens OL, Kanaras AG (2017) Graphene oxide-Upconversion nanoparticle based optical sensors for targeted detection of mRNA biomarkers present in Alzheimer's disease and prostate cancer. ACS Sens 2(1):52–56. https://doi.org/10.1021/acssensors.6b00651

Weller J, Budson A (2018) Current understanding of Alzheimer's disease diagnosis and treatment [version 1; referees: 2 approved]. F1000Research 7(1161). https://doi.org/10.12688/f1000research.14506.1

Yadav A, Faisal M, Subramaniam A, Verma N (2017) Nickel nanoparticle-doped and steam-modified multiscale structure of carbon micro-nanofibers for hydrogen storage: effects of metal, surface texture and operating conditions. Int J Hydrogen Energy 42(9):6104–6117. https://doi.org/10.1016/j.ijhydene.2016.11.070

Yiannopoulou KG, Papageorgiou SG (2013) Current and future treatments for Alzheimer's disease. Ther Adv Neurol Disord 6(1):19–33. https://doi.org/10.1177/1756285612461679

Zäch M, Hägglund C, Chakarov D, Kasemo B (2006) Nanoscience and nanotechnology for advanced energy systems. Curr Opin Solid State Mater Sci 10(3):132–143. https://doi.org/10.1016/j.cossms.2007.04.004

Zhao Y, Tan W, Sheng W, Li X (2015) Identification of biomarkers Associated with Alzheimer's disease by bioinformatics analysis. Am J Alzheimers Dis Other Dement 31(2):163–168. https://doi.org/10.1177/1533317515588181

Oxidative Stress in Alzheimer's Disease: Molecular Hallmarks of Underlying Vulnerability

5

Md. Sahab Uddin and Md. Tanvir Kabir

Abstract

Alzheimer's disease (AD) is the most common neurodegenerative disease characterized by the presence of senile plaques (SPs) and neurofibrillary tangles (NFTs) in the hippocampus and cortex of afflicted patients. In case of AD patients, during the progression of the disease, there is proof that brain tissues of these individuals are exposed to oxidative stress (OS). In AD, advanced glycation end products (AGEs) exist in amyloid plaques; moreover, accelerated oxidation of glycated proteins might cause its extracellular accumulation. AGEs have also found to take part in neuronal death, initiating production of free radical and therefore increasing OS. In case of progression of AD, emerging proof has shown that OS plays a key role. Nevertheless, the processes that ultimately cause disturbance of redox balance and also the sources of the free radicals are still unclear. Likewise, an excessive amount of reactive oxygen species (ROS) might be produced from processes including dysfunction of mitochondria and/or abnormal transition metal accumulation, though the redox imbalance seems to be promoted by the aberrant accumulation of amyloid ß (Aβ) and tau proteins. For Aβ- and tau-mediated neurotoxicity, the resulted OS has been associated. There is also the availability of evidence that OS might increase the aggregation and production of Aβ and also assist the polymerization as well as phosphorylation of tau, therefore creating a malicious cycle that stimulates the progression and even initiation of Alzheimer's. This chapter represents the critical pathogenic mechanism of OS and AD.

Md. S. Uddin (✉)
Department of Pharmacy, Southeast University, Dhaka, Bangladesh

Pharmakon Neuroscience Research Network, Dhaka, Bangladesh
e-mail: msu-neuropharma@hotmail.com

Md. T. Kabir
Department of Pharmacy, BRAC University, Dhaka, Bangladesh

G. M. Ashraf, A. Alexiou (eds.), *Biological, Diagnostic and Therapeutic Advances in Alzheimer's Disease*, https://doi.org/10.1007/978-981-13-9636-6_5

Keywords
Oxidative stress · Alzheimer's disease · Amyloid ß · Tau · Mitochondria dysfunction · Metal dyshomeostasis

Abbreviations

AD Alzheimer's disease
SPs senile plaques
NFTs neurofibrillary tangles
OS oxidative stress

5.1 Introduction

Oxidative balance is now regarded as a significant issue to understand the Alzheimer's disease (AD) pathogenesis. AD brain examination has proved a great extent of oxidative damage, connected with both normal appearing pyramidal neurons as well as in hallmark pathologies such as senile plaques (SPs) and neurofibrillary tangles (NFTs) (Smith et al. 2000; Uddin et al. 2016a, 2018a). Worldwide, the population of the elderly people continues to rise, and henceforth AD prevalence has significantly increased. Furthermore, among the elderly, AD is now considered as one of the foremost causes of death and also disability (Uddin et al. 2019a). Although significant advancement has been observed in case of AD research, the precise pathogenesis and causes of AD are yet to be fully understood (Uddin et al. 2019b). At present, there is no treatment available which can effectively treat the disease (Uddin et al. 2019c).

The occurrences of neuronal loss, NFTs, and SPs are the key AD pathological features of AD brains (Uddin and Rashid 2019). SPs are chiefly composed of Aβ peptide that is generated via proteolytic cleavage of the transmembrane amyloid precursor protein (APP) (Uddin and Haque 2016). Arrays of paired helical filaments (PHFs) structures are the key structural features of NFTs; furthermore, they primarily contain self-aggregated hyperphosphorylated tau, which is regarded as a multifunctional protein found to be involved in microtubule stabilization and assembly (Uddin et al. 2016b). Growing evidence has exhibited that occurrence of widespread OS is a feature of AD brains, along with the well-known pathology of NFTs and SPs (Praticò 2008; Uddin et al. 2016c). Furthermore, in AD brains, it has been found that the levels of 3-nitrotyrosine and protein carbonyls, which are produced from protein oxidation, and markers of oxidative damage to deoxyribonucleic acid (DNA) and ribonucleic acid (RNA), for instance, 8-hydroxyguanosine (8-OHG) and 8-hydroxydeoxyguanosine (8-OHdG) are increased (Nunomura et al. 1999). F2-isoprostanes (F2-IsoPs), malondialdehyde (MDA), and 4-hydroxy-2-nonenal

(HNE) are the products of lipid peroxidation which are also found to rise in multiple areas of the brain and cerebrospinal fluid of people containing mild cognitive impairment (MCI) or AD (Praticò and Sung 2004; Lovell et al. 1995).

In case of peripheral tissues and also central nervous system of AD patients, along with the buildup of free radical damage, it has been observed that changes in the expression or actions of antioxidant enzymes including catalase and superoxide dismutase (SOD) have been detected (Uddin et al. 2016a; Marcus et al. 1998; Padurariu et al. 2010). In addition, in case of MCI and AD brains, the augmented oxidative damage to proteins and lipids and the diminished antioxidant enzyme and glutathione actions are more localized to the synapses and also correspond with the extent of the disease, signifying contribution of OS in the synaptic loss which is related to AD (Ansari and Scheff 2010; Kasapoglu and Ozben 2001). Essentially, it needs to be noted that most of the aforesaid studies exhibit increased OS in MCI, which is suggested as an intermediate state among dementia and normal aging, signifying that the OS damage in Alzheimer's might take place preceding the onset of the disease. Moreover, during the development and initiation of AD, these previously mentioned findings collectively recommend that OS might be one of the initial changes that take place.

OS is regarded as a proximal event in AD pathogenesis. Nevertheless, the exact processes through which redox balance is changed in the disease still stay obscure (Zhu et al. 2004; Uddin et al. 2018b, 2016d). OS is intimately associated with a number of crucial pathological mechanisms found in AD. Therefore, the purpose of this chapter is to emphasize the impact of OS in the pathogenesis of AD including neurotoxicity induced by Aβ, tau pathology, metal dyshomeostasis, and dysfunction of mitochondria.

5.2 Oxidative Stress in Alzheimer's Disease

There are an enormous number of proofs that, during the disease advancement, brain tissue in AD individuals is introduced to OS (Fig. 5.1). As OS is categorized by a disproportion in the radical production of antioxidative defense and ROS, both of them are regarded to have a significant contribution to cognitive decline and process of age-associated neurodegeneration (Gsell et al. 1996; Sayre et al. 2008, 2005).

5.2.1 Lipid Oxidation in Alzheimer's

In AD brains, ROS has found to cause modifications of lipids, and there is a strong link among NFTs, amyloid plaques, antioxidant enzymes, and lipid peroxides (Lovell et al. 1995). Furthermore, in case of AD brains, a number of breakdown products of OS, such as F2-IsoPs, MDA, acrolein, and HNE, have been detected and compared to age-matched controls (Mark et al. 1997; Markesbery and Lovell 1998). Interestingly, it has been found that HNE has the ability to alter proteins, resulting

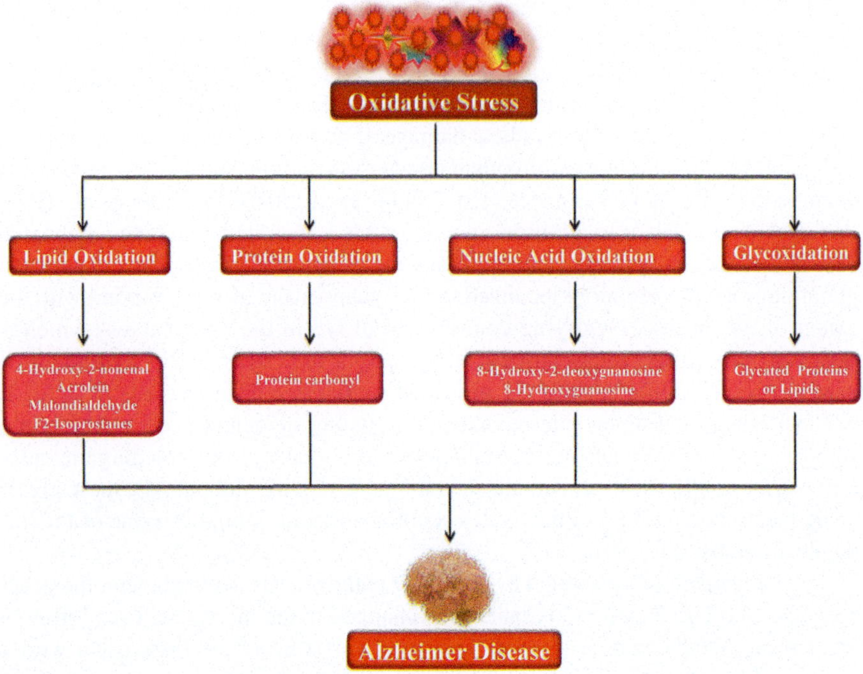

Fig. 5.1 Pathogenic mechanisms by which oxidative stress triggers Alzheimer's pathogenesis.

in a group of effects, such as dysregulation of intracellular calcium (Ca^{2+}) signaling, activation of kinases, sodium-potassium adenosine triphosphatase pump (Na^+/K^+-ATPase) inhibition, glutamate transporter inhibition, and neuronal glucose inhibition that eventually can lead to induction of an apoptotic cascade process (Keller et al. 1997; Mattson and Chan 2003). Since NFTs have adducts of HNE and MDA (i.e., an extremely reactive products of lipid peroxidation), NFTs are found to contain the footprints of oxidative membrane damage.

In addition, dystrophic neurites of SPs that contain filaments of NFTs exhibit higher membrane damage as compared to those that have a deficiency of filaments (Serrano-Pozo et al. 2011). There is rising evidence which recommends that bifunctional HNE is considered as the main cytotoxic products of lipid peroxidation. In AD cases, after lipid peroxidation, a 2-pentylpyrrole modification of lysine is the only currently recognized "advanced" (stable end product) adduct that generates from the alteration of proteins via HNE. Henceforth, the aforesaid results along with the current evidence exhibit that HNE is cytotoxic to neurons, and also it damages the action of membrane proteins including the neuronal glucose transporter 3 (GLUT3), which specify HNE as a distinctive marker that can lead to neurodegeneration in Alzheimer's (Bruce-Keller et al. 1998; Chen and Kong 2012).

5.2.2 Protein Oxidation in Alzheimer's

Oxidation of protein side chains arbitrated by ROS has been studied (Davies 2005), and it is found to result in the protein-based carbonyl's generation or in the hydroxyl group introduction. Furthermore, through oxidizing amino acid residue side-chain hydroxyls into an aldehyde or ketone derivatives, carbonyl groups are presented in proteins (Berlett and Stadtman 1997; Wong et al. 2010). Carbonylation of proteins can be caused via a number of oxidative pathways (Dalle-Donne et al. 2014). Interestingly, the introduction of carbonyl groups in proteins can be mediated through direct oxidation of threonine residues, proline, arginine, and lysine or due to the cleavage of peptide bonds via oxidation of glutamyl residues or via α-amidation pathway (Weng et al. 2017).

ROS has also been found to react with other molecules, including sugars (i.e., glycoxidation), DNA (i.e., DNA oxidation), and lipids (i.e., oxidation of lipid), leading to the production of reactive aldehydes and carbonyl derivatives, which can sequentially cause a reaction with proteins and can ultimately generate protein-bound carbonyls (Rahman et al. 2017). On the other hand, protein carbonylation measurement is supposed to be a decent assessment for the degree of oxidative damage of proteins linked with a range of conditions of OS, physiological disorders, neurodegeneration, AD, and aging (Korolainen et al. 2007; Smith et al. 1998).

5.2.3 Nucleic Acid Oxidation in Alzheimer's

DNA bases are found to be susceptible to OS-mediated damage, including nitration, protein carbonylation, and hydroxylation (Gabbita et al. 1998; Collins et al. 1996; Lovell and Markesbery 2007). In case of AD, it has been noticed that ROS of the brain triggers Ca^{2+} influx, through glutamate receptors, and can ultimately trigger a response which is excitotoxic in nature, and this can result in cell death (Mattson and Chan 2003). In addition, oxygen can react with unregulated redox-active metals to generate ROS (White et al. 2006). RNA and DNA oxidations are discernable by augmented levels of 8-OHD and 8-OHdG (Nunomura et al. 1999, 2001). Moreover, these aforementioned markers have been contained in NFTs and Aβ plaques (Mecocci et al. 1994).

In AD, raised levels of DNA strand breaks have been observed (Moreira et al. 2008). Although they were initially regarded as a part of apoptosis, it is now well-known that oxidative damage is accountable for DNA strand breaks. Furthermore, in case of AD, the aforesaid phenomenon is unswerving with the augmented free carbonyls in the nuclei of glia and neurons. On the other hand, in AD brains, the stimulation of heme oxygenase-1 (i.e., an enzyme that catalyzes the conversion of heme to bilirubin) is raised and also found to be strongly connected with NFTs (Choi et al. 2000; Schipper et al. 2009; Barone et al. 2012).

5.2.4 Glycoxidation in Alzheimer's

Advanced glycation end products (AGEs), which are proinflammatory molecules and powerful neurotoxins, are generated through a nonenzymatic reaction of sugars along with long-lived protein accumulations (Gkogkolou and Böhm 2012). With a free amino acid group of proteins to produce a labile Schiff base, protein glycation initiates as a nonenzymatic process along with the unconstrained condensation of aldehyde or ketone groups of sugars. Furthermore, the aforesaid phenomenon is in line with the classical reaction defined by Maillard in 1912 (Smith et al. 1994). Subsequently, a series of reactions take place to generate AGEs, which are found to be containing irreversibly cross-linked heterogeneous protein aggregates. On the other hand, extensive covalent protein cross-linking-mediated insolubility of Aβ plaques is also supported by the growing evidence (Smith et al. 1996). AGEs are regarded as one of the processes through which long-lived proteins can be cross-linked (Ortwerth and Olesen 1988; Prabhakaram and Ortwerth 1994). The buildup of extracellular AGEs has been confirmed in coronas of classic plaques and primitive plaques, different cortical areas, and SPs. A very high extent of colocalization of AGEs with ApoE has been demonstrated through immunohistochemical studies (Li and Dickson 1997).

In AD, enhanced oxidation of glycated proteins (i.e., glycoxidation) causes extracellular AGE accumulation (Münch et al. 1998). In addition, accumulations of intracellular proteins including Lewy bodies, NFTs of individuals with Hirano bodies, and Parkinson's disease are also cross-connected by AGEs (Loske et al. 2000), which might clarify their resistance to proteases and insolubility in detergents. The key constituent of the NFTs, microtubule-linked protein tau (MAP-tau), has been revealed to be individual to the formation of intracellular AGEs (Gella and Durany 2009; Hagino and Kobayashi 2018). On the other hand, MAP-tau can be interestingly glycated in vitro, decreasing its capacity to bind to microtubules. Moreover, in the tubulin-binding region, MAP-tau isolated from brains of AD individuals is glycated, which can lead to β-sheet fibril formation (González et al. 1998; Ledesma et al. 1998). Interestingly, various experiments have exposed the existence of AGEs along with two key AD proteins such as MAP-tau and Aβ (Vitek et al. 1994; Younessi and Yoonessi 2011; Yan et al. 1994; Emendato et al. 2018). This previously mentioned finding strengthens the debate that AGEs have a contribution in the process of the AD pathogenesis (Colaco and Harrington 1994; Smith et al. 1995). Additionally, it has been observed that free radicals can evidently promote the Aβ cross-linking formation and can also play a role in glycation processes (Mattson et al. 1995; Uddin and Upaganlawar 2019).

5.3 Oxidative Stress and Pathological Hallmarks of Alzheimer's Disease

5.3.1 Oxidative Stress in Aβ Pathology

Aβ is generated through the sequential proteolytic cleavages of APP. One of the proteolytic enzymes is known as beta-secretase, otherwise recognized as beta-site APP-cleaving enzyme 1 (BACE1). The other membrane-bound protease is gamma-secretase, which is a multiprotein complex containing presenilin enhancer protein 2, anterior pharynx-defective 1 (APH-1), presenilin (PSEN), and nicastrin (NCT) (Querfurth and LaFerla 2010; Steiner 2004). Moreover, it has been found that APP at the N-terminal end is cleaved by BACE1, generating a 99-amino acid APP C-terminal fragment (CTF99), which is additionally cleaved within the transmembrane domain by gamma-secretase, which eventually can result in the Aβ peptide release (Querfurth and LaFerla 2010; Walsh and Selkoe 2007). On the other hand, from the cleavages through beta- and gamma-secretases, various peptides of varying lengths can be produced. In addition, among them, more profusely generated 40-amino acid form of Aβ (Aβ40) is much less toxic than the 42-amino acid form of Aβ (Aβ42); this phenomenon is perhaps due to Aβ42's rapid self-assembly into oligomers (Walsh and Selkoe 2007; Yankner and Lu 2009). Actually, evidence from several studies recommend that Aβ oligomers are considered as the most neurotoxic; furthermore, in case of AD, their levels relate with the extent of the cognitive decline (Jongbloed et al. 2015; Ferreira et al. 2015). Fascinatingly, the generation of toxic Aβ peptides can be precluded by the APP cleavage by a 3rd enzyme known as alpha-secretase (Esler and Wolfe 2001). Enhanced production and/or lowered Aβ peptide clearance can cause a buildup of Aβ, which can trigger diverse cell signaling pathways which ultimately can result in weakening of cognitive function, neuronal loss, and synaptic degeneration (Yankner and Lu 2009; Roberson and Mucke 2006; Hsiao et al. 1996).

Numerous studies have associated OS in Aβ-mediated neurotoxicity (Mattson 1997; Cheignon et al. 2018). Aβ treatment might enhance the levels of hydrogen peroxide (H_2O_2) and lipid peroxides, which is revealed by in vitro experiments by means of cell models (Behl et al. 1994; Klein and Ackerman 2003). Reliably, enhanced H_2O_2 and nitric oxide generation plus increased oxidative changes of lipids and proteins were linked with the age-related buildup of Aβ in several AD transgenic mouse models containing mutants of *PSEN1* and *APP*. Collectively, the aforementioned findings endorse that Aβ can encourage OS (Matsuoka et al. 2001; Apelt et al. 2004). ROS stimulation via soluble Aβ oligomers necessitated the N-methyl-D-aspartate (NMDA) receptor activation and also was linked with a fast upsurge in the levels of neuronal Ca^{2+} in hippocampal neuronal cell cultures. Together, these findings signify a probable contribution of soluble Aβ oligomers as proximal neurotoxins and also the contribution of OS in the neuronal loss and synaptic damage stimulated via soluble Aβ oligomers (De Felice et al. 2007; Zhang et al. 2016). Steadily, natural antioxidants, for instance, green tea catechins, curcumin, and EGb 761 (i.e., nutraceuticals), can exhibit neuroprotective functions by weakening Aβ-facilitated neuronal apoptosis and ROS generation, which has been confirmed in studies with AD cell and animal models (Cole et al. 2007; Zhao and Zhao 2012).

Besides facilitating Aβ-mediated cytotoxicity, many experiments have recommended that OS can promote Aβ production. In addition, in transgenic mice overexpressing *APP* mutant, faulty antioxidant defense system can cause increased OS and can markedly increase deposition of Aβ (Li et al. 2004; Nishida et al. 2006). In contrast, antioxidants present in diets, for instance, curcumin, lowered Aβ plaque burden and brain Aβ levels and decreased the increment of oxidized proteins (Lim et al. 2001). Moreover, through supplementation containing an antioxidant, the elevated deposition of Aβ and its related initial onset and more severe cognitive dysfunction stimulated by the faulty antioxidant defense system might be improved (Nishida et al. 2006; Billings et al. 2005). Consistent with these results, decrease in oxidation of protein and elevated antioxidant defense ability in brains, while restoring the memory deficit and decreasing Aβ plaque burden, were observed in Tg19959 transgenic mice overexpressing *APP* mutant, due to the overexpression of manganese SOD (MnSOD) (Dumont et al. 2009). In addition, in comparison with control AD mice, in Tg2576 *APP*-overexpressing AD mouse model, removal of cytoplasmic copper/zinc SOD (Cu-Zn-SOD) caused elevation of oligomerization of Aβ while fastening the loss of memory and spatial learning, signifying a probable contribution of oxidative damage in oligomerization of Aβ (Murakami et al. 2011). Collectively, these findings recommend that, for the development and initiation of AD, the augmentation of plaque formation/production of Aβ as well as oligomerization of Aβ via OS is crucial.

Experiments on how OS increases the production of Aβ have exposed that OS inhibits the action of alpha-secretase while supporting the activation and expression of gamma- and beta-secretases, the enzymes which are regarded as vital for the production of Aβ from APP (Oda et al. 2009; Yoo et al. 2010; Quiroz-Baez et al. 2009). The stimulation of *PSEN1* and *BACE1* expression and the gamma-secretase activation through OS were observed to be reliant on the initiation of an OS-stimulated major cell signaling cascade, c-Jun N-terminal kinase (JNK) pathway (Tamagno et al. 2008; Shen et al. 2008). Actually, the 5′ untranslated and promoter region of *BACE* gene has binding sites for a number of transcription factors including the nuclear factor-kappa B (NF-κB) and redox-sensitive activator protein (AP1), activation of which by OS may sequentially increase the expression of *BACE* (Sambamurti et al. 2004). Interestingly, both the rise in *PSEN1* and *BACE1* activity/expression (Fukumoto et al. 2002; Yang et al. 2003) and the JNK signaling cascade activation have been identified (Zhu et al. 2001; Lagalwar et al. 2006). Henceforth, it is probable that the elevated OS in Alzheimer's brains might start the activation of a series of redox-sensitive cell signaling pathways including JNK, which can cause promotion of the expression of *PSEN1* and *BACE1*, which ultimately can increase cognitive function impairment and Aβ production. On the other hand, in case of Aβ-stimulated neuronal apoptosis, JNK has also been associated (Yao et al. 2005). Therefore, via the initiation of redox-sensitive signaling pathways including JNK, OS might increase the production of Aβ as well as can arbitrate neurotoxicity induced by Aβ.

The increase of production of Aβ via OS might be a compensatory reaction to OS. In case of AD individuals, it was observed that neuronal oxidative damage was more severe with minor levels of Aβ deposition or with the disease having shorter duration (Nunomura et al. 2001, 2004). Furthermore, an inverse relationship was observed amid the extents of neuronal oxidative damage to nucleic acids and the extents of intraneuronal Aβ42 in the subiculum and the hippocampus of Alzheimer's brains (Nunomura et al. 2010). Undeniably, evidence has recommended that low nanomolar or picomolar levels of Aβ can be neuroprotective or neurotrophic (Luo et al. 1996; Plant et al. 2003). Collectively, lower Aβ levels might have a contribution in the normal functioning of neuronal cells and the unusual aggregation and accumulation of specific forms of Aβ, which could be elevated by OS and contributing to the pathological development of AD. Better comprehension is needed to recognize the pathological and physiological contribution of Aβ, which might help to develop efficacious techniques for the interventions of AD.

5.3.2 Oxidative Stress in Tau Pathology

Hyperphosphorylated tau protein is considered as the main NFTs constituent, which is also regarded as another distinctive feature of the pathology of AD that associates with cognitive decline and neurodegeneration (Hossain et al. 2019). Irregular tau hyperphosphorylation can impair its ability to promote microtubule congregation and its binding with tubulin, which eventually can result in its self-accumulation into filaments (Zaplatic et al. 2019). On the other hand, in the abnormal tau phosphorylation, several protein phosphatases and protein kinases have been associated, including calcium-calmodulin kinase, protein kinase C, mitogen-activated protein kinase (MAPK), glycogen synthase kinase-3 beta (GSK-3 beta), and cyclin-dependent kinase 5 (Billingsley and Kincaid 1997). In fact, it has been recommended that Aβ accumulation might become evident before the pathology of tau. In addition, that aggregates of Aβ could be one of a series of molecular processes which can result in hyperphosphorylation of tau (Zheng et al. 2002; Gotz et al. 2001). In contrast, it was stated that tau overexpression decreased Golgi-derived vesicles into neurites, neurofilaments, and kinesin-dependent transport of peroxisomes, eventually instigating transport deficiency in primary neuronal cells including the APP trafficking (Stamer et al. 2002). Specifically, the APP transport into dendrites and axons was obstructed, initiating its buildup in the cell body (Stamer et al. 2002; Petersen et al. 2014).

Based on the studied experiments, it was found that OS is interlinked with the pathology of tau. In addition, the cells that overexpress tau protein had augmented vulnerability against OS; this phenomenon takes place probably because of the peroxisome depletion (Stamer et al. 2002; Petersen et al. 2014; Gendron and Petrucelli 2009). The decrease in gene dosage of mitochondrial SOD2 or thioredoxin reductase (TrxR) increased tau-mediated neuronal apoptosis and neurodegenerative histological abnormalities, in a *Drosophila* model of human tauopathies expressing a disease-associated mutant form of human tau (tau R406W) (Dias-Santagata et al. 2007). On the other hand, tau-stimulated neuronal cell death was attenuated by vitamin E treatment or overexpression of these antioxidant enzymes (Dias-Santagata et al. 2007). In addition, the levels of ROS were elevated in cortical neurons in a transgenic rat model expressing a human truncated variant form of tau protein, as compared to control nontransgenic neurons. Furthermore, in the previously mentioned case, it was also noticed that the increased level of ROS was considerably eliminated due to the use of antioxidants, for example, vitamin C (Cente et al. 2006, 2009). These findings recommend that oxidative damage at least partly mediates tau-stimulated neurotoxicity (Dias-Santagata et al. 2007; Bondy and Campbell 2016). The relationship between tau pathology and OS was confirmed in P301L and P301S transgenic mouse models carrying the human tau gene with P301L or P301S mutations, which display hyperphosphorylated tau buildup and also cause the development of neurodegeneration and neurofibrillary tangles (Yoshiyama et al. 2007). In P301L tau transgenic mice, reduced nicotinamide adenine dinucleotide (NADH)-ubiquinone oxidoreductase action along with mitochondrial dysfunction was observed, which was linked with the elevated production of ROS, weakened synthesis of adenosine triphosphate (ATP), and mitochondrial respiration in aged animals (David et al. 2005). Interestingly, signs of the increased OS along with elevated protein carbonyl levels in cortex mitochondria were exhibited by the brains of P301S transgenic mice.

In the aforesaid phenomenon, modifications in the content and activity of mitochondrial enzymes involved in energy metabolism and formation of ROS were also found, further recommending that in case of tau pathology, mitochondrial dysfunction and OS may have significant contribution (Dumont et al. 2011). Unswervingly, reduced lipid peroxidation and markedly elevated level of complex I action while improving behavioral deficits and survival of the mice were observed following administration of P301S mice with coenzyme Q10, a vital component of the electron transport chain and an antioxidant (Elipenahli et al. 2012). In addition, in a triple transgenic mouse model (pR5/APP/PS2), the convergence of pathologies of tau and Aβ on the dysfunction of mitochondria was confirmed, which further shows both tau and Aβ pathologic characteristics of the disease in the brain of the animal (Eckert et al. 2010). Then again, enormous deregulation of 24 proteins was verified through the proteomics analyses of the brain samples, of which one-third were mitochondrial proteins mostly linked to complexes IV and I of the oxidative phosphorylation system (Rhein et al. 2009). Extraordinarily, complex I deregulation was noticed to be reliant on tau, whereas mitochondrial complex IV deregulation was revealed to be reliant on Aβ (Rhein et al. 2009; Eckert et al. 2011). Moreover, the actions of tau and Aβ on functions of mitochondria were observed to be age-associated and synergistic, which can result in the decrease of ATP synthesis and reduction of the mitochondrial respiratory ability, which can eventually lead to neuronal death and synaptic loss (Rhein et al. 2009; Nisbet et al. 2015).

Rising evidence has revealed that OS might have a contribution in polymerization and hyperphosphorylation of tau. In addition, fatty acid oxidation, which is observed to be increased in Alzheimer's brains, was stated to aid tau polymerization and henceforth may act as a probable connection between OS and the formation of the fibrillar pathology in Alzheimer's (Gamblin et al. 2000). Also, decrease in cytoplasmic SOD1 or mitochondrial SOD2 deficiency (Melov et al. 2007) enhanced phosphorylation of tau in Tg2576 AD transgenic mice. Moreover, the aforementioned phenomenon recommends that, in tau hyperphosphorylation, ROS might have a significant contribution (Murakami et al. 2011; Kosik and Shimura 2005).

5.4 Oxidative Stress and Mitochondria Dysfunction in Alzheimer's Disease

Undoubtedly, mitochondria play critical roles in a number of cellular functions including cell survival and death, Ca^{2+} homeostasis, and ATP synthesis. In the meantime, mitochondria are particularly susceptible to OS. In addition, in the cell, the major location of generation of ROS is regarded as the mitochondrial respiratory chain (Harilal et al. 2019). In the pathogenesis of AD, it has been confirmed by numerous studies that dysfunction of mitochondria is a significant factor. Furthermore, in the hippocampal neurons of AD, several metabolic and mitochondrial irregularities have been recognized in comparison with age-matched controls (Zhu et al. 2006; Mutisya et al. 1994). A marked reduction in mitochondria was shown through morphometric analysis of biopsies from AD brains (Mutisya et al. 1994; Hirai et al. 2001; Moreira et al. 2010). Moreover, these irregularities of mitochondrial were found to occur with oxidative damage marked by nitrotyrosine and 8-OHG. This further signifies that, during AD progression, the mitochondria were damaged (Hirai et al. 2001; Wang et al. 2014). Similarly, in the cortical regions of AD brains, an important reduction in the action of mitochondrial cytochrome oxidase (i.e., complex IV) was stated (Mutisya et al. 1994; Kish et al. 1999). Henceforth, lacking this important electron transport enzyme might result in the reduction in energy stores and elevation of ROS production, which ultimately can play a role in the neurodegenerative process (Mutisya et al. 1994; Liu et al. 2017).

In case of AD, evidence recommends that Aβ might directly interrupt the functions of mitochondria and play a role in the energy metabolism deficit and also neuronal death. In addition, Aβ was localized to mitochondria in the brain of transgenic mice and also in neuroblastoma cells in AD people, thereby firmly expressing human mutant *APP* (Caspersen et al. 2005; Manczak et al. 2006). On the other hand, the existence of Aβ in mitochondria was connected with elevated mitochondrial ROS production and weakened mitochondrial metabolism (Manczak et al. 2006; Onyango et al. 2016; Pavlov et al. 2009). Actually, studies in isolated

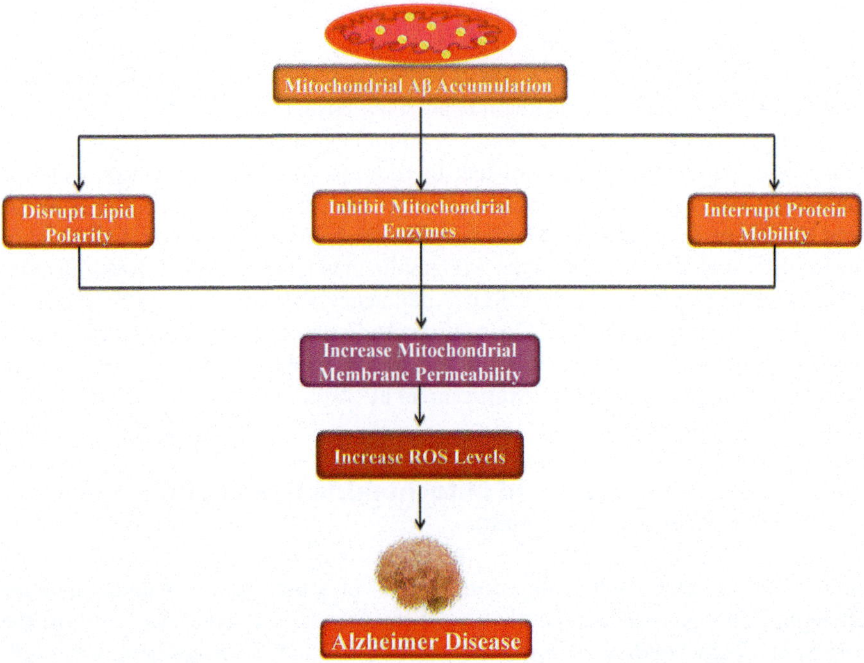

Fig. 5.2 Mitochondrial accumulation of amyloid ß causes compromised mitochondrial function and amplifies reactive oxygen species that lead to Alzheimer's disease. *Aβ* amyloid ß, *ROS* reactive oxygen species

mitochondria revealed that treatment of Aβ could decrease important enzymes of the mitochondria respiratory chain, disrupt protein mobility and lipid polarity, and also result in oxidative injury to mitochondrial membrane, which ultimately can cause rise in cytochrome c release and permeability of mitochondrial membrane (Rodrigues et al. 2001; Casley et al. 2002), which is given in Fig. 5.2. In a double homozygous knock-in mouse model expressing *APP* and *PSEN1* mutants, a principal antioxidant enzyme that protects mitochondria against superoxide, known as MnSOD, was observed to be a target of inactivation and nitration (Anantharaman et al. 2006). The reduced activity of antioxidant defense enzymes including MnSOD might elevate the levels of ROS and compromise the function of mitochondria, taking part in the loss of potential of the mitochondrial membrane and ultimately activation of caspases and even apoptosis (Anantharaman et al. 2006).

Furthermore, Aβ can also change other cellular protective processes against oxidative damage to mitochondria. On the other hand, it is known that uncoupling proteins (UCPs) are a class of mitochondrial anion carrier proteins (Rousset et al. 2004). It has been found that products of lipid peroxidation or ROS can activate UCP3 and UCP2 in order to reduce proton-motive force and decrease ATP production and mitochondrial membrane potential, decreasing generation of ROS from mitochondria and also causing mitochondria uncoupling (Echtay 2007). Thus, the activation and expression of UCPs are regarded as a protective mechanism in response to OS. Interestingly, in AD brains, this protective mechanism seems dysfunctional, where the expression of UCP5, UCP4, and UCP2 is markedly reduced (de la Monte and Wands 2006). The upregulation of levels of UCP4 and UCP2 protein in response to the introduction to the superoxide was revealed. Even though the aforementioned processes are not sufficiently clear, it recommends that buildup of Aβ might be linked to OS (Wu et al. 2009). Likewise, in cells overexpressing *APP* mutant or *APP*, it has been observed that, in response to superoxide treatment, the UCP4- and UCP2-dependent upregulation of mitochondrial free Ca^{2+} was reduced, signifying that the accumulation of Aβ might be linked with mitochondria dysfunction (Manczak et al. 2006; Wu et al. 2009, 2010).

Interestingly, before the appearance of plaques of Aβ, the mitochondria-linked Aβ in conjunction with the rise in H_2O_2 and reduction in cytochrome oxidase action was identified. The aforementioned findings signify the fact that, in the AD pathogenesis, defects in mitochondria can take place earlier. Henceforth, in case of delaying AD, early mitochondrially targeted therapeutic interventions might be useful (Caspersen et al. 2005; Manczak et al. 2006; Kumar and Atamna 2011; Mao and Reddy 2011).

5.5 Oxidative Stress and Metal Homeostasis in Alzheimer's Disease

In the human body, for a broad range of biological processes, copper (Cu), zinc (Zn), and iron (Fe) are the transition metals that play pivotal roles including brain functions and also have noteworthy catalytic roles in many enzymes (Wright and Baccarelli 2007). On the other hand, additional Cu or Fe directly can interact with oxygen to generate hydroxyl radical (•OH), H_2O_2, and superoxide ion (O_2^-), which might lead to OS and a range of biochemical changes that can eventually result in neuronal cell death (Kenche and Barnham 2011). Indeed, rising evidence has revealed that there is an intimate connection between the AD and the disturbance of metal homeostasis (Kenche and Barnham 2011; Uddin and Amran 2018). Moreover, in AD hippocampus and amygdala, unusual levels of Zn, Fe, and Cu have been detected; these mentioned areas are found to exhibit intense histopathologic changes (Deibel et al. 1996). Furthermore, in transgenic mouse models and also in AD patients, the aforementioned transition metals have been found within the deposits of amyloid (Lovell et al. 1998; Zhang et al. 2006). Collectively, these findings recommend that, in AD, abnormal transition metal accumulation might be closely connected with Aβ pathology. Nevertheless, in case of AD, the exact reason and the form of the association of brain metal dyshomeostasis are yet to be well-known.

In AD individuals, the prevalence of transition metals within the deposits of amyloid specifies that transition metals might interact directly with Aβ (Maynard et al. 2005; Duce et al. 2011). Actually, via a tyrosine residue (i.e., Tyr^{10}) and three residues of histidine (i.e., His^6, His^{13}, and His^{14}), both Zn^{2+} and Cu^{2+} can bind to monomers of Aβ, generating conformational alterations in the peptide that encourage its accumulation (Curtain et al. 2001; Hesse et al. 1994). Steadily, in vitro data

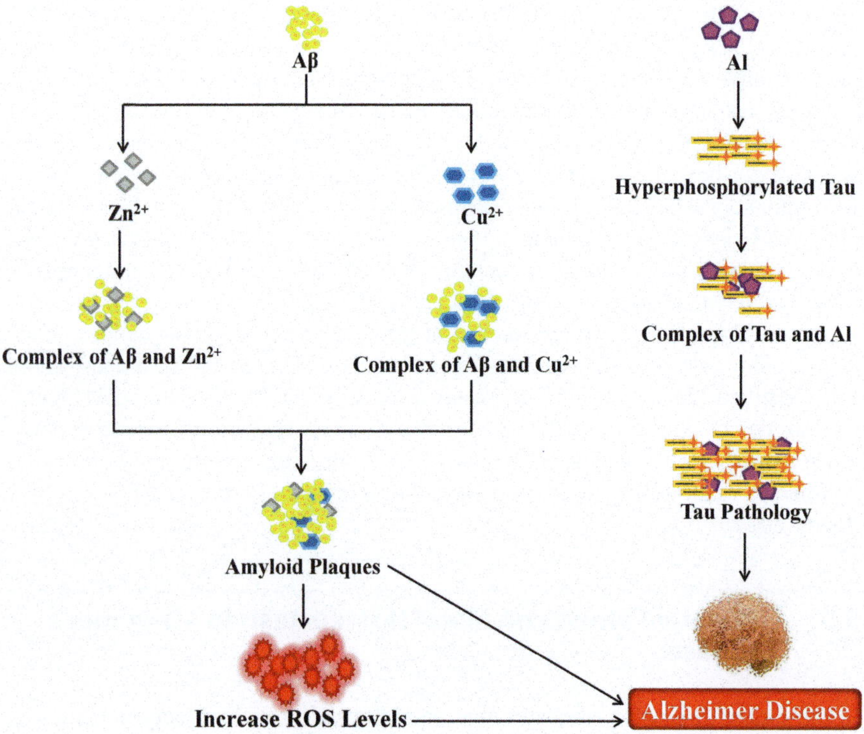

Fig. 5.3 The interaction of zinc and copper with amyloid ß, as well as the impact of aluminum to tau, represents the hallmark pathogenic features that cause Alzheimer's disease. *Aβ* amyloid ß, *Zn^{2+}* zinc ion, *Cu^{2+}* copper ion, *Al* aluminum

revealed that Zn and Cu quickly persuaded the soluble Aβ peptide aggregation (Atwood et al. 1998; Bush et al. 1994), which is given in Fig. 5.3. Thus, in AD brains, the unusual Aβ interaction with metal ions and the disruption in homeostasis of metal might directly be involved in the Aβ deposition process (Atwood et al. 1998; Strausak et al. 2001; Bush 2002). The abnormal interaction between Aβ and transition metals might serve as a cause of ROS generation. Furthermore, aluminum (Al) also serves as a possible modulator of tau pathology (Fig. 5.3) in the context of AD pathogenesis. Then again, a cuproenzyme-like complex is formed due to the high-affinity Aβ binding with Cu^{2+}, which results in the formation of positive charge containing Aβ radical (•$Aβ^+$) (Opazo et al. 2002).

Moreover, oxygen accepts two electrons (e^-) from Cu^+, which leads to the generation of H_2O_2 (Opazo et al. 2002; Huang et al. 1999a; Farooqui 2016); this further sets up circumstances to generate hydroxyl radicals (i.e., Fenton-type reaction) (Huang et al. 1999b; Lynch et al. 2000). Interestingly, subsequent to the donation of e^- to O_2, the radicalized •Cu^{2+} complex might be reinstated to Aβ•Cu^{2+} via e^- transfer from biological reducing agents, including catecholamines, vitamin C, and cholesterol (Sayre et al. 2008; Opazo et al. 2002). The efficiency of H_2O_2 generation is higher for Aβ42 as compared to Aβ40, associating with their cytotoxic action (Huang et al.

1999a, 1999b; Gravina et al. 1995). Like the interaction between copper and Aβ, Fe binding to Aβ can result in the reduction of ferric ion to ferrous ion and can result in H_2O_2 production (Rottkamp et al. 2001; Zhou et al. 2009). These findings recommend that, in AD pathogenesis and Aβ-induced neurotoxicity, ROS produced from the interaction of transition metals with Aβ are the key role players in case of OS. Moreover, it has been found that there is an intimate relationship between the generation and processing of APP and Fe/Cu homeostasis. Fascinatingly, treatment with Fe stimulated Aβ42 release in SH-SY5Y cells, overexpressing the Swedish mutant form of a human *APP* (APPsw) (Wan et al. 2011; Zheng et al. 2009).

Experiments have revealed that, in the cortex and also in the hippocampus of APP/PSEN1 transgenic mice, the levels of metal transporters including divalent metal transporter 1 (DMT1) and Zn transporters are elevated. In addition, like transition metals, in SPs in the cortex of AD brains, these transporters of metals are colocalized with Aβ (Zheng et al. 2009, 2010, 2008). Furthermore, this results in the assumption that, in AD, metal transporters might have significant contribution in the abnormal metal homeostasis. On the other hand, DMT1, otherwise recognized as divalent cation transporter 1 (DCT1) or also as natural resistance-associated macrophage protein 2 (Nramp2), is a recently uncovered proton-coupled metal-ion transport protein accountable for the uptake of a wide range of divalent metal ions, including Zn, Cu, and Fe (Li et al. 2003). Captivatingly, as compared to the control cells, a significant rise in the levels of DMT1 was observed in APPsw cells (Zheng et al. 2009). In addition, it was noticed that, along with the elevated OS and cell toxicity, the intracellular Fe was markedly increased (Wan et al. 2011; Zheng et al. 2009; Oshiro et al. 2000). Instead, silencing of endogenous DMT1 via RNA interference (RNAi) inhibited bivalent ion influx into the cells and also reduced the protein levels of DMT1. These aforesaid findings further signify that the increment of DMT1 may be associated with the disturbance of homeostasis of Fe observed in APPsw cells (Wan et al. 2011; Zheng et al. 2009).

Interestingly, when interacting with Cu, free radicals are found to be generated by APP, which contains a Cu-binding site at N-terminal cysteine-rich region (Barnham et al. 2003). Furthermore, Cu homeostasis has also been found to be modulated by APP. A noteworthy decrease in levels of Cu was caused in Tg2576 transgenic mice due to the APP overexpression (Maynard et al. 2002). Cu levels were found to be markedly increased in the cerebral cortex (i.e., a part of the brain predominantly associated with the AD) in APP knockout mice (Bellingham et al. 2004; White et al. 1999). This results in the assumption that the secreted APP and/or Aβ might encourage Cu efflux or cause prevention of its uptake, therefore decreasing its levels (Maynard et al. 2002; Bush 2003). Despite the fact that the increase in cellular levels of Cu stimulated the departure of APP from the Golgi to a broader distribution all over plasma membrane and the cytoplasm, endogenous APP had a partial colocalization with the Golgi marker GM130, in SH-SY5Y cells [191]. Moreover, the elevation of cell surface APP was caused by an associated reduction in endocytosis and increased exocytosis (Acevedo et al. 2011; Xiao et al. 2012). Henceforth, in Cu efflux pathways, the copper-responsive trafficking of APP is found to be in line with a role for APP (Acevedo et al. 2011, 2014). Therefore, these findings signify the existence of an interdependent relationship between Cu homeostasis and metabolism of APP. Moreover, agitations of either of these might cause a change of the other, which ultimately can promote the free radical generation and Aβ accumulation (White et al. 1999).

Abnormal metal homeostasis has significant contribution in a number of vital features of the pathogenesis of AD including aggregation and production of Aβ and Aβ-mediated OS. Furthermore, as an auspicious therapeutic choice, AD therapy-targeted metal-Aβ interaction is quickly developing (Budimir 2011).

5.6 Conclusion

Though multiple pathophysiologic mechanisms and etiologies are probably involved in AD, OS has been found to have significant contribution in the pathophysiologic process. The neurotoxicity stimulated by abnormal buildup of Aβ and tau proteins might increase generation of Aβ and aggregation along with facilitation of tau polymerization and phosphorylation, which can further cause an increase in various neurotoxic events including production of ROS; therefore, the formation of a vicious cycle takes place that can further stimulate the initiation and progression of AD. Instead, mitochondria dysfunction and/or abnormal transition metal accumulation can cause generation of excessive ROS, which may be due to the combination of the aberrant accumulation of Aβ and tau pathology, and ultimately can result in OS. In AD, this is a noteworthy contribution of the oxidative brain, as the brain is susceptible to OS. Nevertheless, the ability of antioxidant in assuaging AD progression, supported by multicenter trials, signifies the complexity of Alzheimer's. However, to get a well understanding of the disease and treatment strategy, more investigations are required.

References

Acevedo KM, Hung YH, Dalziel AH et al (2011) Copper promotes the trafficking of the amyloid precursor protein. J Biol Chem 286:8252–8262. https://doi.org/10.1074/jbc.M110.128512

Acevedo KM, Opazo CM, Norrish D et al (2014) phosphorylation of amyloid precursor protein at threonine 668 is essential for its copper-responsive trafficking in SH-SY5Y neuroblastoma cells. J Biol Chem 289:11007–11019. https://doi.org/10.1074/jbc.M113.538710

Anantharaman M, Tangpong J, Keller JN et al (2006) Beta-amyloid mediated nitration of manganese superoxide dismutase: implication for oxidative stress in a APPNLH/NLH X PS-1P264L/P264L double knock-in mouse model of Alzheimer's disease. Am J Pathol 168:1608–1618

Ansari MA, Scheff SW (2010) Oxidative stress in the progression of Alzheimer disease in the frontal cortex. J Neuropathol Exp Neurol 69:155–167. https://doi.org/10.1097/NEN.0b013e3181cb5af4

Apelt J, Bigl M, Wunderlich P, Schliebs R (2004) Aging-related increase in oxidative stress correlates with developmental pattern of beta-secretase activity and beta-amyloid plaque formation in transgenic Tg2576 mice with Alzheimer-like pathology. Int J Dev Neurosci 22:475–484. https://doi.org/10.1016/j.ijdevneu.2004.07.006

Atwood CS, Moir RD, Huang X et al (1998) Dramatic aggregation of Alzheimer abeta by Cu(II) is induced by conditions representing physiological acidosis. J Biol Chem 273:12817–12826

Barnham KJ, McKinstry WJ, Multhaup G et al (2003) Structure of the Alzheimer's disease amyloid precursor protein copper binding domain. J Biol Chem 278:17401–17407. https://doi.org/10.1074/jbc.M300629200

Barone E, Di Domenico F, Sultana R et al (2012) Heme oxygenase-1 posttranslational modifications in the brain of subjects with Alzheimer disease and mild cognitive impairment. Free Radic Biol Med 52:2292–2301. https://doi.org/10.1016/j.freeradbiomed.2012.03.020

Behl C, Davis JB, Lesley R, Schubert D (1994) Hydrogen peroxide mediates amyloid beta protein toxicity. Cell 77:817–827

Bellingham SA, Ciccotosto GD, Needham BE et al (2004) Gene knockout of amyloid precursor protein and amyloid precursor-like protein-2 increases cellular copper levels in primary mouse cortical neurons and embryonic fibroblasts. J Neurochem 91:423–428. https://doi.org/10.1111/j.1471-4159.2004.02731.x

Berlett BS, Stadtman ER (1997) Protein oxidation in aging, disease, and oxidative stress. J Biol Chem 272:20313–20316

Billings LM, Oddo S, Green KN et al (2005) Intraneuronal Aβ causes the onset of early Alzheimer's disease-related cognitive deficits in transgenic mice. Neuron 45:675–688. https://doi.org/10.1016/j.neuron.2005.01.040

Billingsley ML, Kincaid RL (1997) Regulated phosphorylation and dephosphorylation of tau protein: effects on microtubule interaction, intracellular trafficking and neurodegeneration. Biochem J 323(Pt 3):577–591

Bondy SC, Campbell A (2016) Inflammation, aging, and oxidative stress. Springer, New York

Bruce-Keller AJ, Li YJ, Lovell MA et al (1998) 4-Hydroxynonenal, a product of lipid peroxidation, damages cholinergic neurons and impairs visuospatial memory in rats. J Neuropathol Exp Neurol 57:257–267

Budimir A (2011) Metal ions, Alzheimer's disease and chelation therapy. Acta Pharma 61:1–14. https://doi.org/10.2478/v10007-011-0006-6

Bush AI (2002) Metal complexing agents as therapies for Alzheimer's disease. Neurobiol Aging 23:1031–1038

Bush AI (2003) The metallobiology of Alzheimer's disease. Trends Neurosci 26:207–214. https://doi.org/10.1016/S0166-2236(03)00067-5

Bush AI, Pettingell WH, Multhaup G et al (1994) Rapid induction of Alzheimer A beta amyloid formation by zinc. Science 265:1464–1467

Casley CS, Canevari L, Land JM et al (2002) Beta-amyloid inhibits integrated mitochondrial respiration and key enzyme activities. J Neurochem 80:91–100

Caspersen C, Wang N, Yao J et al (2005) Mitochondrial Aβ: a potential focal point for neuronal metabolic dysfunction in Alzheimer's disease. FASEB J 19:2040–2041. https://doi.org/10.1096/fj.05-3735fje

Cente M, Filipcik P, Pevalova M, Novak M (2006) Expression of a truncated tau protein induces oxidative stress in a rodent model of tauopathy. Eur J Neurosci 24:1085–1090. https://doi.org/10.1111/j.1460-9568.2006.04986.x

Cente M, Filipcik P, Mandakova S et al (2009) Expression of a truncated human tau protein induces aqueous-phase free radicals in a rat model of tauopathy: implications for targeted antioxidative therapy. J Alzheimers Dis 17:913–920. https://doi.org/10.3233/JAD-2009-1107

Cheignon C, Tomas M, Bonnefont-Rousselot D et al (2018) Oxidative stress and the amyloid beta peptide in Alzheimer's disease. Redox Biol 14:450–464. https://doi.org/10.1016/j.redox.2017.10.014

Chen GC, Kong J (2012) Neural regeneration research. Elsevier (Singapore) Pte. Ltd, Singapore

Choi Y-G, Kim J-I, Lee H-P et al (2000) Induction of heme oxygenase-1 in the brains of scrapie-infected mice. Neurosci Lett 289:173–176. https://doi.org/10.1016/S0304-3940(00)01277-5

Colaco CA, Harrington CR (1994) Glycation: a pathological modification in neuropathies?: a hypothesis. Neuroreport 5:859–861

Cole GM, Teter B, Frautschy SA (2007) Neuroprotective effects of curcumin. In: The molecular targets and therapeutic uses of curcumin in health and disease. Springer, Boston, MA, pp 197–212

Collins AR, Dusinská M, Gedik CM, Stětina R (1996) Oxidative damage to DNA: do we have a reliable biomarker? Environ Health Perspect 104(Suppl 3):465–469. https://doi.org/10.1289/ehp.96104s3465

Curtain CC, Ali F, Volitakis I et al (2001) Alzheimer's Disease amyloid-β binds copper and zinc to generate an allosterically ordered membrane-penetrating structure containing superoxide dismutase-like subunits. J Biol Chem 276:20466–20473. https://doi.org/10.1074/jbc.M100175200

Dalle-Donne I, Aldini G, Carini M et al (2014) Protein carbonylation, cellular dysfunction, and disease progression. J Cell Mol Med 10:389–406

David DC, Hauptmann S, Scherping I et al (2005) Proteomic and functional analyses reveal a mitochondrial dysfunction in P301L Tau transgenic mice. J Biol Chem 280:23802–23814. https://doi.org/10.1074/jbc.M500356200

Davies MJ (2005) The oxidative environment and protein damage. Biochim Biophys Acta, Proteins Proteomics 1703:93–109. https://doi.org/10.1016/j.bbapap.2004.08.007

De Felice FG, Velasco PT, Lambert MP et al (2007) Aβ oligomers induce neuronal oxidative stress through an *N*-methyl-D-aspartate receptor-dependent mechanism that is blocked by the Alzheimer drug memantine. J Biol Chem 282:11590–11601. https://doi.org/10.1074/jbc.M607483200

de la Monte SM, Wands JR (2006) Molecular indices of oxidative stress and mitochondrial dysfunction occur early and often progress with severity of Alzheimer's disease. J Alzheimers Dis 9:167–181

Deibel MA, Ehmann WD, Markesbery WR (1996) Copper, iron, and zinc imbalances in severely degenerated brain regions in Alzheimer's disease: possible relation to oxidative stress. J Neurol Sci 143:137–142

Dias-Santagata D, Fulga TA, Duttaroy A, Feany MB (2007) Oxidative stress mediates tau-induced neurodegeneration in Drosophila. J Clin Invest 117:236–245. https://doi.org/10.1172/JCI28769

Duce JA, Bush AI, Adlard PA (2011) Role of amyloid-β–metal interactions in Alzheimer's disease. Future Neurol 6:641–659. https://doi.org/10.2217/fnl.11.43

Dumont M, Wille E, Stack C et al (2009) Reduction of oxidative stress, amyloid deposition, and memory deficit by manganese superoxide dismutase overexpression in a transgenic mouse model of Alzheimer's disease. FASEB J 23:2459–2466. https://doi.org/10.1096/fj.09-132928

Dumont M, Stack C, Elipenahli C et al (2011) Behavioral deficit, oxidative stress, and mitochondrial dysfunction precede tau pathology in P301S transgenic mice. FASEB J 25:4063–4072. https://doi.org/10.1096/fj.11-186650

Echtay K (2007) Mitochondrial uncoupling proteins--what is their physiological role? Free Radic Biol Med 43:1351–1371. https://doi.org/10.1016/j.freeradbiomed.2007.08.011

Eckert A, Schulz KL, Rhein V, Götz J (2010) Convergence of amyloid-β and tau pathologies on mitochondria in vivo. Mol Neurobiol 41:107–114. https://doi.org/10.1007/s12035-010-8109-5

Eckert A, Schmitt K, Götz J (2011) Mitochondrial dysfunction – the beginning of the end in Alzheimer's disease? Separate and synergistic modes of tau and amyloid-β toxicity. Alzheimers Res Ther 3:15. https://doi.org/10.1186/alzrt74

Elipenahli C, Stack C, Jainuddin S et al (2012) Behavioral improvement after chronic administration of coenzyme Q10 in P301S transgenic mice. J Alzheimers Dis 28:173–182. https://doi.org/10.3233/JAD-2011-111190

Emendato A, Milordini G, Zacco E et al (2018) Glycation affects fibril formation of Aβ peptides. J Biol Chem 293:13100–13111. https://doi.org/10.1074/jbc.RA118.002275

Esler WP, Wolfe MS (2001) A portrait of Alzheimer secretases--new features and familiar faces. Science (80-) 293:1449–1454. https://doi.org/10.1126/science.1064638

Farooqui AA (2016) Therapeutic potentials of curcumin for Alzheimer disease. Springer, Cham

Ferreira ST, Lourenco MV, Oliveira MM, De Felice FG (2015) Soluble amyloid-Î² oligomers as synaptotoxins leading to cognitive impairment in Alzheimerâ€™s disease. Front Cell Neurosci 9:191. https://doi.org/10.3389/fncel.2015.00191

Fukumoto H, Cheung BS, Hyman BT, Irizarry MC (2002) Beta-secretase protein and activity are increased in the neocortex in Alzheimer disease. Arch Neurol 59:1381–1389

Gabbita SP, Lovell MA, Markesbery WR (1998) Increased nuclear DNA oxidation in the brain in Alzheimer's disease. J Neurochem 71:2034–2040

Gamblin TC, King ME, Kuret J et al (2000) Oxidative regulation of fatty acid-induced tau polymerization. Biochemistry 39:14203–14210

Gella A, Durany N (2009) Oxidative stress in Alzheimer disease. Cell Adhes Migr 3:88–93

Gendron TF, Petrucelli L (2009) The role of tau in neurodegeneration. Mol Neurodegener 4:13. https://doi.org/10.1186/1750-1326-4-13

Gkogkolou P, Böhm M (2012) Advanced glycation end products. Dermato-Endocrinology 4:259–270. https://doi.org/10.4161/derm.22028

González C, Farías G, Maccioni RB (1998) Modification of tau to an Alzheimer's type protein interferes with its interaction with microtubules. Cell Mol Biol (Noisy-le-grand) 44:1117–1127

Gotz J, Chen F, van Dorpe J, Nitsch RM (2001) Formation of neurofibrillary tangles in P301L Tau transgenic mice induced by Abeta 42 fibrils. Science (80-) 293:1491–1495. https://doi.org/10.1126/science.1062097

Gravina SA, Ho L, Eckman CB et al (1995) Amyloid beta protein (A beta) in Alzheimer's disease brain. Biochemical and immunocytochemical analysis with antibodies specific for forms ending at A beta 40 or A beta 42(43). J Biol Chem 270:7013–7016

Gsell W, Strein I, Riederer P (1996) The neurochemistry of Alzheimer type, vascular type and mixed type dementias compared. J Neural Transm Suppl 47:73–101

Hagino N, Kobayashi S (2018) Mind in Aging – Nobuyoshi Hagino, Shinjiro Kobayashi – Google Books. https://books.google.com.bd/books?hl=en&lr=&id=RoO9DgAAQBAJ&oi=fnd&pg=PP1&dq=N.+Hagino,+S.+Kobayashi,+Mind+in+Aging.+Scientific+Research+Publishing,+Inc.+(USA,+2017).&ots=U4wit92ivJ&sig=NuiTXLdWzjA1iVYNa_x0kwK2VPA&redir_esc=y#v=onepage&q=N. Hagino%2C. Accessed 14 Nov 2018

Harilal S, Jose J, Parambi DGT, Kumar R, Mathew GE, Uddin MS et al (2019) Advancements in nanotherapeutics for Alzheimer's disease: current perspectives. J Pharm Pharmacol. https://doi.org/10.1111/jphp.13132

Hesse L, Beher D, Masters CL, Multhaup G (1994) The βA4 amyloid precursor protein binding to copper. FEBS Lett 349:109–116. https://doi.org/10.1016/0014-5793(94)00658-X

Hirai K, Aliev G, Nunomura A et al (2001) Mitochondrial abnormalities in Alzheimer's disease. J Neurosci 21:3017–3023

Hossain MF, Uddin MS, Uddin GMS et al (2019) Melatonin in Alzheimer's disease: a latent endogenous regulator of neurogenesis to mitigate Alzheimer's neuropathology. Mol Neurobiol. https://doi.org/10.1007/s12035-019-01660-3

Hsiao K, Chapman P, Nilsen S et al (1996) Correlative memory deficits, Abeta elevation, and amyloid plaques in transgenic mice. Science 274:99–102

Huang X, Cuajungco MP, Atwood CS et al (1999a) Cu(II) potentiation of alzheimer abeta neurotoxicity. Correlation with cell-free hydrogen peroxide production and metal reduction. J Biol Chem 274:37111–37116

Huang X, Atwood CS, Hartshorn MA et al (1999b) The Aβ peptide of Alzheimer's disease directly produces hydrogen peroxide through metal ion reduction. Biochemistry 38:7609–7616. https://doi.org/10.1021/bi990438f

Jongbloed W, Bruggink KA, Kester MI et al (2015) Amyloid-β oligomers relate to cognitive decline in Alzheimer's disease. J Alzheimers Dis 45:35–43. https://doi.org/10.3233/JAD-142136

Kasapoglu M, Ozben T (2001) Alterations of antioxidant enzymes and oxidative stress markers in aging. Exp Gerontol 36:209–220

Keller JN, Pang Z, Geddes JW et al (1997) Impairment of glucose and glutamate transport and induction of mitochondrial oxidative stress and dysfunction in synaptosomes by amyloid beta-peptide: role of the lipid peroxidation product 4-hydroxynonenal. J Neurochem 69:273–284

Kenche VB, Barnham KJ (2011) Alzheimer's disease & metals: therapeutic opportunities. Br J Pharmacol 163:211–219. https://doi.org/10.1111/j.1476-5381.2011.01221.x

Kish SJ, Mastrogiacomo F, Guttman M et al (1999) Decreased brain protein levels of cytochrome oxidase subunits in Alzheimer's disease and in hereditary spinocerebellar ataxia disorders: a nonspecific change? J Neurochem 72:700–707

Klein JA, Ackerman SL (2003) Oxidative stress, cell cycle, and neurodegeneration. J Clin Invest 111:785–793. https://doi.org/10.1172/JCI18182

Korolainen MA, Nyman TA, Nyyssonen P et al (2007) Multiplexed proteomic analysis of oxidation and concentrations of cerebrospinal fluid proteins in Alzheimer disease. Clin Chem 53:657–665. https://doi.org/10.1373/clinchem.2006.078014

Kosik KS, Shimura H (2005) Phosphorylated tau and the neurodegenerative foldopathies. Biochim Biophys Acta Mol basis Dis 1739:298–310. https://doi.org/10.1016/J.BBADIS.2004.10.011

Kumar R, Atamna H (2011) Therapeutic approaches to delay the onset of Alzheimer's disease. J Aging Res 2011:1–11. https://doi.org/10.4061/2011/820903

Lagalwar S, Guillozet-Bongaarts AL, Berry RW, Binder LI (2006) Formation of phospho-SAPK/JNK granules in the hippocampus is an early event in Alzheimer disease. J Neuropathol Exp Neurol 65:455–464. https://doi.org/10.1097/01.jnen.0000229236.98124.d8

Ledesma MD, Pérez M, Colaco C, Avila J (1998) Tau glycation is involved in aggregation of the protein but not in the formation of filaments. Cell Mol Biol (Noisy-le-grand) 44:1111–1116

Li YM, Dickson DW (1997) Enhanced binding of advanced glycation endproducts (AGE) by the ApoE4 isoform links the mechanism of plaque deposition in Alzheimer's disease. Neurosci Lett 226:155–158

Li H, Li F, Sun H, Qian ZM (2003) Membrane-inserted conformation of transmembrane domain 4 of divalent-metal transporter. Biochem J 372:757–766. https://doi.org/10.1042/BJ20030075

Li F, Calingasan NY, Yu F et al (2004) Increased plaque burden in brains of APP mutant MnSOD heterozygous knockout mice. J Neurochem 89:1308–1312. https://doi.org/10.1111/j.1471-4159.2004.02455.x

Lim GP, Chu T, Yang F et al (2001) The curry spice curcumin reduces oxidative damage and amyloid pathology in an Alzheimer transgenic mouse. J Neurosci 21:8370–8377

Liu Z, Zhou T, Ziegler AC et al (2017) Oxidative stress in neurodegenerative diseases: from molecular mechanisms to clinical applications. Oxidative Med Cell Longev 2017:1–11. https://doi.org/10.1155/2017/2525967

Loske C, Gerdemann A, Schepl W et al (2000) Transition metal-mediated glycoxidation accelerates cross-linking of beta-amyloid peptide. Eur J Biochem 267:4171–4178

Lovell MA, Markesbery WR (2007) Oxidative DNA damage in mild cognitive impairment and late-stage Alzheimer's disease. Nucleic Acids Res 35:7497–7504. https://doi.org/10.1093/nar/gkm821

Lovell MA, Ehmann WD, Butler SM, Markesbery WR (1995) Elevated thiobarbituric acid-reactive substances and antioxidant enzyme activity in the brain in Alzheimer's disease. Neurology 45:1594–1601

Lovell MA, Robertson JD, Teesdale WJ et al (1998) Copper, iron and zinc in Alzheimer's disease senile plaques. J Neurol Sci 158:47–52

Luo Y, Sunderland T, Roth GS, Wolozin B (1996) Physiological levels of beta-amyloid peptide promote PC12 cell proliferation. Neurosci Lett 217:125–128

Lynch T, Cherny RA, Bush AI (2000) Oxidative processes in Alzheimer's disease: the role of abeta-metal interactions. Exp Gerontol 35:445–451

Manczak M, Anekonda TS, Henson E et al (2006) Mitochondria are a direct site of Aβ accumulation in Alzheimer's disease neurons: implications for free radical generation and oxidative damage in disease progression. Hum Mol Genet 15:1437–1449. https://doi.org/10.1093/hmg/ddl066

Mao P, Reddy PH (2011) Aging and amyloid beta-induced oxidative DNA damage and mitochondrial dysfunction in Alzheimer's disease: implications for early intervention and therapeutics. Biochim Biophys Acta Mol basis Dis 1812:1359–1370. https://doi.org/10.1016/j.bbadis.2011.08.005

Marcus DL, Thomas C, Rodriguez C et al (1998) Increased peroxidation and reduced antioxidant enzyme activity in Alzheimer's Disease. Exp Neurol 150:40–44. https://doi.org/10.1006/exnr.1997.6750

Mark RJ, Lovell MA, Markesbery WR et al (1997) A role for 4-hydroxynonenal, an aldehydic product of lipid peroxidation, in disruption of ion homeostasis and neuronal death induced by amyloid beta-peptide. J Neurochem 68:255–264

Markesbery WR, Lovell MA (1998) Four-hydroxynonenal, a product of lipid peroxidation, is increased in the brain in Alzheimer's disease. Neurobiol Aging 19:33–36

Matsuoka Y, Picciano M, La Francois J, Duff K (2001) Fibrillar beta-amyloid evokes oxidative damage in a transgenic mouse model of Alzheimer's disease. Neuroscience 104:609–613. https://doi.org/10.1016/S0306-4522(01)00115-4

Mattson MP (1997) Cellular actions of beta-amyloid precursor protein and its soluble and fibrillogenic derivatives. Physiol Rev 77:1081–1132. https://doi.org/10.1152/physrev.1997.77.4.1081

Mattson MP, Chan SL (2003) Neuronal and glial calcium signaling in Alzheimer's disease. Cell Calcium 34:385–397

Mattson MP, Carney JW, Butterfield DA (1995) A tombstone in Alzheimer's? Nature 373:481–481. https://doi.org/10.1038/373481a0

Maynard CJ, Cappai R, Volitakis I et al (2002) Overexpression of Alzheimer's disease amyloid-β opposes the age-dependent elevations of brain copper and iron. J Biol Chem 277:44670–44676. https://doi.org/10.1074/jbc.M204379200

Maynard CJ, Bush AI, Masters CL et al (2005) Metals and amyloid-β in Alzheimer's disease. Int J Exp Pathol 86:147–159. https://doi.org/10.1111/j.0959-9673.2005.00434.x

Mecocci P, MacGarvey U, Beal MF (1994) Oxidative damage to mitochondrial DNA is increased in Alzheimer's disease. Ann Neurol 36:747–751. https://doi.org/10.1002/ana.410360510

Melov S, Adlard PA, Morten K et al (2007) Mitochondrial oxidative stress causes hyperphosphorylation of tau. PLoS One 2:e536. https://doi.org/10.1371/journal.pone.0000536

Moreira PI, Nunomura A, Nakamura M et al (2008) Nucleic acid oxidation in Alzheimer disease. Free Radic Biol Med 44:1493–1505. https://doi.org/10.1016/j.freeradbiomed.2008.01.002

Moreira PI, Carvalho C, Zhu X et al (2010) Mitochondrial dysfunction is a trigger of Alzheimer's disease pathophysiology. Biochim Biophys Acta Mol basis Dis 1802:2–10. https://doi.org/10.1016/j.bbadis.2009.10.006

Münch G, Cunningham AM, Riederer P, Braak E (1998) Advanced glycation end products are associated with Hirano bodies in Alzheimer's disease. Brain Res 796:307–310

Murakami K, Murata N, Noda Y et al (2011) SOD1 (copper/zinc superoxide dismutase) deficiency drives amyloid β protein oligomerization and memory loss in mouse model of Alzheimer disease. J Biol Chem 286:44557–44568. https://doi.org/10.1074/jbc.M111.279208

Mutisya EM, Bowling AC, Beal MF (1994) Cortical cytochrome oxidase activity is reduced in Alzheimer's disease. J Neurochem 63:2179–2184

Nisbet RM, Polanco J-C, Ittner LM, Götz J (2015) Tau aggregation and its interplay with amyloid-β. Acta Neuropathol 129:207–220. https://doi.org/10.1007/s00401-014-1371-2

Nishida Y, Yokota T, Takahashi T et al (2006) Deletion of vitamin E enhances phenotype of Alzheimer disease model mouse. Biochem Biophys Res Commun 350:530–536. https://doi.org/10.1016/j.bbrc.2006.09.083

Nunomura A, Perry G, Pappolla MA et al (1999) RNA oxidation is a prominent feature of vulnerable neurons in Alzheimer's disease. J Neurosci 19:1959–1964

Nunomura A, Perry G, Aliev G et al (2001) Oxidative damage is the earliest event in Alzheimer disease. J Neuropathol Exp Neurol 60:759–767

Nunomura A, Chiba S, Lippa CF et al (2004) Neuronal RNA oxidation is a prominent feature of familial Alzheimer's disease. Neurobiol Dis 17:108–113. https://doi.org/10.1016/j.nbd.2004.06.003

Nunomura A, Tamaoki T, Tanaka K et al (2010) Intraneuronal amyloid beta accumulation and oxidative damage to nucleic acids in Alzheimer disease. Neurobiol Dis 37:731–737. https://doi.org/10.1016/j.nbd.2009.12.012

Oda A, Tamaoka A, Araki W (2009) Oxidative stress up-regulates presenilin 1 in lipid rafts in neuronal cells. J Neurosci Res 88. https://doi.org/10.1002/jnr.22271

Onyango IG, Dennis J, Khan SM (2016) Mitochondrial dysfunction in Alzheimer's disease and the rationale for bioenergetics based therapies. Aging Dis 7:201. https://doi.org/10.14336/AD.2015.1007

Opazo C, Huang X, Cherny RA et al (2002) Metalloenzyme-like activity of Alzheimer's disease β-amyloid. J Biol Chem 277:40302–40308. https://doi.org/10.1074/jbc.M206428200

Ortwerth BJ, Olesen PR (1988) Ascorbic acid-induced crosslinking of lens proteins: evidence supporting a Maillard reaction. Biochim Biophys Acta Protein Struct Mol Enzymol 956:10–22. https://doi.org/10.1016/0167-4838(88)90292-0

Oshiro S, Kawahara M, Kuroda Y et al (2000) Glial cells contribute more to iron and aluminum accumulation but are more resistant to oxidative stress than neuronal cells. Biochim Biophys Acta Mol basis Dis 1502:405–414. https://doi.org/10.1016/S0925-4439(00)00065-X

Padurariu M, Ciobica A, Hritcu L et al (2010) Changes of some oxidative stress markers in the serum of patients with mild cognitive impairment and Alzheimer's disease. Neurosci Lett 469:6–10. https://doi.org/10.1016/j.neulet.2009.11.033

Pavlov PF, Petersen CH, Glaser E, Ankarcrona M (2009) Mitochondrial accumulation of APP and Aβ: significance for Alzheimer disease pathogenesis. J Cell Mol Med 13:4137–4145. https://doi.org/10.1111/j.1582-4934.2009.00892.x

Petersen JD, Kaech S, Banker G (2014) Selective microtubule-based transport of dendritic membrane proteins arises in concert with axon specification. J Neurosci 34:4135–4147. https://doi.org/10.1523/JNEUROSCI.3779-13.2014

Plant LD, Boyle JP, Smith IF et al (2003) The production of amyloid beta peptide is a critical requirement for the viability of central neurons. J Neurosci 23:5531–5535

Prabhakaram M, Ortwerth BJ (1994) Determination of glycation crosslinking by the sugar-dependent incorporation of [14C]Lysine into protein. Anal Biochem 216:305–312. https://doi.org/10.1006/abio.1994.1046

Praticò D (2008) Oxidative stress hypothesis in Alzheimer's disease: a reappraisal. Trends Pharmacol Sci 29:609–615. https://doi.org/10.1016/j.tips.2008.09.001

Praticò D, Sung S (2004) Lipid peroxidation and oxidative imbalance: early functional events in Alzheimer's disease. J Alzheimers Dis 6:171–175

Quiroz-Baez R, Rojas E, Arias C (2009) Oxidative stress promotes JNK-dependent amyloidogenic processing of normally expressed human APP by differential modification of α-, β- and γ-secretase expression. Neurochem Int 55:662–670. https://doi.org/10.1016/j.neuint.2009.06.012

Rahman A, Haque A, Uddin MS et al (2017) *In Vitro* screening for antioxidant and anticholinesterase effects of *Uvaria littoralis* Blume: a nootropic phytotherapeutic remedy. J Intellect Disabil – Diagnosis Treat 5:50–60. https://doi.org/10.6000/2292-2598.2017.05.02.3

Rhein V, Song X, Wiesner A et al (2009) Amyloid- and tau synergistically impair the oxidative phosphorylation system in triple transgenic Alzheimer's disease mice. Proc Natl Acad Sci 106:20057–20062. https://doi.org/10.1073/pnas.0905529106

Roberson ED, Mucke L (2006) 100 years and counting: prospects for defeating Alzheimer's disease. Science (80-) 314:781–784. https://doi.org/10.1126/science.1132813

Rodrigues CMP, Solá S, Brito MA et al (2001) Amyloid β-peptide disrupts mitochondrial membrane lipid and protein structure: protective role of tauroursodeoxycholate. Biochem Biophys Res Commun 281:468–474. https://doi.org/10.1006/bbrc.2001.4370

Rottkamp CA, Raina AK, Zhu X et al (2001) Redox-active iron mediates amyloid-beta toxicity. Free Radic Biol Med 30:447–450

Rousset S, Alves-Guerra M-C, Mozo J et al (2004) The biology of mitochondrial uncoupling proteins. Diabetes 53(Suppl 1):S130–S135

Sambamurti K, Kinsey R, Maloney B et al (2004) Gene structure and organization of the human β-secretase (*BACE*) promoter. FASEB J 18:1034–1036. https://doi.org/10.1096/fj.03-1378fje

Sayre LM, Moreira PI, Smith MA, Perry G (2005) Metal ions and oxidative protein modification in neurological disease. Ann Ist Super Sanita 41:143–164

Sayre LM, Perry G, Smith MA (2008) Oxidative stress and neurotoxicity. Chem Res Toxicol 21:172–188. https://doi.org/10.1021/tx700210j

Schipper HM, Song W, Zukor H et al (2009) Heme oxygenase-1 and neurodegeneration: expanding frontiers of engagement. J Neurochem 110:469–485. https://doi.org/10.1111/j.1471-4159.2009.06160.x

Serrano-Pozo A, Frosch MP, Masliah E, Hyman BT (2011) Neuropathological alterations in Alzheimer disease. Cold Spring Harb Perspect Med 1:a006189. https://doi.org/10.1101/cshperspect.a006189

Shen C, Chen Y, Liu H et al (2008) Hydrogen peroxide promotes Abeta production through JNK-dependent activation of gamma-secretase. J Biol Chem 283:17721–17730. https://doi.org/10.1074/jbc.M800013200

Smith MA, Kutty RK, Richey PL et al (1994) Heme oxygenase-1 is associated with the neurofibrillary pathology of Alzheimer's disease. Am J Pathol 145:42–47

Smith MA, Sayre LM, Monnier VM, Perry G (1995) Radical AGEing in Alzheimer's disease. Trends Neurosci 18:172–176

Smith MA, Perry G, Richey PL et al (1996) Oxidative damage in Alzheimer's. Nature 382:120–121. https://doi.org/10.1038/382120b0

Smith MA, Sayre LM, Anderson VE et al (1998) Cytochemical demonstration of oxidative damage in Alzheimer disease by immunochemical enhancement of the carbonyl reaction with 2,4-dinitrophenylhydrazine. J Histochem Cytochem 46:731–735. https://doi.org/10.1177/002215549804600605

Smith MA, Rottkamp CA, Nunomura A et al (2000) Oxidative stress in Alzheimer's disease. Biochim Biophys Acta 1502:139–144

Stamer K, Vogel R, Thies E et al (2002) Tau blocks traffic of organelles, neurofilaments, and APP vesicles in neurons and enhances oxidative stress. J Cell Biol 156:1051–1063. https://doi.org/10.1083/jcb.200108057

Steiner H (2004) Uncovering gamma-secretase. Curr Alzheimer Res 1:175–181

Strausak D, Mercer JF, Dieter HH et al (2001) Copper in disorders with neurological symptoms: Alzheimer's, Menkes, and Wilson diseases. Brain Res Bull 55:175–185

Tamagno E, Guglielmotto M, Aragno M et al (2008) Oxidative stress activates a positive feedback between the gamma- and beta-secretase cleavages of the beta-amyloid precursor protein. J Neurochem 104:683–695. https://doi.org/10.1111/j.1471-4159.2007.05072.x

Uddin MS, Rashid M (2019) Advances in neuropharmacology: drugs and therapeutics. Apple Academic Press, Canada

Uddin MS, Upaganlawar AB (2019) Oxidative stress and antioxidant defense: biomedical value in health and diseases. Nova Science Publishers, USA

Uddin MS, Haque A, Al MA et al (2016) Searching the linkage between high fat diet and Alzheimer's disease: a debatable proof stand for ketogenic diet to alleviate symptoms of Alzheimer's patient with APOE ε4 Allele. J Neurol Neurophysiol 07:1–9. https://doi.org/10.4172/2155-9562.1000397

Uddin MS, Al MA, Hossain MS et al (2016a) Neuroprotective effect of *Phyllanthus acidus* L. on learning and memory impairment in scopolamine-induced animal model of dementia and oxidative stress: natural wonder for regulating the development and progression of Alzheimer's disease. Adv Alzheimer's Dis 05:53–72. https://doi.org/10.4236/aad.2016.52005

Uddin MS, Al MA, Hossain MS et al (2016b) Exploring the effect of *Phyllanthus emblica* L. on cognitive performance, brain antioxidant markers and acetylcholinesterase activity in rats: promising natural gift for the mitigation of Alzheimer's disease. Ann Neurosci 23:218–229. https://doi.org/10.1159/000449482

Uddin MS, Al MA, Iqbal MA et al (2016c) Analyzing nootropic effect of *Phyllanthus reticulatus* Poir. on cognitive functions, brain antioxidant enzymes and acetylcholinesterase activity against aluminium-induced Alzheimer's model in rats: applicable for controlling the risk factors of Alzheimer's disease. Adv Alzheimer's Dis 05:87–102. https://doi.org/10.4236/aad.2016.53007

Uddin MS, Nasrullah M, Hossain MS et al (2016d) Evaluation of nootropic activity of *Persicaria flaccida* on cognitive performance, brain antioxidant markers and acetylcholinesterase activ-

ity in rats: implication for the management of Alzheimer's disease. Am J Psychiatry Neurosci 4(2):26–37. https://doi.org/10.11648/j.ajpn.20160402.12
Uddin MS, Amran MS (2018) Handbook of research on critical examinations of neurodegenerative disorders. Hershey, IGI Global
Uddin MS, Stachowiak A, Al MA et al (2018a) Autophagy and Alzheimer's disease: from molecular mechanisms to therapeutic implications. Front Aging Neurosci 10:1–18. https://doi.org/10.3389/fnagi.2018.00004
Uddin MS, Al Mamun A, Kabir MT et al (2018b) Nootropic and anti-Alzheimer's actions of medicinal plants: molecular insight into therapeutic potential to alleviate Alzheimer's neuropathology. Mol Neurobiol:1–20. https://doi.org/10.1007/s12035-018-1420-2
Uddin MS, Al MA, Labu ZK et al (2019a) Autophagic dysfunction in Alzheimer's disease: cellular and molecular mechanistic approaches to halt Alzheimer's pathogenesis. J Cell Physiol 234(6):8094–8112. https://doi.org/10.1002/jcp.27588
Uddin MS, Kabir MT, Al Mamun A et al (2019b) *APOE* and Alzheimer's disease: evidence mounts that targeting *APOE4* may combat Alzheimer's pathogenesis. Mol Neurobiol 56(4):2450–2465. https://doi.org/10.1007/s12035-018-1237-z
Uddin MS, Al MA, Takeda S et al (2019c) Analyzing the chance of developing dementia among geriatric people: a cross-sectional pilot study in Bangladesh. Psychogeriatrics 19(2):87–94. https://doi.org/10.1111/psyg.12368
Vitek MP, Bhattacharya K, Glendening JM et al (1994) Advanced glycation end products contribute to amyloidosis in Alzheimer disease. Proc Natl Acad Sci U S A 91:4766–4770
Walsh DM, Selkoe DJ (2007) Aβ Oligomers – a decade of discovery. J Neurochem 101:1172–1184. https://doi.org/10.1111/j.1471-4159.2006.04426.x
Wan L, Nie G, Zhang J et al (2011) β-Amyloid peptide increases levels of iron content and oxidative stress in human cell and Caenorhabditis elegans models of Alzheimer disease. Free Radic Biol Med 50:122–129. https://doi.org/10.1016/j.freeradbiomed.2010.10.707
Wang X, Wang W, Li L et al (2014) Oxidative stress and mitochondrial dysfunction in Alzheimer's disease. Biochim Biophys Acta Mol basis Dis 1842:1240–1247. https://doi.org/10.1016/j.bbadis.2013.10.015
Weng S-L, Huang K-Y, Kaunang FJ et al (2017) Investigation and identification of protein carbonylation sites based on position-specific amino acid composition and physicochemical features. BMC Bioinf 18:66. https://doi.org/10.1186/s12859-017-1472-8
White AR, Reyes R, Mercer JF et al (1999) Copper levels are increased in the cerebral cortex and liver of APP and APLP2 knockout mice. Brain Res 842:439–444
White AR, Barnham KJ, Bush AI (2006) Metal homeostasis in Alzheimer's disease. Expert Rev Neurother 6:711–722. https://doi.org/10.1586/14737175.6.5.711
Wong CM, Marcocci L, Liu L, Suzuki YJ (2010) Cell signaling by protein carbonylation and decarbonylation. Antioxid Redox Signal 12:393–404. https://doi.org/10.1089/ars.2009.2805
Wright RO, Baccarelli A (2007) Metals and neurotoxicology. J Nutr 137:2809–2813. https://doi.org/10.1093/jn/137.12.2809
Wu Z, Zhang J, Zhao B (2009) Superoxide anion regulates the mitochondrial free Ca $^{2+}$ through uncoupling proteins. Antioxid Redox Signal 11:1805–1818. https://doi.org/10.1089/ars.2009.2427
Wu Z, Zhao Y, Zhao B (2010) Superoxide anion, uncoupling proteins and Alzheimer's disease. J Clin Biochem Nutr 46:187–194. https://doi.org/10.3164/jcbn.09-104-2
Xiao Q, Gil S-C, Yan P et al (2012) Role of phosphatidylinositol Clathrin assembly lymphoid-myeloid leukemia (PICALM) in intracellular amyloid precursor protein (APP) processing and amyloid plaque pathogenesis. J Biol Chem 287:21279–21289. https://doi.org/10.1074/jbc.M111.338376
Yan SD, Chen X, Schmidt AM et al (1994) Glycated tau protein in Alzheimer disease: a mechanism for induction of oxidant stress. Proc Natl Acad Sci U S A 91:7787–7791
Yang L-B, Lindholm K, Yan R et al (2003) Elevated β-secretase expression and enzymatic activity detected in sporadic Alzheimer disease. Nat Med 9:3–4. https://doi.org/10.1038/nm0103-3

Yankner BA, Lu T (2009) Amyloid beta-protein toxicity and the pathogenesis of Alzheimer disease. J Biol Chem 284:4755–4759. https://doi.org/10.1074/jbc.R800018200

Yao M, Nguyen T-V V, Pike CJ (2005) Beta-amyloid-induced neuronal apoptosis involves c-Jun N-terminal kinase-dependent downregulation of Bcl-w. J Neurosci 25:1149–1158. https://doi.org/10.1523/JNEUROSCI.4736-04.2005

Yoo M-H, Gu X, Xu X-M et al (2010) Delineating the role of glutathione peroxidase 4 in protecting cells against lipid hydroperoxide damage and in Alzheimer's disease. Antioxid Redox Signal 12:819–827. https://doi.org/10.1089/ars.2009.2891

Yoshiyama Y, Higuchi M, Zhang B et al (2007) Synapse loss and microglial activation precede tangles in a P301S Tauopathy mouse model. Neuron 53:337–351. https://doi.org/10.1016/j.neuron.2007.01.010

Younessi P, Yoonessi A (2011) Advanced glycation end-products and their receptor-mediated roles: inflammation and oxidative stress. Iran J Med Sci 36:154–166

Zaplatic E, Bule M, Shah SZA, Uddin MS et al (2019) Molecular mechanisms underlying protective role of quercetin in attenuating Alzheimer's disease. Life Sci 224:109–119. https://doi.org/10.1016/j.lfs.2019.03.055

Zhang J, Liu Q, Chen Q et al (2006) Nicotine attenuates β-amyloid-induced neurotoxicity by regulating metal homeostasis. FASEB J 20:1212–1214. https://doi.org/10.1096/fj.05-5214fje

Zhang L-H, Wang X, Stoltenberg M et al (2008) Abundant expression of zinc transporters in the amyloid plaques of Alzheimer's disease brain. Brain Res Bull 77:55–60. https://doi.org/10.1016/j.brainresbull.2008.03.014

Zhang L-H, Wang X, Zheng Z-H et al (2010) Altered expression and distribution of zinc transporters in APP/PS1 transgenic mouse brain. Neurobiol Aging 31:74–87. https://doi.org/10.1016/j.neurobiolaging.2008.02.018

Zhang Y, Li P, Feng J, Wu M (2016) Dysfunction of NMDA receptors in Alzheimer's disease. Neurol Sci 37:1039–1047. https://doi.org/10.1007/s10072-016-2546-5

Zhao Y, Zhao B (2012) Natural antioxidants in prevention and management of Alzheimer's disease. Front Biosci (Elite Ed) 4:794–808

Zheng W-H, Bastianetto S, Mennicken F et al (2002) Amyloid beta peptide induces tau phosphorylation and loss of cholinergic neurons in rat primary septal cultures. Neuroscience 115:201–211

Zheng W, Xin N, Chi Z-H et al (2009) Divalent metal transporter 1 is involved in amyloid precursor protein processing and Aβ generation. FASEB J 23:4207–4217. https://doi.org/10.1096/fj.09-135749

Zhou Q, Lam PY, Han D, Cadenas E (2009) Activation of c-Jun-N-terminal kinase and decline of mitochondrial pyruvate dehydrogenase activity during brain aging. FEBS Lett 583:1132–1140. https://doi.org/10.1016/j.febslet.2009.02.043

Zhu X, Raina AK, Rottkamp CA et al (2001) Activation and redistribution of c-jun N-terminal kinase/stress activated protein kinase in degenerating neurons in Alzheimer's disease. J Neurochem 76:435–441

Zhu X, Raina AK, Perry G, Smith MA (2004) Alzheimer's disease: the two-hit hypothesis. Lancet Neurol 3:219–226. https://doi.org/10.1016/S1474-4422(04)00707-0

Zhu X, Perry G, Moreira PI et al (2006) Mitochondrial abnormalities and oxidative imbalance in Alzheimer disease. J Alzheimers Dis 9:147–153

Strengthen Alzheimer's Awareness Through Biomusic

6

Panagiota Simou, Pavlos Vakalos, Athanasios Alexiou, Stylianos Chatzichronis, Mahmoud A. Ali, Ioannis Haranas, Ioannis Gkigkitzis, Abdul Hafeez, Asma Perveen, and Ghulam Md Ashraf

Abstract

A large number of neurodegenerative diseases, the age-related dementia, and the human frailty are caused, among others, from non-folding proteins and the aggregation of them leading to neuronal death. In the case of Alzheimer's disease, a significant number of proteins like amyloid beta and tau seem to form unordered and problematic structures, leading through unknown mechanisms to pathological conditions and implications. Additionally, few studies have already shown the possibility of composing music sculptures retrieved from biological data, either for music audience or for cases of art therapy. In this study, we use a novel algorithm in order to present complex sound sculptures based on a hypothetical computerized interaction of amyloid beta and tau. The designed algorithm is completely adjustable, regarding several sound parameters, recognizing and emphasizing common

P. Simou
Novel Global Community Educational Foundation, Hebersham, NSW, Australia

P. Vakalos
Independent Musician-Researcher, Thessaloniki, Greece

A. Alexiou (✉)
Novel Global Community Educational Foundation, Hebersham, NSW, Australia

AFNP Med, Wien, Austria
e-mail: alexiou@ngcef.net

S. Chatzichronis
Novel Global Community Educational Foundation, Hebersham, NSW, Australia

Department of Informatics and Telecommunications, Bioinformatics Program, National and Kapodistrian University of Athens, Athens, Greece

M. A. Ali
Biotechnology Program, Faculty of Agriculture, Cairo University, Giza, Egypt

G. M. Ashraf, A. Alexiou (eds.), *Biological, Diagnostic and Therapeutic Advances in Alzheimer's Disease*, https://doi.org/10.1007/978-981-13-9636-6_6

patterns between the tested proteins for the generation of acoustic stimulus to the audience, and revealing Alzheimer's complexity and tragedy.

Keywords
Alzheimer's disease · Amino acids · Amyloid beta protein · Art therapy · Musical instrument digital interface · Music neuroscience · Neuroscience education · Tau protein

6.1 Introduction

A large number of neurodegenerative diseases in humans result from non-folding proteins and the aggregation of them. Unfolded proteins are believed to be the primary cause of Alzheimer's disease (AD), Parkinson's disease (PD), Huntington's disease (HD), cystic fibrosis, and many other neurodegenerative disorders (Mantzavinos and Alexiou 2017; Alexiou et al. 2018). In the case of AD, it is well known that the overexpression of amyloid beta (Aβ) may be a consequence of neuronal apoptosis and causes alteration in mitochondrial functions, mitochondrial fragmentation, increase in reactive oxygen species and ATP production, and reduced mitochondrial membrane potential (Wang et al. 2008; Lustbader et al. 2004). The most known neuropathological lesions of AD include neurofibrillary tangles and Aβ plaques that are strongly related to cognitive decline. While paired helical filaments are one of the principal constituents in AD pathology, we identify as their main component the tubule-associated hyperphosphorylated tau protein (Brion et al. 1986; Grundke-Iqbal et al. 1986; Grundke-Iqbal et al. 1988; Gold et al. 2015). There is evidence that several peptide regions of Aβ and tau interact with each other, explaining in a way the lack of co-localization of neurofibrillary tangles and senile plaques in AD (Pérez et al. 2004; Smith et al. 1995).

I. Haranas
Dept. of Physics and Computer Science, Wilfrid Laurier University, Waterloo, ON, Canada

I. Gkigkitzis
NOVA, Department of Mathematics, Annandale, VA, USA

A. Hafeez
Glocal School of Pharmacy, Glocal University, Saharanpur, India

A. Perveen
Glocal School of Life Sciences, Glocal University, Saharanpur, India

G. M. Ashraf
King Fahd Medical Research Center, King Abdulaziz University, Jeddah, Saudi Arabia

Department of Medical Laboratory Technology, Faculty of Applied Medical Sciences, King Abdulaziz University, Jeddah, Saudi Arabia
e-mail: gashraf@kau.edu.sa

However, should biology be considered as an inspiration for the artists? It is referred that artists are already involved in biological sciences and adopt a new way of expressing genetic information and biological data or present human body through art masterpieces or BioArt projects (Simou et al. 2013). In the case of biomusic, composers attempt to extract audio material using gene code, muscle and nerve movement, neurobiological data, fMRI results or electroencephalogram activity of the human brain, electrocardiogram, galvanic skin response, and respiration (Wald 1983; Rosling and Kitchen 1993; Rosenboom 1976, 1997, 2003, 2014; Minciacchi 2003; Kverno et al. 2009; Wu et al. 2009; Paul et al. 2015; Bukowska et al. 2015; Deuel et al. 2017). Obviously, music can be composed as an alternative representation of biological data, as a novel artistic project, or even as a therapeutic method. Nevertheless, the significant challenge is to combine inspiration and real human data in order to compose emotionally charged music, fully configurable and adjustable to different audio techniques and methods. It is known that scientists and artists have already developed artworks as a medium of communication in education and in diagnosis and treatment of mental disorders (Ahmed et al. 2006, Websites for Arts in Dementia and AD Management, Simou et al. 2015). In this chapter, we generated audio sequences, by transforming amino acids to musical instrument digital interface (midi) notes, and correlate pairs of audio samples in association with Aβ and tau protein sequences. According to latest studies (Vialatte et al. 2012), there are cases where audio representations of electroencephalographic multichannel signals are characterized as clinically valuable for medical practitioners and neuroscientists and a new innovative way of representing brain signals and biological data. Therefore, the proposed algorithm correlates randomly amino acid sequences to musical instrument digital interface notes, in order to produce sounds and compose music that is fully customizable and adjustable to different selections and pairing of data with notes. The composed music targets the hundreds of millions of people that in someway are related to an AD patient, to raise the awareness and increase the sensitization of the social bodies that are actively involved in the management of the disease.

6.2 Material and Methods

6.2.1 Computational Biology Analysis

The primary information is extracted from the UniProt database and the correspondent FASTA format for the amyloid beta A4 protein (Fig. 6.1) and the microtubule-associated tau protein, respectively (Fig. 6.2). Amyloid beta is an integral membrane protein that is expressed in many tissues and is concentrated in the synapses of neurons. Its primary function is unclear, although it is considered to regulate the formation of synapses in neural plasticity and iron export. It is known as the precursor molecule whose proteolysis produces beta-amyloid peptide whose fibrillar form is the major component of the amyloid plaques found in the brains of patients with AD (Selkoe 2001).

```
MLPGLALLLLAAWTARALEVPTDGNAGLLAEPQIAMFCGRLNMHMNVQNGKWDSDPSGTKTCIDTKEGIL
QYCQEVYPELQITNVVEANQPVTIQNWCKRGRKQCKTHPHFVIPYRCLVGEFVSDALLVPDKCKFLHQER
MDVCETHLHWHTVAKETCSEKSTNLHDYGMLLPCGIDKFRGVEFVCCPLAEESDNVDSADAEEDDSDVWW
GGADTDYADGSEDKVVEVAEEEEVAEVEEEEADDDEDDEDGDEVEEEAEEPYEEATERTTSIATTTTTTT
ESVEEVVREVCSEQAETGPCRAMISRWYFDVTEGKCAPFFYGGCGGNRNNFDTEEYCMAVCGSAMSQSLL
KTTQEPLARDPVKLPTTAASTPDAVDKYLETPGDENEHAHFQKAKERLEAKHRERMSQVMREWEEAERQA
KNLPKADKKAVIQHFQEKVESLEQEAANERQQLVETHMARVEAMLNDRRRLALENYITALQAVPPRPRHV
FNMLKKYVRAEQKDRQHTLKHFEHVRMVDPKKAAQIRSQVMTHLRVIYERMNQSLSLLYNVPAVAEEIQD
EVDELLQKEQNYSDDVLANMISEPRISYGNDALMPSLTETKTTVELLPVNGEFSLDDLQPWHSFGADSVP
ANTENEVEPVDARPAADRGLTTRPGSGLTNIKTEEISEVKMDAEFRHDSGYEVHHQKLVFFAEDVGSNKG
AIIGLMVGGVVIATVIVITLVMLKKKQYTSIHHGVVEVDAAVTPEERHLSKMQQNGYENPTYKFFEQMQN
```

Fig. 6.1 >gi|112927|sp|P05067.3|A4_HUMAN, (UniProtKB: locus A4_HUMAN, accession P05067) (Kang et al. 1987)

```
MAEPRQEFEVMEDHAGTYGLGDRKDQGGYTMHQDQEGDTDAGLKESPLQTPTEDGSEEPGSETSDAKSTP
TAEDVTAPLVDEGAPGKQAAAQPHTEIPEGTTAEEAGIGDTPSLEDEAAGHVTQEPESGKVVQEGFLREP
GPPGLSHQLMSGMPGAPLLPEGPREATRQPSGTGPEDTEGGRHAPELLKHQLLGDLHQEGPPLKGAGGKE
RPGSKEEVDEDRDVDESSPQDSPPSKASPAQDGRPPQTAAREATSIPGFPAEGAIPLPVDFLSKVSTEIP
ASEPDGPSVGRAKGQDAPLEFTFHVEITPNVQKEQAHSEEHLGRAAFPGAPGEGPEARGPSLGEDTKEAD
LPEPSEKQPAAAPRGKPVSRVPQLKARMVSKSKDGTGSDDKKAKTSTRSSAKTLKNRPCLSPKHPTPGSS
DPLIQPSSPAVCPEPPSSPKYVSSVTSRTGSSGAKEMKLKGADGKTKIATPRGAAPPGQKGQANATRIPA
KTPPAPKTPPSSGEPPKSGDRSGYSSPGSPGTPGSRSRTPSLPTPPTREPKKVAVVRTPPKSPSSAKSRL
QTAPVPMPDLKNVKSKIGSTENLKHQPGGGKVQIINKKLDLSNVQSKCGSKDNIKHVPGGGSVQIVYKPV
DLSKVTSKCGSLGNIHHKPGGGQVEVKSEKLDFKDRVQSKIGSLDNITHVPGGGNKKIETHKLTFRENAK
AKTDHGAEIVYKSPVVSGDTSPRHLSNVSSTGSIDMVDSPQLATLADEVSASLAKQGL
```

Fig. 6.2 >gi|334302961|sp|P10636.5|TAU_HUMAN (UniProtKB: locus TAU_HUMAN, accession P10636) (Goedert et al. 1988)

Additionally, tau proteins stabilize microtubules, and they are rich in neurons of the central nervous system. Pathological disorders and dementia of the nervous system such as AD can be triggered when tau proteins are defective and the microtubules are not properly stabilized (Rosenmann et al. 2012). The basic computational analysis of these two proteins was executed using the CLC Workbench software (CLC Bio Qiagen 2018) in order to calculate the amino acid distribution (Fig. 6.3), to identify protein patterns with hidden Markov models and any other similarities that could be translated to audio reference.

In general, proteins, according to their quaternary structure, have a different arrangement and configuration in the space and consist of a different amino acid sequence. However, the computational analysis resulted in several common patterns, which may demonstrate certain properties and suggested common functionalities or related mutations in several cases of AD (McLaurin and Fraser 2000). These common biological patterns consist of different amino acid sequences that are commonly found in the two tested proteins and were carefully chosen to produce sounds, by simultaneously playing in various audio channels. No traditional musical styles or contents were used, in order to produce stylistically independent sounds based on raw, acoustic features (Rosenboom 2014). These sound sculptures could be mixed with other biological data as well, modeled with alternative instrumental and notes

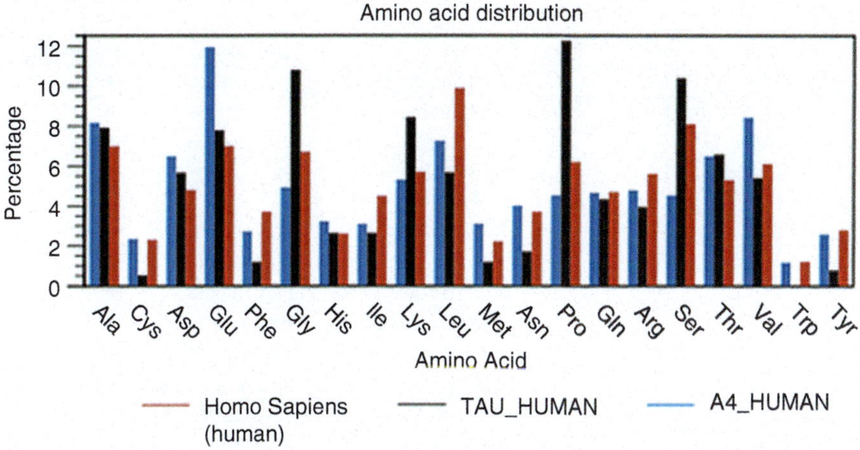

Fig. 6.3 Aβ and tau amino acid distribution comparison

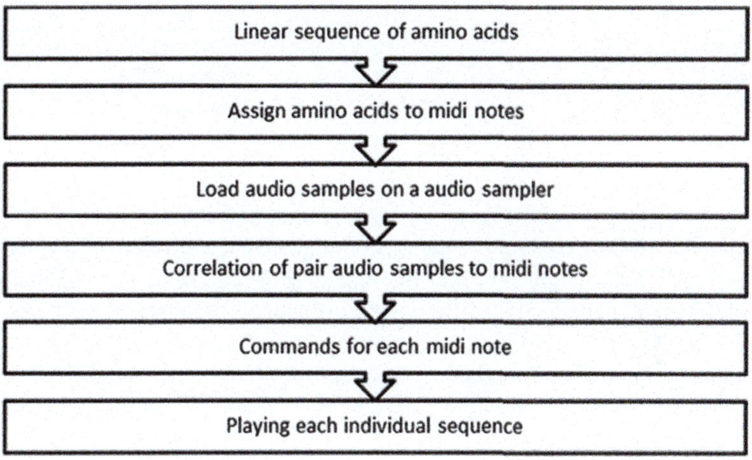

Fig. 6.4 The biomusic algorithm

selection and used for personalized neurological treatment like the NASESE Software (Chatzichronis et al. 2019).

6.2.2 Duet for Amyloid Beta A4 and Tau

The basic idea for the proposed biomusic algorithm can be described as follows (Fig. 6.4): Initially, we randomly assigned the amino acids from the two protein sequences on different midi notes, for example, (alanine) ➔ C-2, cysteine ➔ C#-2, etc. Then, several audio samples were chosen and uploaded on an audio sampler.

Table 6.1 Example 1 – Simple tones

Amino acids	Midi mapping	Samples	Commands	Sequencing
Alanine (A)	C-2	65.5Hz sine wave	Play sample 1	➔
Cysteine (C)	C#-2	73.4Hz sine wave	Play sample 2	➔
Aspartic Acid (D)	D-2	77.8Hz sine wave	Play sample 3	➔
Glutamic Acid (E)	D#-2	87.3Hz sine wave	Play sample 4	➔
Phenylanine (F)	E-2	98Hz sine wave	Play sample 5	➔
Glysine (G)	F-2	103.8Hz sine wave	Play sample 6	➔
Histidine (H)	F#-2	116.5Hz sine wave	Play sample 7	➔
Isoleusine (I)	G-2	130.8Hz sine wave	Play sample 8	➔
Lysine (K)	G#-2	146.8Hz sine wave	Play sample 9	➔
Leucine (L)	A-2	155.5Hz sine wave	Play sample 10	➔
Methionine (M)	A#-2	174.6Hz sine wave	Play sample 11	➔
Asparagine (N)	B-2	196Hz sine wave	Play sample 12	➔
Proline (P)	C-1	207.6Hz sine wave	Play sample 13	➔
Glutamine (Q)	C#-1	233.8Hz sine wave	Play sample 14	➔
Arginine (R)	D-1	261.6Hz sine wave	Play sample 15	➔
Serine (S)	D#-1	293.6Hz sine wave	Play sample 16	➔
Threonine (T)	E-1	311.1 Hz sine wave	Play sample 17	➔
Valine (V)	F-1	349.2Hz sine wave	Play sample 18	➔
Tryptophan (W)	F#-1	392Hz sine wave	Play sample 19	➔
Tyrosine (Y)	G-1	415.3Hz sine wave	Play sample 20	➔

The audio samples were correlated on specific midi notes, for example, sample1 ➔ C-2, sample2 ➔ C#2, etc., and sequential commands were given to every midi note (play sample1, play sample2, or play nothing, or play reverse sample1). Finally, two midi files were assigned in two different audio channels (channel 1 for Aβ and channel 2 for tau) to play together, simulating the proteins' interaction. We have to mention that in the future and for other composers as well, different samples can be used every time, simple or more complex, depending on the variety of the commands and the chosen audio samples.

In the alzheimers.edu.gr, three audio files (.mp3) can be heard, in respect to the different types of audio samples and the commands in the midi sequencer. In the first example (Table 6.1), simple audio samples were chosen (different frequencies in sine waves on c minor) and commands (play sample1, play sample2, etc.) in order to produce an initial flat and nonlinear sound (tones.mp3).

In the second case (Table 6.2), more complex audio samples were chosen (cello for amyloid beta A4 on c minor and drums for tau) and commands (play sample1, reverse sample2, or play nothing, etc.) in order to produce a more instrumental and asynchronous sound (cello and drums.mp3).

In the third case (Table 6.3), multilayer audio samples were used. Even though the same simple command for each sample (play sample) was given, complex sounds (like well-known audio and voices) were applied in less than 3-second duration and after sound processing (duet for Aβ and tau.mp3). The two samples

Table 6.2 Example 2 – Cello and drums

Amino acids	Midi mapping	Samples Amyloid beta A4	Samples Tau	Commands for Amyloid beta A4	Commands for Tau	Sequencing
Alanine (A)	C-2	Cello C2-sample1	kick1-sample 1	Play sample 1	Play sample 1	➔
Cysteine (C)	C#-2	Cello D2-sample2	snare-sample 2	Play sample 2	Play sample 2	➔
Aspartic Acid (D)	D-2	Cello D#2-sample3	tom-sample 3	Play sample 3	Play sample 3	➔
Glutamic Acid (E)	D#-2	Cello F2-sample4	ride-sample 4	Play sample 4	Play sample 4	➔
Phenylanine (F)	E-2	Cello G2-sample5	clap-sample 5	Play sample 5	Play sample 5	➔
Glysine (G)	F-2	Cello G#2-sample6	kick2-sample 6	Play sample 6	Play sample 6	➔
Histidine (H)	F#-2	Cello A#2-sample7	kick3-sample 7	Play sample 7	Play sample 7	➔
Isoleusine (I)	G-2	Cello C3-sample8	snare2-sample 8	Play sample 8	Play sample 8	➔
Lysine (K)	G#-2			Play sample 1 reverse	Play sample 1 reverse	➔
Leucine (L)	A-2			Play sample 2 reverse	Play sample 2 reverse	➔
Methionine (M)	A#-2			Play sample 3 reverse	Play sample 3 reverse	➔
Asparagine (N)	B-2			Play sample 4 reverse	Play sample 4 reverse	➔
Proline (P)	C-1			Play sample 5 reverse	Play sample 5 reverse	➔
Glutamine (Q)	C#-1			Play sample 6 reverse	Play sample 6 reverse	➔
Arginine (R)	D-1			Play sample 7 reverse	Play sample 7 reverse	➔
Serine (S)	D#-1			Play sample 8 reverse	Play sample 8 reverse	➔
Threonine (T)	E-1			Play nothing	Play nothing	➔

(continued)

Table 6.2 (continued)

Amino acids	Midi mapping	Samples Amyloid beta A4	Samples Tau	Commands for Amyloid beta A4	Commands for Tau	Sequencing
Valine (V)	F-1			Play nothing	Play nothing	→
Tryptophan (W)	F#-1			Play nothing	Play nothing	→
Tyrosine (Y)	G-1			Play nothing	Play nothing	→

categories produce a complex multi-audio experience, simulating the lesions of the AD due to these protein interactions.

Even though the produced sounds are not considered yet for music treatment of neurodegeneration or rehabilitation, they can be adjusted to different midi mapping on biological sequences with the application of alternative audio samples and other commands. Besides, it is well known that there are a number of advantages to perceive the brain via an auditory music, while humans have the ability to distinguish between several simultaneous voices or instruments even in a noisy environment (Jovanov et al. 1999).

6.3 Discussion

In this study, we presented for the first time sounds derived for the artificial interaction of two proteins related to AD progression, the Aβ and the tau. A novel algorithm translated amino acid distribution to midi notes, extracting reconfigurable sounds in terms of midi notes and commands. While science and art are the most common types of knowledge and emotional expression, bioartists and researchers nowadays focus on the way that art can manipulate human existence and inspire social life (Simou et al. 2013; Rosenboom 2014). DNA sequences or brain activities can be translated and processed via alternative representations such as discrete mathematics and combinatorics, stochastic programming, image analysis through expert systems, or even artistic media like music and colors. Music is a unique auditory art form (Peretz 2006) with an unbound set of principles and rules regarding aesthetics, adaptation, and harmonization in customs of human cultures (Simou et al. 2013). Considering the rapidly increasing population of AD patients worldwide and the disturbing expectations for the next decades, we believe that every effort on the familiarization, prognosis, diagnosis, or even treatment of neurodegeneration diseases can be acceptable, in respect to bioethics and scientific reliability. Even though music neuroscience (Rosenboom 2014) is still considered as a newly established method of understanding and

Table 6.3 Example 3 – Duet for Aβ and tau

Amino acids	Midi mapping	Samples Amyloid beta A4	Samples Tau	Commands for Amyloid beta A4	Commands for Tau	Sequencing
Alanine (A)	C-2	Pad-sample1	Orthodox Monks chant-sample1	Play sample1	Play sample1	➔
Cysteine (C)	C#-2	Kick-sample2	Drum-sample2	Play sample2	Play sample2	➔
Aspartic Acid (D)	D-2	Lil Kim voice-sample3	W. Ryder voice-sample3	Play sample3	Play sample3	➔
Glutamic Acid (E)	D#-2	Gong-sample4	Voice-sample4	Play sample4	Play sample4	➔
Phenylanine (F)	E-2	S. Korea president voice-sample5	Hammond-sample5	Play sample5	Play sample5	➔
Glysine (G)	F-2	Voice-sample6	Voice-sample6	Play sample6	Play sample6	➔
Histidine (H)	F#-2	Pad-sample7	Piano-sample7	Play sample7	Play sample7	➔
Isoleusine (I)	G-2	Pad-sample8	Voice-sample8	Play sample8	Play sample8	➔
Lysine (K)	G#-2	Voice-sample9	Drum-sample9	Play sample9	Play sample9	➔
Leucine (L)	A-2	Rhythm-sample10	Drum-sample10	Play sample10	Play sample10	➔
Methionine (M)	A#-2	Pad-sample11	Drum-sample11	Play sample11	Play sample11	➔
Asparagine (N)	B-2	Pad-sample12	Hyakunin fairy tale voice-sample12	Play sample12	Play sample12	➔
Proline (P)	C-1	Crying woman voice-sample13	Pad-sample13	Play sample13	Play sample13	➔
Glutamine (Q)	C#-1	Drum-sample14	Drum-sample14	Play sample14	Play sample14	➔
Arginine (R)	D-1	Drum-sample15	Drum-sample15	Play sample15	Play sample15	➔
Serine (S)	D#-1	Drum-sample16	Drum-sample16	Play sample16	Play sample16	➔
Threonine (T)	E-1	Drum-sample17	Voice-sample17	Play sample17	Play sample17	➔

(continued)

Table 6.3 (continued)

Amino acids	Midi mapping	Samples Amyloid beta A4	Samples Tau	Commands for Amyloid beta A4	Commands for Tau	Sequencing
Valine (V)	F-1	Drum-sample18	Crying woman voice-sample18	Play sample18	Play sample18	➜
Tryptophan (W)	F#-1	Drum-sample19	Pad-sample19	Play sample19	Play sample19	➜
Tyrosine (Y)	G-1	Drum-sample20	Pad-sample20	Flay sample20	Play sample20	➜

modeling brain activity, it can be part of dialogue education (Vella 2008) and an experiential way of understanding the AD.

Conflict of Interest The authors declare no conflict of interest.

References

Ahmed SH et al (2006) Essay: healing through art therapy in disaster settings. Lancet 368:S28–S29

Alexiou A, Soursou G, Chatzichronis S et al (2018) Role of GTPases in the regulation of mitochondrial dynamics in Alzheimer's disease and CNS-related disorders. Mol Neurobiol 56:4530. https://doi.org/10.1007/s12035-018-1397-x

Brion JP, Flament-Durand J, Dustin P (1986) Alzheimer's disease and tau proteins (letter). Lancet 2(8515):1098

Bukowska AA, Krezalek P, Mirek E, Bujas P, Marchewka A (2015) Neurologic music therapy training for mobility and stability rehabilitation with Parkinson's disease – a pilot study. Front Hum Neurosci 9:710. https://doi.org/10.3389/fnhum.2015.00710

Chatzichronis S, Alexiou A, Simou P, Mantzavinos V, Tsiamis V et al (2019) Neurocognitive assessment software for enrichment sensory environments. J Proteomics Bioinform 12:018–028. https://doi.org/10.4172/0974-276X.1000492

CLC Main Workbench website: http://www.clcbio.com/

Deuel T, Pampin J, Sundstrom J, Darvas F (2017) The encephalophone: a novel musical biofeedback device using conscious control of electroencephalogram (EEG). Front Hum Neurosci 11. https://doi.org/10.3389/fnhum.2017.00213

Goedert M, Wischik CM, Crowther RA, Walker JE, Klug A (1988) Cloning and sequencing of the cDNA encoding a core protein of the paired helical filament of Alzheimer disease: identification as the microtubule-associated protein tau. Proc Natl Acad Sci U S A 85(11):4051–4055

Gold G, Herrmann F, Bouras C, Kövari E, Xekardaki A (2015) Decrease in amyloid deposition in aging brain-an autopsy study. Adv Exp Med Biol 821:9–10. https://doi.org/10.1007/978-3-319-08939-3_5

Grundke-Iqbal I, Iqbal K, Quinlan M, Tung YC, Zaidi MS, Wisniewski HM (1986) Microtubule-associated protein tau. a component of Alzheimer paired helical filaments. J Biol Chem 261:6084–6089

Grundke-Iqbal I, Vorbrodt AW, Iqbal K, Tung YC, Wang GP, Wisniewski HM (1988) Microtubule-associated polypeptides tau are altered in Alzheimer paired helical filaments. Brain Res 464:43–52

Jovanov E, Wegner K, Radivojevic V, Starcevic D, Quinn MS, Karron DB (1999) Tactical audio and acoustic rendering in biomedical applications. IEEE Trans Inf Technol Biomed 3(2):109–118
Kang J, Lemaire HG, Unterbeck A, Salbaum JM, Masters CL, Grzeschik KH, Multhaup G, Beyreuther K, Muller-Hill B (1987) The precursor of Alzheimer's disease amyloid A4 protein resembles a cell-surface receptor. Nature 325(6106):733–736
Kverno KS, Black BS, Nolan MT, Rabins PV (2009) Research on treating neuropsychiatric symptoms of advanced dementia with non-pharmacological strategies, 1998–2008: a systematic literature review. Int Psychogeriatr 21:825–843
Lustbader JW, Cirilli M, Lin C, Xu JW, Takuma K, Wang N, Caspersen C, Chen X, Pollak S, Chaney M, Trinchese F, Liu S, Gunn-Moore F, Lue LF, Walker JW, Kuppsamy P, Zewier ZL, Arancio O, Stern D, Yan SS, Wu H (2004) ABAD directly links Aβ to mitochondrial toxicity in Alzheimer's disease. Science 304:448–452
Mantzavinos V, Alexiou A (2017) Biomarkers for Alzheimer's disease diagnosis. Curr Alzheimer Res 14:1149–1154. https://doi.org/10.2174/1567205014666170203125942
McLaurin J, Fraser PE (2000) Effect of amino-acid substitutions on Alzheimer's amyloid-β peptide–glycosaminoglycan interactions. Eur J Biochem 267:6353–6361
Minciacchi D (2003) Translation from neurobiological data to music parameters. Ann N Y Acad Sci 999:282–301. https://doi.org/10.1196/annals.1284.039
http://afnpmed.com/asd-and-dementia/ (2019) Neurocognitive Assessment Software for Enrichment Sensory Environments (NASESE)
Paul A et al (2015) The effect of sung speech on socio-communicative responsiveness in children with autism spectrum disorders. Front Hum Neurosci 9:555. https://doi.org/10.3389/fnhum.2015.00555
Peretz I (2006) The nature of music from a biological perspective. Cognition 100(1):1–32
Pérez M, Cuadros R, Benítez MJ, Jiménez JS (2004) Interaction of Alzheimer's disease amyloid beta peptide fragment 25-35 with tau protein, and with a tau peptide containing the microtubule binding domain. J Alzheimers Dis 6(5):461–467
Rosenboom D (1976) Biofeedback and the arts, results of early experiments. Aesthetic Research Centre of Canada Publications, Vancouver
Rosenboom D (1997) Extended musical interface with the human nervous system: assessment and prospectus. Available online at: http://www.davidrosenboom.com/media/extended-musical-interface-human-nervous-system-assessment-and-prospectus. [Original (1990). San Francisco: Leonardo Monograph Series,1]
Rosenboom D (2014) Active imaginative listening-a neuromusical critique. Front Neurosci 8:251. https://doi.org/10.3389/fnins.2014.00251
Rosenboom D (2003) In: Avazini G, Faienza C, Minciacchi D (eds) Propositional music from extended musical interface with the human nervous system, The neurosciences and music, vol 999. Annals of the New York Academy of Sciences, NewYork, pp 263–271
Rosenmann H, Blum D, Kayed R, Ittner LM (2012) Tau protein: function and pathology. Int J Alzheimers Dis 2012:707482
Rosling LK, Kitchen J (1993) Music and drawing with institutionalized elderly. Act Adapt Aging 17:27–38. https://doi.org/10.1300/J016v17n02_03
Selkoe DJ (2001) Alzheimer's disease: genes, proteins, and therapy. Physiol Rev 81:741–766
Simou P, Tiligadis K, Alexiou A (2015) Artificial humanoid for the elderly people. In: Vlamos P, Alexiou A (eds) GeNeDis 2014, advances in experimental medicine and biology, vol 821. Springer, Switzerland. https://doi.org/10.1007/978-3-319-08939-3_7
Simou P, Tiligadis K, Alexiou A (2013) Exploring artificial intelligence utilizing bioart. 9th artificial intelligence applications and innovations conference, 1st workshop on ethics and philosophy in artificial intelligence. In: Papadopoulos H et al (eds) AIAI 2013, IFIP AICT 412, ©IFIP International Federation for Information Processing 2013. Springer, Heidelberg/Paphos, pp 687–692
Smith MA, Siedlak SL, Richey PL, Mulvihill P, Ghiso J, Frangione B, Tagliavini F, Giaccone G, Bugiani O, Praprotnik D (1995) Tau protein directly interacts with the amyloid beta-protein precursor: implications for Alzheimer's disease. Nat Med 1(4):365–369

Vella J (2008) On teaching and learning: putting the principles and practices of into action. Jossey-Bass, San Francisco

Vialatte FB, Dauwels J, Musha T, Cichocki A (2012) Audio representations of multi-channel EEG: a new tool for diagnosis of brain disorders. Am J Neurodegener Dis 1(3):292–304

Wald J (1983) Alzheimer's disease and the role of art therapy in its treatment. Am J Art Ther 22:57–64

Wang X, Su B, Siedlak SL, Moreira PI, Fujioka H, Wang Y, Casadesus G, Zhu X (2008) Amyloid-beta overproduction causes abnormal mitochondrial dynamics via differential modulation of mitochondrial fission/fusion proteins. Proc Natl Acad Sci U S A 105(49):19318–19323

Websites for arts in dementia and AD management:

Alzheimer's poetry project. www.alzpoetry.com

Arts4Dementia. www.arts4dementia.org.uk

Arts for dementia. www.artsfordementia.com

Creative dementia arts network. www.creativedementia.org

Dementia & imagination. www.dementiaandimagination.org.uk/art-and-dementia

Dementia arts. www.dementiaarts.com

Meet Mre at MoMA. www.moma.org

Music & memory. www.musicandmemory.org

The Society for the Arts in Dementia Care, www.cecd-society.org

Wu D, Liu Y, Yao DZ (2009) Listen to the brain in real time – the Chengdu brainwave music. J Electrochem Sci Technol China 7:1

Diagnosis of Alzheimer's Disease Using Brain Imaging: State of the Art

7

Atif Shah, Kamal Niaz, Moataz Ahmed, and Reem Bunyan

Abstract

Alzheimer's disease (AD) is one of the prominent diseases in elderly people which leads to language impairment, disorientation, memory loss, and eventually death. Despite the severity of the disease, there is no such drug reported to control, reduce, or stop the progression of AD. The neuroimages played a crucial role in tracking the progression of AD using biomarkers which help the physicians to diagnose the disease more accurately. In this study, we investigate the effectiveness of structural and functional neuroimaging modularities which are used in the state-of-the-art methods to diagnose AD. The finding shows that most of the studies prioritize magnetic resonance imaging techniques (MRIT) solely or combined with other neuroimaging modularities to achieve better performance. Studies also founded that only few public datasets are available, and the most widely used public dataset is Alzheimer's Disease Neuroimaging Initiative.

Keywords

Medical imaging · MRI · Alzheimer's disease · CAD · Features extraction

A. Shah (✉) · M. Ahmed
Department of Information and Computer Science, King Fahd University of Petroleum and Minerals, Dhahran, Kingdom of Saudi Arabia
e-mail: g201408760@kfupm.edu.sa

K. Niaz
Department of Pharmacology and Toxicology, Faculty of Bio-Sciences, Cholistan University of Veterinary and Animal Sciences (CUVAS), Bahawalpur, Pakistan

R. Bunyan
Neurosciences Center, King Fahd Specialist Hospital Dammam, Dammam, Kingdom of Saudi Arabia

G. M. Ashraf, A. Alexiou (eds.), *Biological, Diagnostic and Therapeutic Advances in Alzheimer's Disease*, https://doi.org/10.1007/978-981-13-9636-6_7

Abbreviations

3D	Three dimensions
ACC	Accuracy
AD	Alzheimer's disease
ADNI	Alzheimer's Disease Neuroimaging Initiative
AUC	Area under the curve
Aβ	Amyloid-β
CAD	Computer-aided diagnosis
CNN	Convolutional neural network
CR	Creatine
CSF	Cerebrospinal fluid
CT	Computed tomography
DL	Deep learning
DTI	Diffusion tensor imaging
DWI	Diffusion-weighted imaging
FC	Functional connectivity
FDG	Fluorodeoxyglucose
fMRI	Functional magnetic resonance imaging
FN	False-negative
FP	False-positive
GP-LR	Gaussian process logistic regression
ICA	Independent component analysis
LLE	Local linear embedding
MCI	Mild cognitive impairment
MCI-A	Amnestic MCI
MCI-C	MCI converted
MCI-NC	MCI non-converted
MCI-P	Progressive MCI
MCI-S	Stable MCI
MRI	Magnetic resonance imaging
MRS	Magnetic resonance spectroscopy
NAA	N-Acetyl aspartate
PCA	Principal component analysis
PET	Positron emission tomography
RBF	Radial basis function
re-fMRI	Resting-state fMRI
RFE	Recursive feature elimination
ROC	Receiver operating characteristic
ROS	Reactive oxygen species
SEN	Sensitivity
SPE	Specificity
SPECT	Single photon emission computer tomography
SVM	Support vector machine
T1-w	T1-weighted

T2-w	T2-weighted
TN	True-negative
TP	True-positive
VBM	Voxel-based measure
WM	White matter
WMH	White matter hyperintensities

7.1 Introduction

Alzheimer's disease (AD) is the utmost common cause of cognitive impairment usually in elderly people; around 11% of people at the age of 65 and above suffered from AD (Alzheimer's Association 2014). AD progression leads to memory loss, disorientation, and language impairment and eventually leads to death, which increased to 123% between 2000 and 2015 (Alzheimer's disease facts and figures 2018). It is well understood that the prevalence and severity of AD are high, still there is no drug or treatment that has been reported to reduce the risk or completely stop AD progression. The impact of AD can be understood via the progressive brain changes associated with AD. To achieve this, brain imaging played a profound role in track changes in the follow-up scans using different imaging techniques and transformed AD research and clinical practices. Brain imaging provides informative biomarker even before clinical symptoms have appeared. These biomarkers help in the diagnostic decision and disease assessment and improve diagnosis. The use of a biomarker for AD will also be helpful for clinical trials because of its enrichment strategies or tough inclusion criteria.

Imaging biomarker played an important role in AD detection and its early stage of mild cognitive impairment (MCI) using structural and functional neuroimaging. Recent diagnostic criteria (McKhann et al. 2011) suggest that neuroimaging biomarkers perform better in monitoring AD progression and its early detection than other biomarkers. Different imaging modularities have been used, including computed tomography (CT), magnetic resonance imaging (MRI), functional magnetic resonance imaging (fMRI), positron emission tomography (PET), single photon emission computer tomography (SPECT), diffusion-weighted imaging (DTI), and magnetic resonance spectroscopy (MRS), for biomarkers for detecting AD and predicting its progression. These biomarkers used features such as voxel-wise tissue probability (Kloppel et al. 2008; Liu et al. 2014; Tong et al. 2014), cortical volumes (Wolz et al. 2011; Zhang et al. 2011; Zhou et al. 2011), and thickness (Eskildsen et al. 2013; de Magalhaes Oliveira et al. 2010; Querbes et al. 2009) for classification of AD and MCI.

In this study, we investigate various structural and functional brain imaging techniques to detect AD from HC or MCI individuals. The AD can be detected via its progressive behavior which is illustrated in Fig. 7.1, showing the healthy-control/normal-control (CH/CN), MCI, and AD patient MRI with 2-year follow-up images and their differences. The biomarkers which are automatically extracted and

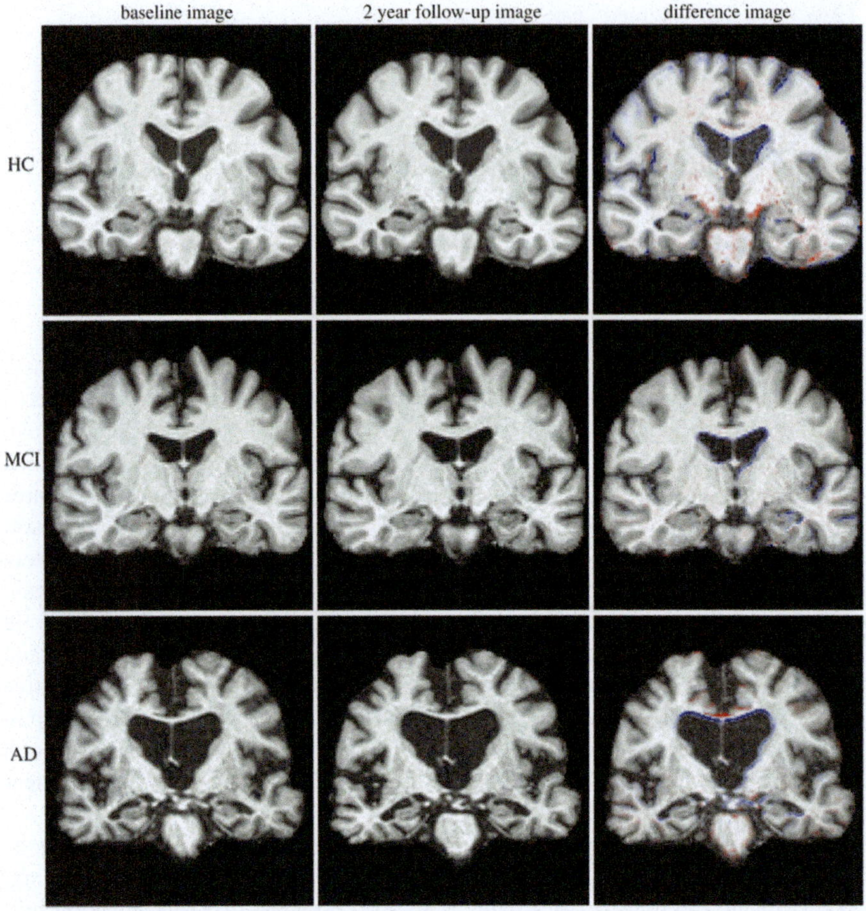

Fig. 7.1 MRI shows the extracted brain of HC, MCI, and AD patient with 2-year follow-up and their differences (Ledig et al. 2018). (The image is reproduced under the Creative Commons license)

classify AD, HC, and MCI patients are called computer-aided diagnosis (CAD). In this chapter, we discussed the relationship of AD with brain imaging, various types of structure and function neuroimaging techniques, and the observations along with conclusion and future directions.

7.2 Relationship of Brain Imaging with AD

AD is a frequent form of neurodegenerative disorder across the world, which has become a public health problem. And imaging plays an important role in the diagnosis of AD and dementia with a special focus on CT, MRI, fMRI, and PET (Johnson et al. 2012). Many of the studies revealed that computer imaging is converted from minor exclusionary role to central position which provides information of temporal and spatial revolution in AD (Selkoe et al. 2012; Krstic and Knuesel 2013). Epidemiological studies indicated that aging and hereditary predisposition are the two key risk factors for AD (Omar et al. 2017). On the other hand, pathological studies of AD have characterized extracellular aggregation of senile plaques (SPs) and development of intracellular neurofibrillary tangles and lesions of cholinergic neurons. Also, molecular pathology such as amyloid deposits can be visualized via imaging in AD (Johnson et al. 2012; Wu et al. 2011). In the central cholinergic system, numerous neurotransmitters and neuronal pathways function together in learning and memory. Thus, the cognitive impairment in AD is related to functional loss in the central cholinergic system (Palle and Neerati 2017). Moreover, aggregation of amyloid-β (Aβ) peptide, which is referred to as amyloid hypothesis, is found to cause synaptic dysfunction and neurodegeneration (van Dyck 2018). These amyloid plaques can't be seen through structural MRI which cannot detect histopathological trademarks of AD (Johnson et al. 2012). Other pathological processes are also reported in relation to AD such as neuroinflammation, impairment of cerebral circulation, altered synaptic function, and cerebral amyloid angiopathy. Disruption of default network activities during sleeping and resting stage has been revealed via PET imaging (Sperling et al. 2009; Hedden et al. 2009). Hence, these pathological processes that are identified to cause AD are considered important drug targets toward AD drug search. However, still further multi-modality studies are needed to explore the correlationship between MRI, fMRI, fluorodeoxyglucose (FDG)-PET, and PET imaging techniques and AD (Jack et al. 2010; Ghanemi 2015). Considerable number of evidence indicates that dietary control can minimize the risk of developing AD. Animal model studies have also shown a decreasing high-calorie diet to be neuroprotective by reducing Aβ accumulation (Hartman et al. 2006). Generally, flavonoids and in particular quercetin are important compounds for the development of AD therapeutics since it can protect the neurons against oxidative agents and excitotoxicity through regulating cell death mechanisms. FDG-PET is the vital biomarker to detect expression of AD-specific genes, mitochondrial dysfunction, neuropathy, oxidative stress, glial excitation, and synapse loss (Johnson et al. 2012) (Ansari et al. 2009). Oxidative stress reflects an imbalance between reactive oxygen species (ROS) production and biological system's antioxidant defense mechanisms which act by detoxifying or repairing the reactive intermediates, thereby causing damage. These damages can be seen via PET, ADNI FDG data analysis, fMRI, and MRI (Johnson et al. 2012; Chen et al. 2010; Dong et al. 2014). ROS act as the neurotransmitters and excitatory amino acids in the brain and neuronal tissue. In addition to that, the brain itself represents a substantial source of oxidative stress, as its metabolism serves as a "factory" of ROS which attacks glial cells and neurons, resulting in neuronal damage which may lead in oxidative injury and programmed cell death by apoptosis which is evaluated via computer imaging (Attwell et al. 2010; Hanisch and Kettenmann 2007; Iadecola 2004; Bezzi and Volterra 2001). All these studies revealed that brain imaging/computer imaging is an important tool to explore the histopathological condition of AD. MRI and CT are used to illustrate structural neuroimaging and SPECT, PET, FMRI, and DTI for functional neuroimaging, respectively, which represent the CAD system as shown in Fig. 7.2.

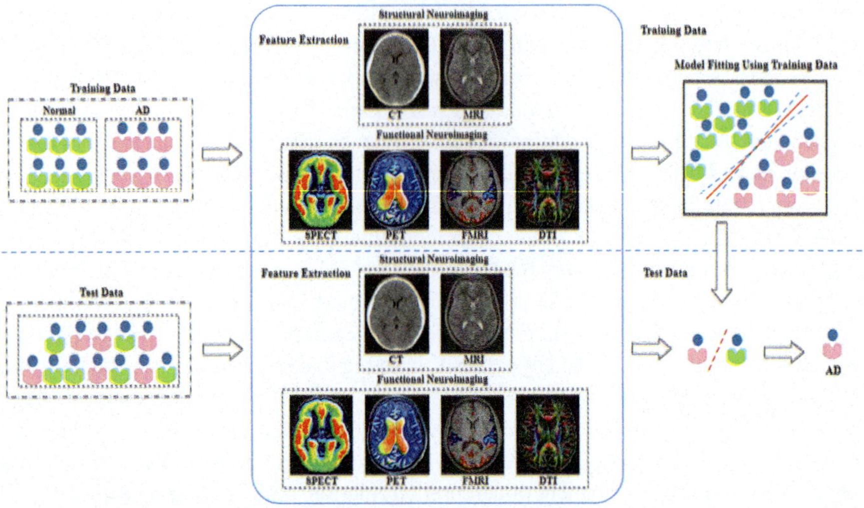

Fig. 7.2 An abstract level of AD classification framework using neuroimaging

7.3 Brain Imaging in AD Diagnosis

7.3.1 Structural Neuroimaging

7.3.1.1 Magnetic Resonance Imaging

Magnetic resonance imaging (MRI) is a medical imaging technique used to measure the hydrogen atoms' radiofrequency energies. During the static magnetic field, the nuclei with higher energy align against the field, while the lower energy aligns with the field. However, with a specific frequency, the low-energy nuclei absorb energies and align against the field. With the discontinuation of radiation, the MRI scanner detects the emitted energies, and the nuclei return to their lower-energy state. Such changes in energy levels can provide brain structures in the form of detailed images: T1-weighted (T1-w) and T2-weighted (T2-w). T1-w distinguishes gray and white matter, while T2-w characterizes white matter hyperintensities.

MRI is one of the major imaging techniques used to evaluate AD in clinical studies (Byun et al. 2017; Boutet et al. 2014; McEvoy and Brewer 2010) as well as in CAD methods, where computers detect AD automatically using machine learning algorithms (Kloppel et al. 2008; Liu et al. 2014; Tong et al. 2014; Wolz et al. 2011).

Beheshti and Demirel (2016) proposed a feature ranking-based method to classify AD and healthy-control (HC) individuals. Voxel-based morphometry was used to compare the global and local differences, a significant difference was extracted called a volume of interest (VOI), each voxel in VOI is used as a feature, t-test score was used to rank the features, and fisher criterion was used to select the dominant features with support vector machine (SVM) classifier. The work extended (Beheshti et al. 2017) where the same author used the feature ranking based on a genetic algorithm to optimize the feature selection process. In their another work, feature ranking was used with classification error to identify AD patients (Beheshti et al. 2016). Seven different ranking methods were used which include information gain, statistical dependency, t-test score, Pearson correlation coefficient, Gini index, Fisher's criterion, and mutual information. The classification error was used to select the discriminatory features during the training phase. SVM was used as a classifier with tenfold cross-validation.

Eskildsen et al. (2015) proposed a novel method to predict AD using T1-w MRI. The mutual information fusion was used to identify the robust features which were stable for the period of 3 years. Five features were used to classify AD and MCI subjects which include left and right hippocampi grading, the right anterior part of the parahippocampal gyrus, and the cortical thickness of the left precuneus and left superior temporal sulcus. These features were used in linear discriminant classifier using leave-one-out cross-validation to predict AD and MCI subjects.

Cuingnet et al. (2011) used ten different techniques for CN, AD, and MCI classification. These methods include five voxel-based methods, two methods using the hippocampus, and three methods based on cortical thickness. The results were compared: CN vs AD, CN vs MCI, and MCI non-converted vs MCI converted. This work also evaluated the time complexity of hyper-parameter tuning and feature selection. Westman et al. (2012) used the combination of MRI and cerebrospinal fluid (CSF) measures to detect AD and MCI conversion. In this study, 34 cortical thickness, 34 subcortical volumetric, and 3 CSF measures with a total of 699 (AD = 187, MCI = 287 and CTL = 225) measures were used for classification. The experiments were performed on AD vs HC, MCI vs HC, and MCI vs AD prediction.

Liu et al. (2013) used an unsupervised method, the local linear embedding (LLE) algorithm. The algorithm is used to transform the MRI data to local linear space with few dimensions. These features were used to train a classifier for AD prediction. The work demonstrated that LLE feature performed better in classification than using direct features. Moradi et al. (2015) used the semi-supervised method to detect the conversion of MCI to AD in MCI subjects. The low-density separation method was used to extract features and integrated with cognitive and age measures to improve the performance. The results showed that the combination of these features improved the area under the curve (AUC).

Sørensen et al. (2017) used cortical thickness hippocampal texture, shape, and volumetry as features to detect CN, AD, and MCI patients. The work secured first place in CAD-dementia challenge where it evaluated on two datasets, Alzheimer's Disease Neuroimaging Initiative (ADNI) and Australian Imaging Biomarkers ad Lifestyle Study of Ageing (AIBL), with tenfold cross-validation. Linear discriminant analysis was used for feature reduction.

7.3.1.2 CT

CT is a diagnostic imaging test which draws the detailed images of internal organs such as bone, vessels, and soft tissues. It is a computer-processed combinations of many X-ray imaging procedure where the rotated beam of X-rays is pointed to the body to make cross-sectional images, also called slices, which can be visualized in three dimensions (3D). The attenuation of the body tissues varies which is why the highest attenuation appears in white such as bones, while other tissues appear in black. That's why CT has the ability to visualize fluid, gas, and soft tissues. However, it cannot differentiate the gray and white matter (G and WM). To detect AD in CT images, boundary detection is one of the important methods which can increase the detection rate. Al-Jibory and El-Zaart (2013) proposed a method based on Weibull distribution instead of Gaussian distribution to detect boundary detection. The advantages of this method are to address more challenging problem where the distribution has both symmetric and asymmetric shapes.

7.3.2 Functional Neuroimaging

7.3.2.1 SPECT

SPECT is a type of nuclear imaging which is used to view how the blood flows in brain arteries and veins. Radioactive materials are used before SPECT scans, called tracer. These materials emit a single photon which is detected by the machines and translated to two-dimensional cross sections. These cross sections are then converted to 3D brain scans (Schuckit 1992). In comparison with PET, SPECT cannot determine glucose metabolism and also have lower resolution in imaging deep structures. Tabei et al. (2017) used SPECT and white matter hyperintensity (WMH) to predict cognitive decline, which is one of the major problems in AD patients. The study was conducted on 182 patients which concludes that regional cerebral blood flow and WMH are the main parameters that affect the cognitive function in AD patients.

7.3.2.2 PET

PET also uses tracers which emit positron. When a positron encounters with the electron, it gets destroyed and releases a photon which travels in opposite direction. The scanner detects the arrival of two photons simultaneously and determines the travel direction of photons. The scanner then uses this information to contract the PET image just like SPECT (Strauss 1986). However, the PET images have good spatial resolution and are used widely because of lower scanner cost and more availability of PET tracers.

Azmi et al. (2017) used the voxel intensity values of PET images as features to detect AD and HC individuals. The voxel intensities feature consists of mean voxel intensity, slice-based intensity, and global mean voxel intensity. Neural network with nodes 1 and 10 was used for the classification with z-scores. Silveira and Marques (2010) used voxel intensity with a total of 150 features with a weak classifier which combines the outputs to make a complex classifier. Each time the classifier was boosted via re-weighting which improves the performance. The work used AdaBoost algorithm, and for evaluation on selected features, SVM with radial basis function (RBF) kernel was used with tenfold cross-validation. Garali et al. (2015) extracted features from the region of interest and used separation power factor method to select the best 21 regions for further classification which improve the performance and reduce the time complexity. SVM and random forest were used as classifiers. Gray et al. (2012) used multiregional features extracted from PET images. The statistical analysis t-test was conducted to evaluate the proposed method, and two-class SVM was used for classification.

Darsana and Abraham (2016) applied the multistructure registration approach using both PET and MRI images to detect AD and MCI individuals. Asim et al. (2018) applied multi-model and multi-atlas-based method for AD detection using MRI and PET images. Voxel-based measure (VBM), GM, and WM maps were extracted from MRI, while cerebral metabolic rates of glucose were extracted from PET images and used as features. Principal component analysis (PCA) was used to reduce the image features with help of four major steps; normalize image data, calculate covariance matrix from the image data, perform single data decomposition (SVD) and find the projection image data to the new basis with reduced features. SVM with RBF kernel was used for leave-one-out cross-validation. Ota et al. (2015) used Loni probabilistic brain atlas and automated anatomical labeling to extract relative cerebral metabolic rate and gray matter density features. The features were fed to SVM-recursive feature elimination (SVM-RFE) for feature reduction to avoid overfitting and for classification with leave-one-out cross-validation.

7.3.2.3 fMRI

fMRI is used to detect brain activity and changes in blood oxygenation and flow which arise due to the initiation of the activity, also called hemodynamics. The active area of the brain consumes more oxygen which increases blood demand. fMRI produce the activation maps which are represented in different color codes, showing the strength and involvement of specific brain region in a particular mental process. The advantage of this imaging technique is that it does not use any radiation just like CT and PET.

Long et al. (2016) used to detect MCI using SVM-based methods with the multi-level characteristic of MRI. The MCI is the transition phase from CH/HC to AD and an important step for early therapeutic interventions. Hurst exponents, gray matter density, the amplitude of low-frequency fluctuations, and regional homogeneity were used as the main features for SVM with leave-one-out cross-validation. Challis et al. (2015) utilized Gaussian process logistic regression (GP-LR) with SVM using fMRI images to detect CH vs amnestic MCI (MCI-A) and MCI-A vs AD. The number of features was selected via Kendall tau correlation coefficient ranking, and features normalized via PCA and the GP-LR parameters were optimized to increase the accuracy. Leave-one-out cross-validation was used for GP-LR and SVM models.

Khazaee et al. (2017) used local and global graph measure features with SVM and naïve Bayes classifier. Fisher algorithm was used for feature selection based on the discrimination ability, followed by forward-sequential feature selection and tenfold cross-validation for robust classification. Using both the classifiers with optimal features, Bayes classifier performed better using all patient data and achieved higher AUC in HC vs AD classification. Hojjati et al. (2017) extracted graph measures of segregation and integration using graph theory. The study focused on MCI patients where they classify MCI converter (MCI-C) from MCI non-converter (MCI-NC). Global and local graph measures were calculated, which ended with 913 features. Feature selection method was used to sort the features, and sequential feature collection was used to find the optimal features for classification. SVM with leave one out cross-validation was used along the statistical analysis.

Schouten et al. (2016) combined various modalities of MRI and fMRI to classify AD and HC individuals. Various features were extracted including anatomical, diffusion, and functional connectivity features with the elastic net classifier using tenfold cross-validation. In another study (de Vos et al. 2018), eight measures were extracted including functional connectivity (FC) dynamics, FC metric, FC states, graph properties, FC for each voxel, FC for each hippocampus and low-frequency fluctuation, eigenvector centrality, and the combination of all these features. Agosta et al. (2012) explored the salience, default mode, and executive networks to detect the resting-state abnormality in MCI-A and HC. The experiments were conducted on very few images which is one of the limitations of this study. Li et al. (2018) used a transfer learning approach to improve the detection rate on a smaller dataset. ADNI dataset was used for training, and then the trained weights were transferred via adaptation method for Tongji dataset classification which consists of only 12 AD and 14 CN individual's data. The results improved by 30% than that using the smaller dataset.

7.3.2.4 DTI

DTI is the extension of diffusion-weighted imaging which provides data about WM region orientation. The measurements are based on Brownian motion and water molecules. DTI provides quantitative information to track the magnitude, orientation, and anisotropy of brain WM regions with the help of computer algorithms.

Ahmed et al. (2017) developed a CAD system by extracting features from T1-w MRI and mean diffusivity from DTI images. The AD disease-related signature was generated which was used in multiple kernel learning approach for classification with tenfold cross-validation. Kantarci et al. (2017) used DTI images in clinical perspective to analyze the white matter integrity and pathologic staging in AD. The voxel-based and atlas-based analyses were used for the evaluation of neurofibrillary tangle stages. Schouten et al. (2017) used voxel tensor measures, graph measures, and independent component analysis (ICA)-clustered measures. All features were used for classification independently and also as a fusion vector via elastic net classification with tenfold cross-validation. The results demonstrate that fractional anisotropy with ICA performed better than other measures.

7.3.2.5 MRS

MRS is considered as a noninvasive diagnostic test to estimate the chemical changes in the brain. MRS compares the normal and abnormal tissues, while MRI is used to detect the anatomical location of abnormal tissues. Most commonly used MRS is proton spectroscopy which analyzes the proton molecules. MRS is used to quantify metabolites such as alanine, N-acetyl-aspartate (NAA), creatine (CR), amino acids, and myoinositol.

Proton MRS was used by Wang et al. (2015) to identify the cerebral metabolite changes in AD patients, which shows that NAA was significantly reduced in bilateral hippocampus and posterior cingulate. NAA/CR ratio also decreased in posterior cingulate. Other studies also discovered the reduction of NAA/CR in AD patients (Schuff et al. 1997; Adalsteinsson et al. 2000).

7.4 Observations

In this study, we observed that most of the biomarkers are extracted using MRI, fMRI, and PET, while the rest of the images such as DTI are used by very few studies as shown in Table 7.1. It is well understood that MRI, fMRI, and PET images separately and in combination provide more details to extract biomarkers for AD with better classification rate as shown in Table 7.1. However, it's unfair to compare the ACC, SEN, and specificity (SPE) of MRI, fMRI, PET, and DTI listed papers because they are not using the same classification pairs or the same datasets, i.e., Beheshti and Demirel (2016) classify CN and AD using MRIs of ADNI dataset, while Garali et al. (2015) classify CN and AD using PET but with a private dataset. Similarly, Silveira and Marques (2010) used PET and ADNI dataset but did the classification of AD and MCI subjects. Table 7.1 also shows that only a few studies used unsupervised or semi-supervised methods (Liu et al. 2013; Moradi et al. 2015), while the rest of the methods used a supervised method with naïve Bayes (Khazaee et al. 2017), neural network (Azmi et al. 2017), and SVM (Long et al. 2016; Hojjati et al. 2017) classifiers.

Deep learning (DL) is another supervised method which recently got attention due to its state-of-the-art performance in image processing and computer vision. DL also cultivated the area of medical imaging (Klang 2018; Lakhani et al. 2018); the auto-encoder, deep neural networks, and the most widely used 2D and 3D convolutional neural networks (CNN) are focused on detection, segmentation, registrations, content-based image retrieval, image reconstruction, and generation (Litjens et al. 2017). However, in the area of AD detection, DL work is limited and it may be due to the small amount of data. DL algorithms such as CNN need a large amount of data, resources, and time to train the model and tune the hyperparameters. To all such issues, one solution is transfer learning, train the model on large dataset, and transfer the trained weights for small dataset image classification as adopted by Li et al. (2018) for AD small dataset classification. Most of the studies conducted the experiments on ADNI (Wyman et al. 2013) dataset which is the largest AD dataset publicly available. According to ADNI, a total of 800 participants including 200 CH/HC subjects, 400 MCI individuals, and 200 AD subjects are recruited from 50 different places from the United States and Canada.

The evaluation metrics for biomarker used in AD are accuracy, SPE, and sensitivity (SEN) which shows the true-positive (TP) rate and true-negative (TN) rate, respectively, while AUC shows the classifier stability, and receiver operating characteristic (ROC) shows the classifier diagnostic capability. These metrics can be

Table 7.1 Summary of AD detection methods using brain imaging

References	Images	Methods	Participants			Results (%)	Datasets
			Healthy-control (HC)/normal-control (CN)	Alzheimer's disease (AD)	Mild cognitive impairment (MCI)		
Beheshti and Demirel (2016)	MRI	Feature ranking-based approach	68	68	–	ACC = 96.32 SEN = 94.11 SPE = 98.52 AUC = 0.9993	ADNI (Wyman et al. 2013)
Beheshti et al. (2017)	MRI	Feature ranking + genetic algorithm	162	160	MCI-S = 65 MCI-P = 71	ACC = 93.01 SEN = 89.13 SPE = 96.80 AUC = 0.9351	ADNI
Eskildsen et al. (2015)	MRI (T1-w)	Measurements of structural pathologic patterns	231	198	405	ACC = 71.9 SEN = 69.6 SPE = 73.6 AUC = 0.763	ADNI
Beheshti et al. (2016)	MRI	Feature ranking + classification error	130	130	–	ACC = 92.38 SEN = 91.07 SPE = 93.89 AUC = 0.96	ADNI
Cuingnet et al. (2011)	MRI	Comparative analysis of 10 methods	81	69	MCI-NC = 67 MCI-C = 37	For CN vs AD: SEN = 81 SPE = 95	ADNI
Westman et al. (2012)	MRI	MRI measures with combination of CSF measures	111	96	162	ACC = 91.8 SEN = 88.5 SPE = 94.6 AUC = 0.958	ADNI
Liu et al. (2013)	MRI	LLE	137	96	MCI-S = 93 MCI-C = 97	ACC = 68.0 SEN = 80.0 SPE = 56.0 P = 0.007	ADNI

Moradi et al. (2015)	MRI	Low-density separation	231	200	MCI-P = 164 MCI-S = 100	AUC = 0.902	ADNI
Sørensen et al. (2017)	MRI (T1-w)	Cortical thickness, hippocampal shape, texture, and volumetry	ADNI: 169 AIBL: 88 CAD-dementia:12	ADNI: 101 AIBL: 28 CAD-dementia:9	ADNI: 234 AIBL: 29 CAD-dementia:9	ACC = 63.0 AUC = 0.832	ADNI, AIBL (Ellis et al. 2009), CAD-dementia (Bron et al. 2015)
Azmi et al. (2017)	PET	Voxel intensities + neural network	219	126	–	Nural-net 10 Nodes: SEN = 83.0 SPE = 94.0 Nural-Net 1 node: ACC = 90.0 SEN = 81.0 SPE = 95.0	ADNI
Gray et al. (2012)	PET	Multiregional analysis	54	50	MCI-S = 64 MCI-P = 53	ACC = 88.4 SEN = 83.2 SPE = 93.6 AUC = 0.94	ADNI
Silveira and Marques (2010)	PET	Voxel intensities + AdaBoost	81	74	113	AD detection: ACC = 90.97 MCI detection: ACC = 79.63	ADNI
Garali et al. (2015)	PET	Region-based features	61	81	–	ACC = 95.07	"La-Timone" University Hospital, in the Nuclear Medicine Department (Marseille, France)

(continued)

Table 7.1 (continued)

References	Images	Methods	Participants			Results (%)	Datasets
			Healthy-control (HC)/normal-control (CN)	Alzheimer's disease (AD)	Mild cognitive impairment (MCI)		
Darsana and Abraham (2016)	PET, MRI	Multi-structure registration approach	133	166	151	ACC = 96.4	ADNI
Asim et al. (2018)	PET, MRI	Multimodal and multi-atlas-based approach	100	100	100	CN vs AD: ACC = 94.00 SEN = 95.0 SPE = 93.0	ADNI
Ota et al. (2015)	PET, MRI	Relative cerebral metabolic rate + gray matter density + SVM-RFE	–	40	MCI-A = 40	AUC = 0.75	SEAD-J (Ota et al. 2014)
Zhu et al. 2015 (Dong et al. 2014)	PET, MRI	Sparse multitasking learning framework	52	51	MCI-C = 43 MCI-NC = 56	AD vs NC ACC = 95.7 SEN = 96.6 SPE = 98.2 AUC = 0.981	ADNI
Long et al. (2016)	fMRI	SVM-based method with multilevel characteristic of MRI	33	–	29	ACC = 96.77 SEN = 93.10 SPE = 100	Private dataset

Challis et al. (2015)	re-fMRI	Gaussian process logistic regression + SVM	39	27	MCI-A = 50	CN vs MCI-A ACC = 75% SEN = 80 SPE = 70 AUC = 0.7 MCI-A vs AD ACC = 97% SEN = 90.0 SPE = 90.0 AUC = 0.89	Private Dataset
Hojjati et al. (2017)	re-fMRI	Graph theory + SVM	–	–	MCI-C = 18 MCI-NC = 62	MCI-C vs MCI-NC ROC = 91.4 SEN = 83.24 SPE = 90.1 AUC = 0.95	ADNI
Khazaee et al. (2017)	re-fMRI	Directed graph measures + naïve Bayes classifier	45	34	89	Overall ACC = 93.3 HC vs MCI AUC = 0.92 MCI vs AD AUC = 0.89 HC vs AD AUC = 0.94	ADNI
Agosta et al. (2012)	re-fMRI	Default mode + fontal network + salience networks	13	13	MCI-A = 12	Default mode network I: F = 19.52 P < 0.001 Executive Network: F = 11.19 P < 0.001	Outpatient Memory Clinic of the IRCCS Centro San Giovanni di Dio, Brescia, Italy.

(continued)

Table 7.1 (continued)

References	Images	Methods	Participants			Results (%)	Datasets
			Healthy-control (HC)/normal-control (CN)	Alzheimer's disease (AD)	Mild cognitive impairment (MCI)		
de Vos et al. (2018)	re-fMRI	Functional connectivity + independent component analysis	173	77	–	ACC = 79.0 SEN = 86.0 SPE = 71.0 AUC = 0.85	PRODEM (Seiler et al. 2012)
Schouten et al. (2016)	MRI, re-fMRI	Multi-modalities + elastic net classifier	173	77	–	ACC = 93.0 SEN = 81.6 SPE = 95.6 AUC = 0.971	PRODEM
Li et al. (2018)	fMRI	Transfer learning	ADNI:175 Tongii:14	ADNI:117 Tongii:12	–	With adaptation: ACC = 84.6 SEN = 92.0 SPE = 79.0 AUC = 0.80	ADNI, Tongji Hospital at Wuhan
Ahmed et al. (2017)	DTI, MRI	Multimodal approach + multiple kernel learning	52	45	58	AD vs NC ACC = 90.2 SEN = 82.92 SPE = 97.2	ADNI
Schouten et al. (2017)	DTI	Voxel-wise measures + elastic net classification	173	77	–	ACC = 85.1 SEN = 86.8 SPE = 84.4 AUC = 0.92 ± 0.018	PRODEM

Some of the notations used in this table are accuracy (ACC), sensitivity (SEN), specificity (SPE), area under the curve (AUC), F-score (F), and p-value (P) extracted from a confusion matrix which consists of TP, TN, false-positive (FP), and false-negative (FN). The mathematical definitions are as follows:

$$\text{Accuracy} = \frac{|\text{TN}|+|\text{TP}|}{|\text{TN}|+|\text{TP}|+|\text{FP}|+|\text{FN}|}$$

$$\text{Sensitivity} = \frac{|\text{TP}|}{|\text{TP}|+|\text{FN}|}$$

$$\text{Specificity} = \frac{|\text{TN}|}{|\text{TN}|+|\text{FP}|}$$

7.5 Conclusion and Future Perspective

AD, one of the leading neurodegenerative disorders, is characterized by brain damage and cognitive impairment in elderly people. Due to the severity of the disease, prevention and/or regression of AD is a challenge to overcome. To track its progression, medical imaging played a vital role from decades, but still, there is room for improvement. The brain images are used to extract the biomarkers which help the doctors in the identification of AD. In this study, we presented various brain image modularities where the research focused more on MRI due to its clarity, availability, and correctly exposing the biomarkers than other imaging techniques (CT, SPECT, PET, FMRI, and DTI).

In future effort, researchers should consider more datasets with a large amount of imaging data for experimentation to improve the AD detection. Researchers should not only focus on imaging but also take clinical symptoms into consideration and used as features which can help in better identification of AD. DL is a new area which has been enriching in medical imaging. However, DL and transfer learning still have not been exploited to the extent in CAD systems for AD.

Acknowledgments The authors wish to acknowledge King Fahd University of Petroleum and Minerals (KFUPM) for utilizing the various facilities in carrying out this research. Other authors of the manuscript thank and acknowledge their respective universities and institutes as well.

Conflict of Interest There is no conflict of interest.

References

2018 Alzheimer's disease facts and figures (2018) Alzheimer's Dement 14(3):367–429

Adalsteinsson E, Sullivan EV, Kleinhans N, Spielman DM, Pfefferbaum A (2000) Longitudinal decline of the neuronal marker N-acetyl aspartate in Alzheimer's disease. Lancet (London, England) 355(9216):1696–1697

Agosta F, Pievani M, Geroldi C, Copetti M, Frisoni GB, Filippi M (2012) Resting state fMRI in Alzheimer's disease: beyond the default mode network. Neurobiol Aging 33(8):1564–1578

Al-Jibory WK, El-Zaart A (2013) Edge detection for diagnosis early Alzheimer's disease by using Weibull distribution. In: 2013 25th International Conference on Microelectronics (ICM), pp 1–5

Alzheimer's Association (2014) 2014 Alzheimer's disease facts and figures. Alzheimers Dement 10(2):e47–e92

Ansari MA, Abdul HM, Joshi G, Opii WO, Butterfield DA (2009) Protective effect of quercetin in primary neurons against Aβ(1–42): relevance to Alzheimer's disease. J Nutr Biochem 20(4):269–275

Asim Y, Raza B, Malik AK, Rathore S, Hussain L, Iftikhar MA (2018) A multi-modal, multi-atlas-based approach for Alzheimer detection via machine learning. Int J Imaging Syst Technol 28(2):113–123

Attwell D, Buchan AM, Charpak S, Lauritzen M, MacVicar BA, Newman EA (2010) Glial and neuronal control of brain blood flow. Nature 468(7321):232–243

Azmi MH, Saripan MI, Nordin AJ, Ahmad Saad FF, Abdul Aziz SA, Wan Adnan WA (2017) 18 F-FDG PET brain images as features for Alzheimer classification. Radiat Phys Chem 137:135–143

Beheshti I, Demirel H (2016) Feature-ranking-based Alzheimer's disease classification from structural MRI. Magn Reson Imaging 34(3):252–263

Beheshti I, Demirel H, Farokhian F, Yang C, Matsuda H (2016) Structural MRI-based detection of Alzheimer's disease using feature ranking and classification error. Comput Methods Prog Biomed 137:177–193

Beheshti I, Demirel H, Matsuda H (2017) Classification of Alzheimer's disease and prediction of mild cognitive impairment-to-Alzheimer's conversion from structural magnetic resource imaging using feature ranking and a genetic algorithm. Comput Biol Med 83:109–119

Ben Ahmed O, Benois-Pineau J, Allard M, Catheline G, Ben Amar C (2017) Recognition of Alzheimer's disease and mild cognitive impairment with multimodal image-derived biomarkers and Multiple Kernel learning. Neurocomputing 220:98–110

Bezzi P, Volterra A (2001) A neuron-glia signalling network in the active brain. Curr Opin Neurobiol 11(3):387–394

Boutet C et al (2014) Detection of volume loss in hippocampal layers in Alzheimer's disease using 7 T MRI: a feasibility study. NeuroImage Clin 5:341–348

Bron EE et al (2015) Standardized evaluation of algorithms for computer-aided diagnosis of dementia based on structural MRI: The CAD dementia challenge. NeuroImage 111:562–579

Byun MS et al (2017) Korean brain aging study for the early diagnosis and prediction of Alzheimer's disease: methodology and baseline sample characteristics. Psychiatry Investig 14(6):851

Challis E, Hurley P, Serra L, Bozzali M, Oliver S, Cercignani M (2015) Gaussian process classification of Alzheimer's disease and mild cognitive impairment from resting-state fMRI. NeuroImage 112:232–243

Chen K et al (2010) Twelve-month metabolic declines in probable Alzheimer's disease and amnestic mild cognitive impairment assessed using an empirically pre-defined statistical region-of-interest: findings from the Alzheimer's disease neuroimaging initiative. NeuroImage 51(2):654–664

Cuingnet R et al (2011) Automatic classification of patients with Alzheimer's disease from structural MRI: a comparison of ten methods using the ADNI database. NeuroImage 56(2):766–781

Darsana S, Abraham L (2016) Multistructure brain registration using multimodal neuroimaging for the detection of Alzheimer's disease. In: 2016 International Conference on Communication Systems and Networks (ComNet), pp 221–226

de Magalhaes Oliveira PP, Nitrini R, Busatto G, Buchpiguel C, Sato JR, Amaro E (2010) Use of SVM methods with surface-based cortical and volumetric subcortical measurements to detect Alzheimer's disease. J Alzheimers Dis 19(4):1263–1272

de Vos F et al (2018) A comprehensive analysis of resting state fMRI measures to classify individual patients with Alzheimer's disease. NeuroImage 167:62–72

Dong Y et al (2014) Protective effect of quercetin against oxidative stress and brain edema in an experimental rat model of subarachnoid hemorrhage. Int J Med Sci 11(3):282–290

Ellis KA et al (2009) The Australian Imaging, Biomarkers and Lifestyle (AIBL) study of aging: methodology and baseline characteristics of 1112 individuals recruited for a longitudinal study of Alzheimer's disease. Int Psychogeriatr 21(4):672

Eskildsen SF, Coupé P, García-Lorenzo D, Fonov V, Pruessner JC, Collins DL (2013) Prediction of Alzheimer's disease in subjects with mild cognitive impairment from the ADNI cohort using patterns of cortical thinning. NeuroImage 65:511–521

Eskildsen SF, Coupé P, Fonov VS, Pruessner JC, Collins DL (2015) Structural imaging biomarkers of Alzheimer's disease: predicting disease progression. Neurobiol Aging 36:S23–S31

Garali I, Adel M, Bourennane S, Guedj E (2015) Region-based brain selection and classification on pet images for Alzheimer's disease computer aided diagnosis. In: 2015 IEEE International Conference on Image Processing (ICIP), pp 1473–1477

Ghanemi A (2015) Alzheimer's disease therapies: Selected advances and future perspectives. Alexandria J Med 51(1):1–3

Gray KR, Wolz R, Heckemann RA, Aljabar P, Hammers A, Rueckert D (2012) Multi-region analysis of longitudinal FDG-PET for the classification of Alzheimer's disease. NeuroImage 60(1):221–229

Hanisch U-K, Kettenmann H (2007) Microglia: active sensor and versatile effector cells in the normal and pathologic brain. Nat Neurosci 10(11):1387–1394

Hartman RE et al (2006) Pomegranate juice decreases amyloid load and improves behavior in a mouse model of Alzheimer's disease. Neurobiol Dis 24(3):506–515

Hedden T et al (2009) Disruption of functional connectivity in clinically normal older adults harboring amyloid burden. J Neurosci 29(40):12686–12694

Hojjati SH, Ebrahimzadeh A, Khazaee A, Babajani-Feremi A (2017) Predicting conversion from MCI to AD using resting-state fMRI, graph theoretical approach and SVM. J Neurosci Methods 282:69–80

Iadecola C (2004) Neurovascular regulation in the normal brain and in Alzheimer's disease. Nat Rev Neurosci 5(5):347–360

Jack CR et al (2010) Hypothetical model of dynamic biomarkers of the Alzheimer's pathological cascade. Lancet Neurol 9(1):119–128

Johnson KA, Fox NC, Sperling RA, Klunk WE (2012) Brain imaging in Alzheimer disease. Cold Spring Harb Perspect Med 2(4):a006213–a006213

Kantarci K et al (2017) White-matter integrity on DTI and the pathologic staging of Alzheimer's disease. Neurobiol Aging 56:172–179

Khazaee A, Ebrahimzadeh A, Babajani-Feremi A, Alzheimer's Disease Neuroimaging Initiative (2017) Classification of patients with MCI and AD from healthy controls using directed graph measures of resting-state fMRI. Behav Brain Res 322(Pt B):339–350

Klang E (2018) Deep learning and medical imaging. J Thorac Dis 10(3):1325–1328

Kloppel S et al (2008) Automatic classification of MR scans in Alzheimer's disease. *Brain* 131(3):681–689

Krstic D, Knuesel I (2013) Deciphering the mechanism underlying late-onset Alzheimer disease. Nat Rev Neurol 9(1):25–34

Lakhani P, Gray DL, Pett CR, Nagy P, Shih G (2018) Hello world deep learning in medical imaging. J Digit Imaging 31(3):283–289

Ledig C, Schuh A, Guerrero R, Heckemann RA, Rueckert D (2018) Structural brain imaging in Alzheimer's disease and mild cognitive impairment: biomarker analysis and shared morphometry database. Sci Rep 8(1):11258

Li W, Zhao Y, Chen X, Xiao Y, Qin Y (2018) Detecting Alzheimer's disease on small dataset: a knowledge transfer perspective. IEEE J Biomed Heal Informatics:1–1

Litjens G et al (2017) A survey on deep learning in medical image analysis. Med Image Anal 42:60–88

Liu X, Tosun D, Weiner MW, Schuff N (2013) Locally linear embedding (LLE) for MRI based Alzheimer's disease classification. NeuroImage 83:148–157

Liu M, Zhang D, Shen D (2014) Hierarchical fusion of features and classifier decisions for Alzheimer's disease diagnosis. Hum Brain Mapp 35(4):1305–1319

Long Z et al (2016) A support vector machine-based method to identify mild cognitive impairment with multi-level characteristics of magnetic resonance imaging. Neuroscience 331:169–176

McEvoy LK, Brewer JB (2010) Quantitative structural MRI for early detection of Alzheimer's disease. Expert Rev Neurother 10(11):1675–1688

McKhann GM et al (2011) The diagnosis of dementia due to Alzheimer's disease: recommendations from the National Institute on Aging-Alzheimer's Association workgroups on diagnostic guidelines for Alzheimer's disease. Alzheimers Dement 7(3):263–269

Moradi E, Pepe A, Gaser C, Huttunen H, Tohka J (2015) Machine learning framework for early MRI-based Alzheimer's conversion prediction in MCI subjects. NeuroImage 104:398–412

Omar SH, Scott CJ, Hamlin AS, Obied HK (2017) The protective role of plant biophenols in mechanisms of Alzheimer's disease. J Nutr Biochem 47:1–20

Ota K, Oishi N, Ito K, Fukuyama H (2014) A comparison of three brain atlases for MCI prediction. J Neurosci Methods 221:139–150

Ota K, Oishi N, Ito K, Fukuyama H (2015) Effects of imaging modalities, brain atlases and feature selection on prediction of Alzheimer's disease. J Neurosci Methods 256:168–183

Palle S, Neerati P (2017) Quercetin nanoparticles attenuates scopolamine induced spatial memory deficits and pathological damages in rats. Bull Fac Pharmacy, Cairo Univ 55(1):101–106

Querbes O et al (2009) Early diagnosis of Alzheimer's disease using cortical thickness: impact of cognitive reserve. Brain 132(8):2036–2047

Schouten TM et al (2016) Combining anatomical, diffusion, and resting state functional magnetic resonance imaging for individual classification of mild and moderate Alzheimer's disease. NeuroImage Clin 11:46–51

Schouten TM et al (2017) Individual classification of Alzheimer's disease with diffusion magnetic resonance imaging. NeuroImage 152:476–481

Schuckit MA (1992) An introduction and overview to clinical applications of neuroSPECT in psychiatry. J Clin Psychiatry 53(Suppl):3–6

Schuff N et al (1997) Changes of hippocampal N-acetyl aspartate and volume in Alzheimer's disease. A proton MR spectroscopic imaging and MRI study. Neurology 49(6):1513–1521

Seiler S et al (2012) Driving Cessation and dementia: results of the prospective registry on dementia in Austria (PRODEM). PLoS One 7(12):e52710

Selkoe D, Mandelkow E, Holtzman D (2012) Deciphering Alzheimer disease. Cold Spring Harb Perspect Med 2(1):a011460–a011460

Silveira M, Marques J (2010) Boosting Alzheimer disease diagnosis using PET images. In: 2010 20th International Conference on Pattern Recognition, pp 2556–2559

Sørensen L et al (2017) Differential diagnosis of mild cognitive impairment and Alzheimer's disease using structural MRI cortical thickness, hippocampal shape, hippocampal texture, and volumetry. NeuroImage Clin 13:470–482

Sperling RA et al (2009) Amyloid deposition is associated with impaired default network function in older persons without dementia. Neuron 63(2):178–188

Strauss HW (1986) Positron emission tomography and autoradiography: principles and applications for the brain and heart. Phelps ME, Mazziotta JC, Schelbert HR (eds) Raven Press, New York, 704 Pages, Illustrated, $89.50 ISBN: 0-88167-118-5. Clin Cardiol 9(5):233–233

Tabei K, Kida H, Hosoya T, Satoh M, Tomimoto H (2017) Prediction of cognitive decline from white matter hyperintensity and single-photon emission computed tomography in Alzheimer's disease. Front Neurol 8:408

Tong T, Wolz R, Gao Q, Guerrero R, Hajnal JV, Rueckert D (2014) Multiple instance learning for classification of dementia in brain MRI. Med Image Anal 18(5):808–818

van Dyck CH (2018) Anti-amyloid-β monoclonal antibodies for Alzheimer's disease: pitfalls and promise. Biol Psychiatry 83(4):311–319

Wang H et al (2015) Magnetic resonance spectroscopy in Alzheimer's disease: systematic review and meta-analysis. J Alzheimers Dis 46(4):1049–1070

Westman E, Muehlboeck J-S, Simmons A (2012) Combining MRI and CSF measures for classification of Alzheimer's disease and prediction of mild cognitive impairment conversion. NeuroImage 62(1):229–238

Wolz R et al (2011) Multi-method analysis of MRI images in early diagnostics of Alzheimer's disease. PLoS One 6(10):e25446

Wu T-Y, Chen C-P, Jinn T-R (2011) Traditional Chinese medicines and Alzheimer's disease. Taiwan J Obstet Gynecol 50(2):131–135

Wyman BT et al (2013) Standardization of analysis sets for reporting results from ADNI MRI data. Alzheimers Dement 9(3):332–337

Zhang D, Wang Y, Zhou L, Yuan H, Shen D (2011) Multimodal classification of Alzheimer's disease and mild cognitive impairment. NeuroImage 55(3):856–867

Zhou L, Wang Y, Li Y, Yap P-T, Shen D (2011) Hierarchical anatomical brain networks for MCI prediction: revisiting volumetric measures. PLoS One 6(7):e21935

A Review of the Relationship Between Gut Microbiota and Memory

8

Amira Benmelouka, Ahmed M. Sherif, and Mahmoud Ahmed Ebada

Abstract

Alzheimer's disease (AD) became a public health problem due to its increasing incidence and the dangerous sequences that affect society. Like many other neurodegenerative disorders, the complete mechanism of AD is not known yet. Scientists think that AD is caused by the combination of various factors related to the environment and the genetic basis. Many studies have demonstrated the role of human gut microbiota in different physiological and pathological pathways affecting many distant organs particularly the nervous system. Researchers observed a variety of disturbances of the intestinal microbiota homeostasis in many conditions such as diet changes, aging, probiotic and antibiotic administration. The study of this homeostasis and its different characteristics can help us understand the various mechanisms by which it influences the gut-brain axis. It also helps to find solutions to many health problems including neurodegenerative diseases.

Keywords

Alzheimer's disease · Cognition · Dementia · Gut microbiota · Memory · Neurodegeneration · Neuroinflammation

A. Benmelouka
University of Algiers, Sidi M'Hamed, Algeria

A. M. Sherif · M. A. Ebada (✉)
Faculty of Medicine, Zagazig University, Zagazig, El-Sharkia, Egypt
e-mail: mahmoud011467@medicine.zu.edu.eg

G. M. Ashraf, A. Alexiou (eds.), *Biological, Diagnostic and Therapeutic Advances in Alzheimer's Disease*, https://doi.org/10.1007/978-981-13-9636-6_8

8.1 Introduction

Medicine development and technology advancement contribute to the promotion of life expectancy and the maturation of the population. According to many United Nations reports, the phenomenon of people aging will increase in the next years. Scientists stated that the global number of the population aging greater than 60 years had attained 962 million in 2017. They also expect that this number will be more than its double in the period between 2017 and 2050 (Hu et al. 2016; United Nations 2017). Despite all the improvements of healthcare systems, age-related neurodegenerative disabilities especially Alzheimer's disease and dementia continue to be a public health problem that threats both the world economy and social stability. That is why developing strategies and medications that can help prevent or delay the occurrence of such conditions is becoming an urgent concern for both medical scientists and social experts. Recent data indicate that gut microbiota can have a critical impact on the neuro-hormonal communicating system between the human gut and brain neurons (commonly known by gut-brain axis) (Collins et al. 2012; Cryan and O'Mahony 2011). Furthermore, many studies have suggested the possible association between microbiota-gut-brain communication and the modulation of the same behavioral functions and memory mechanisms that can be impaired in neuroinflammatory processes (Foster and Neufeld 2018; Aziz et al. 2013). In this review, we summarized relevant studies which discussed a possible link between human intestinal bacteria and cognitive dysfunction in Alzheimer's disease.

8.2 Background on Alzheimer's Disease

Alzheimer disease (AD) is a common incurable, progressive neurodegeneration of the human brain which is characterized by cognitive impairment and declined memory abilities. In AD, we find specific neuropathological hallmarks which are represented by the deposition of extracellular β-amyloid (Aβ), senile plaques (SP), and intracellular neurofibrillary tangles (NFT) that result from the tau proteins hyperphosphorylation in the neurons (Reitz et al. 2011). This neurodegenerative disease may arise from complex interactions between some human genetic factors including amyloid precursor protein (APP) gene mutations, presenilin genes mutations, and apolipoprotein E allele and the exposure to many environmental conditions such as nutrition habits, infections, and chemical agents (Hu et al. 2016; Xu and Wang 2016).

Alzheimer's disease forms can be classified, according to the age of onset, into early-onset Alzheimer's disease (EOAD) which occurs before 65 years old and late-onset Alzheimer's disease (LOAD) in which the clinical manifestations start after the age of 65 years (Kowalski et al. 2019). Many hypotheses have been generated to figure out the pathogenesis of AD types (Agahi et al. 2018). First, we have the amyloids cascade theory, in which we think that the alteration of APP gene expression may lead to Aβ synthesis and amyloid plaques formation. These plaques are believed to play a role in cerebral inflammation, in neurodegeneration and as a result in many cognitive disorders (Dominicé and Lehto 1991). Amyloid cascade hypothesis is more likely to be applicable in familial (autosomal dominant) forms of AD in which there are mutations of APP, PS1, and PS2 genes. The second hypothesis is the mitochondrial cascade theory (Swerdlow et al. 2010; Swerdlow and Khan 2004). In this theory, it is believed that altered brain mitochondria are capable of initiating AD pathogenesis by influencing tau proteins hyperphosphorylation and APP production. Mitochondrial cascade hypothesis consists of the observed alterations in mitochondrial functions that can develop during aging. It helps explain the sporadic cases of AD that are so far more frequent than genetic forms (Bekris et al. 2011).

According to the Alzheimer's association, AD is the most famous and common form of dementia representing 60–80% of all dementia. The incidence of AD is rising every year. This serious augmentation contributes to the degradation of elderly life quality, causing heavy burdens to caregivers, to patient families, and the whole society. Alzheimer's disease engenders high costs including direct medical costs (e.g., nursing and medication), direct nonmedical costs (e.g., in-home daycare), and indirect costs which are related to people's productivity (Meek et al. 1998).

8.3 The Diversity of Gut Microbiota

Gut microbiota can contribute to maintaining host health by its implications on immune response against pathogens and in many human metabolism pathways (Cho and Blaser 2012). Intestinal microbiota metagenome consists of trillions of microbial organisms. Recent data indicate that microbiota genes are about 100-fold as human genes (Gill et al. 2006). This illustrates the diversity of microbiota components which is controlled by both genetic (host genes) and environmental factors including unbalanced diet, medications, stress situations, and geographic conditions such as the altitude (De Filippo et al. 2010; Hufeldt et al. 2010; Taichi and Suzuki 2014). This diversity may be the primary determinant of human genetic individuality and many human metabolic pathways (Aziz et al. 2013). To clarify this point, we will take the example of the human genome; scientists discovered only about 26,600 protein-coding genes during the Human Genome Project. However, this number is not enough to explain the complexity of human biochemistry. This led the scientists to think more about the place of human symbiotic microbes' genome in human genetic complexity. According to Human Microbiome Project Consortium 2012, the human gut microbiota consists of about 4,000,000 genes; this makes the human genetic complexity closer to 4,026,600 genes (Kobayashi et al. 2017; Qin et al. 2010).

Another relevant point is that of the pathological modification of the gut microbiome composition. Many combined factors and circumstances including diet disturbances with increased sugar and fat consumption can affect healthy microbiota and change its phylogenetic structure. Such kind of modifications can lead to gut dysbiosis (Proctor et al. 2017) and thus to the perturbation of the normal physiology of gut-brain axis which results in increased inflammation of the hippocampus and consequently decreasing planning abilities and mental flexibility and altered memory functions (Solas et al. 2017).

8.4 Aging and Modification of Gut Microbiota

Aging has been recognized as the most common risk factor of Alzheimer's disease and dementia. We can explain this association with the alterations of physiological functions in the elderly. Many studies explored some changes in intestinal microbiota with aging confirming that beneficial intestinal microbes (e.g., lactobacilli, bifidobacteria) can be decreased in old people. This may contribute to the generation of many diseases especially inflammatory reactions, immune system impairments, and age-related mental health problems. Other bacterial species modifications can correlate positively with age. Among them, we can find *Oscillospira*, *Odoribacter*, and *Butyricimonas* (Salazar et al. 2017).

Moreover, with age, the gastrointestinal tract becomes progressively colonized by more opportunistic bacteria that belong to the Proteobacteria group (Buford 2017). It was observed in the elderly that there was a loss of some genes like those coding short-chain fatty acid production in human microbiota and the decrease seen in their saccharolytic functions. In addition to that, microbiome proteolytic activities were accentuated compared with the intestinal metagenome of young people (Rampelli et al. 2013). All these changes are due to various factors such as dietary factors (poor nutrition, changes in nutrition habits in old people) and modifications of both immune system (immunosenescence and inflammation-aging process) and gastrointestinal system (gastric motility, chemical modifications in the gut tube) in elder people (Buford 2017; Tiihonen et al. 2010; Woodmansey 2007).

8.5 Diet, Microbiota, and Neurodegeneration

It is suggested that there is an association between altered gut microbiota composition, nutrition habits and some chronic diseases in older people (Claesson et al. 2012). Modified gut microbiota may cause a change in host behaviors by controlling brain functions (Sampson and Mazmanian 2015; Gareau 2014). Now researchers know that the exposition of microbiota to high-fat concentrations may lead to alterations in exploratory, cognitive, and stereotypical behaviors, concluding that diet and pharmacological modulation of gut microbiome might help reduce the impact of obesity on the neurodegeneration phenomenon (Proctor et al. 2017; Bruce-Keller et al. 2015). Additionally, people with inflammatory bowel disease (a health condition in which the microbiota is altered (Nishida et al. 2018) and in which there is impaired digestion and absorption of nutrients and electrolytes in the gastrointestinal tube (Peuhkuri et al. 2016; Priyamvada et al. 2016; Semrin et al. 2016) may have low verbal memory levels (Collins et al. 2012). More recent findings illustrate the consequences of omega-3 use on both intestinal microbiota and brain tissue. For example, a recent study showed that omega-3 FAs could be considered as a probiotic since they have a critical role in modulating microbiota composition by raising the number of bacteria producing butyrate (La Rosa et al. 2018). Scientists have also remarked that, in ApoE epsilon four noncarriers, the regular consumption of fish and omega-3 rich oil is associated with a significant decrease in the risk of dementia (Barberger-Gateau et al. 2007).

8.6 Gut Microbiota, Neurotransmitters, and Central Nervous System Receptors

Increasing studies confirm the role of microbiota in various CNS affections and behavioral disorders such as autism, multiple sclerosis, Alzheimer's disease, Parkinson's disease, and anxiety. Gut microbiota can regulate bidirectional communication in the Brain-Gut axis (Cryan and O'Mahony 2011; Bravo et al. 2012; Cryan and Dinan 2012; Borre et al. 2014). For instance, some gut bacteria like bifidobacterium and lactobacillus can produce a CNS inhibitory neurotransmitter called γ-aminobutyric acid (GABA) from glutamate. This neurotransmitter is involved in different neurological illnesses such as depression anxiety and many cognitive impairments that are seen in Alzheimer's disease and dementia (Foster and Neufeld 2018; Aziz et al. 2013; Barrett et al. 2012; Hornig 2013).

Another example is the relation between the NMDA receptor and some toxins secreted by altered gut bacteria. N methyl-D-aspartate (NMDA) receptor is an ionotropic glutamate receptor that is known to influence brain cognition through regulating synaptic plasticity. This receptor can be targeted by a neurotoxin called beta-N-methylamino-L-alanine (BMAA) which is an oxidative stress-inducing neurotoxin which is produced by cyanobacteria in the intestinal microbiome. BMAA can gain CNS tissue by crossing the blood-brain barrier. In susceptible people who cannot prevent the accumulation of BMAA in their central nervous system, the interaction between BMAA and NMDA receptor may lead to the same neurological dysfunction. Scientists confirmed that BMAA could trigger the formation of neurological hallmarks of Alzheimer's disease (neurofibrillary tangles NFT and β amyloid plaques). This indicates that the secretion of BMAA toxin by intestinal microbiota can play the leading role in neurodegeneration in human beings (Smith et al. 1992; Nunn and Ponnusamy 2009; Bradley 2009; Banack et al. 2010; Lakhan et al. 2013; Brenner 2013; Cox et al. 2016).

8.7 More Evidence About Cyanobacteria and Neurodegeneration in Older People

In older people, the intestinal epithelial barrier permeability may increase under the influence of microRNA. This increased permeability may lead to more cyanobacteria-generated neurotoxins in the brain tissue. Among these toxins, we have saxitoxin and anatoxin-α which can contribute to the development of many neurodegenerative illnesses (Tran and Greenwood-Van 2013).

8.8 Altered Microbiota in Liver Disease and Cognitive Impairment

Liver cirrhosis is characterized by increased bacterial species of *Enterobacteriaceae*, *Alcaligenaceae*, and *Fusobacteriaceae* and decreased bacterial species of Ruminococcaceae and Lachnospiraceae in fecal microbiota. These alterations in microbiota were also correlated with cognitive dysfunction and endotoxemia (Bajaj et al. 2011).

8.9 Microglial Dysfunction and Intestinal Microbiota

Microglial dysfunction is observed in many neurodegenerative and neuroinflammatory diseases such as Alzheimer's disease and multiple sclerosis (MS). This association is explained in a big part by nitric oxide (NO) pathways modifications which may be enhanced by altered intestinal flora. Many gut bacterial species can produce a significant amount of amyloids, LPS, and NO. These molecules may migrate to the nervous tissue and activate microglia inducing prolonged inflammation. This phenomenon may be exacerbated in older people who have increased blood-brain barrier and gastrointestinal epithelium permeability. When microglial cells become activated by NO and amyloids that come from the gut, it can start elaborating high concentrations of APP and NO. This can form a vicious cycle (Tse 2017; Asiimwe et al. 2016; Pistollato et al. 2016; Heneka et al. 2015; Banks 2008; Wang et al. 2018) resulting in the deposition of vast amounts of brain amyloid plaques. Moreover, in a recent article, Rothhammer et al. reported that tryptophan metabolites that the gut microbiome elaborates can modulate the expression of TGFα and VEGF-B by microglia ameliorating the inflammatory processes in the brain (Rothhammer et al. 2018). This consumption of tryptophan is associated with tryptophan levels' depletion in the serum and with cognitive impairments in senile dementia (Mendelsohn et al. 2009) (Rampelli et al. 2013).

8.10 Gut Microbiota and Amyloids Formation

Aβ is a transmembrane protein that comes from the cleavage of a type 1 glycoprotein called amyloid precursor protein (APP) by β- and γ-secretases. Aβ is associated with many cellular processes including intracellular transport, neuronal development, microglia maturation, and protection against microbial infections (Strooper 2010; Chen et al. 2017; Kumar et al. 2016). This peptide is a crucial element in Alzheimer's disease. Its excessive production causes its aggregation and the formation of oligomers, protofibrils, and fibrils whose deposition results in the apparition of senile plaques (Kowalski et al. 2019). Bacteria and fungi (as well as gut microbiota) can produce amyloid proteins that are involved in biofilm formation, bacterial invasion, host adhesion, and bacterial resistance (Larsen et al. 2007) (Larsen et al. 2008; Hufnagel et al. 2013). Bacterial amyloids interact with innate immune cells using the same mechanisms that are involved in recognizing Aβ (Tükel et al. 2010). This leads us to think that gut bacterial amyloids proteins may also activate brain microglia and cause neuroinflammation resulting in the priming of neurodegenerative mechanisms (Chen et al. 2016). In Chen et al. study, it was demonstrated that exposure to a functional bacterial amyloid protein named Curli (a bacterial amyloid that is produced by many Gram-negative bacteria and by *Escherichia coli*) led to alpha-synuclein aggregation and its deposition in the gut and brain neurons. This deposition was accompanied by excessive expression of TLR2, IL-6, and TNF in rat brains (Chen et al. 2016). Aβ in Alzheimer's disease can be considered as prion-like proteinaceous nucleating particles since it can self-propagate when the concentration beats a certain threshold in the brain tissue (Walker et al. 2016). It is possible that bacterial amyloids can have some prion protein characteristics and causes the cross-seeding of amyloidogenic proteins using molecular mimicry properties that involve many molecules such as TLR 2/1, CD14, and NFκB (Friedland 2015). Interestingly, nutrients can modify intestinal microbiota components and consequently might affect bacterial amyloids proteins production or aggregation (Friedland 2015; Zhou et al. 2012; Scott et al. 2013). This indicates that the use of active molecules like probiotics and antibiotics can be an excellent future strategy that would aim to attenuate chronic neuroinflammation observed in AD and different neurodegenerative conditions (Pistollato et al. 2016).

8.11 Lipopolysaccharides and Neuroinflammation

Neuroinflammatory effects of lipopolysaccharides have been described by many researchers. On February 2000, a study reported the inflammatory reaction caused by chronic injection of lipopolysaccharides into the fourth ventricle. This immuno-inflammatory process consisted in the activation of astrocytes and the increase of microglia number and activity in the dentate gyrus and CA3 region of the hippocampus. The researcher also highlighted the elevation of IL1β, TNFα, and β amyloid precursor protein mRNAs (Hauss-Wegrzyniak et al. 2000). Kahn et al. also described in their study an increase of central Aβ after seven consecutive peritoneal infusions of 250 μg/kg of lipopolysaccharides (Distrutti et al. 2014). These findings were confirmed by many subsequent articles (Asti and Gioglio 2014; Zhao et al. 2017a).

Furthermore, this endotoxin is thought to be involved in conformational changes of prion proteins to a β sheet structure that can aggregate and induce neurological damages and behavioral impairment symptoms (Saleem et al. 2014; Ladner-Keay et al. 2016). *Bacteroides fragilis* lipopolysaccharides are recognized by many microglial receptors (CD14, TLR4, and TLR2). This recognition gives rise to the expression of the proinflammatory transcription factor NF/κB (P50/P65) complex (Lukiw 2016). The role of CD14 (lipopolysaccharides receptor in microglial cells) in the generation of neurotoxicity has been elucidated by many studies (Liu et al. 2005; Fassbender et al. 2004). Other different outcomes demonstrated the significant association between lipopolysaccharides, neuroinflammation, and neurodegeneration processes by the discovery of lipopolysaccharides in postmortem AD brain tissue especially in the hippocampus (Zhao et al. 2017b). All this accumulated evidence and elucidated mechanisms highlight the pivotal role that gastrointestinal microbiome-derived lipopolysaccharides in amyloidogenesis and the development of human neurological impairments (Zhao et al. 2017a; Zhao et al. 2017b; Zhao et al. 2017c).

8.12 Germ-free Experiments and Cognitive Dysfunction

Recent experiments on germ-free animals demonstrated that gut microbiota has an influence on the enteric nervous system and the brain (Biagi et al. 2013), and showed that germ-free animals (GFA) are more susceptible to developing spatial and working memory troubles and many psychiatric disorders like depression (Gareau et al. 2011). Many possible mechanisms of gut-brain communication can affect normal brain development in GFA. For instance, germ-free mice have decreased expression of NGFI-A mRNA which is a synaptic plasticity-related gene (Heijitz et al. 2011).

Moreover, Braniste et al. confirmed that, in germ-free mice, the brain-blood barrier (BBB) could be more permeable to many substances. They also observed that after the exposition of these GFA to a pathogen-free microbiota, the BBB permeability decreased due to the upregulation of tight junction proteins (Braniste et al. 2014). These modifications help explain in part the susceptibility of GFA to many neurotoxins. Furthermore, scientists could detect the hypermyelinated neuronal axons in the prefrontal axons of germ-free mice. This hypermyelination was reversed by the administration of conventional microbiota in mice. These findings may illustrate how gut microbiota affects brain maturation and repair (Hoban et al. 2016; Erny et al. 2017).

Brain-derived neurotrophic factor (known as BDNF) promotes learning and synaptic remodeling. Experiments discovered the depletion of this factor in both the hippocampal and the cortical regions of germ-free animals. This reduction may be correlated to increased anxiety and with progressive cognitive disorders (Collins et al. 2012; Ozawa et al. 2013; Lu et al. 2013).

8.13 Probiotics and Cognition

Probiotics are involved in many anti-inflammatory mechanisms. They can positively modulate immunosenescence and inflammation-aging processes. They are also suggested to improve longevity (Biagi et al. 2013; Pérez Martinez et al. 2014; Akbari et al. 2016). We will give some evidence to illustrate the effects of probiotic treatment on behavior and cognitive functions. First, the consumption of *Bifidobacterium breve* strain A1 can enhance the cognitive impairment in AD mice. It positively affects both working and long-term memory. It also reduces the toxicity of Aβ and reestablishes the regular gene expression in the hippocampus (Kobayashi et al. 2017). SLAB51 is a probiotic that is capable of modifying gut microbiota. The use of SLAB51 probiotic in AD mice helped reduce pro-inflammatory cytokines (IL1α, IL2, IL12, IL17, and IFNγ) and increase cytokines that play a crucial role in inflammatory response downregulation (IL4, IL6, G-CSF). The experiments have also shown that SLAB51 improved cognitive impairment in AD mice. Ghrelin, leptin, and GLP-1 are gut hormones that have neuroprotective effects. They may decrease with aging. The administration of SLAB51 results in a significant increase in the concentration of these hormones in the plasma. This probiotic can regulate apoptotic activity and restore autophagy mechanisms that are altered in amyloid plaques formation in AD (Bonfili et al. 2017). Another probiotic composed of 8 Gram-positive bacterial species (VSL#3) was observed to reduce age-related synaptic plasticity deficits by attenuating long-term potentiation (LTP) deficits (Distrutti et al. 2014). A recent study documented the administration of three probiotic strains (*Lactobacillus rhamnosus*, *Bifidobacterium lactis*, and *Bifidobacterium longum*) could modulate the immunological processes by favoring the macrophage M2 type and causing an increase in anti-inflammatory cytokine IL-10 levels. They also reported, in the same study, that this serobioma plays a role in pro-inflammatory cytokines downregulation (IL1β and IL6 concentrations were decreased by 70% and 80%, respectively) (Gareau et al. 2011). Probiotics are also thought to prevent and reverse memory dysfunction in acute stress situations caused by bacterial infection (Gareau et al. 2011). As we all know, diabetes may be a serious risk factor for developing cognitive functions disorders. Many studies suggested that probiotics could improve impaired cognitive levels in mice with brain dysfunction due to diabetes mellitus (Crane et al. 2013; Davari et al. 2013; Ohara 2016). Lastly, in a new double-blinded clinical trial, the outcomes revealed that there was no effect of the probiotic composition that has been used on both AD symptoms and biomarkers. The researchers explained these results by the interference of many possible factors including the severity of the disease, the probiotic formulation, and its administration timing (Agahi et al. 2018). In summary, all these studies open our eyes to the future possibilities of probiotics use as a prophylactic treatment of inflammatory and immunological manifestations that result of gut-brain axis disturbances in humans.

8.14 Antibiotics, Gut Microbiota, and Neuroinflammation

Gut microbiota products may initiate microglial cascades, and activate an inflammatory response in the CNS, which results in increased neurotoxicity and decreased amyloid clearance (Kowalski et al. 2019). This led researchers to think about the modulation of gut microbiota by antimicrobial agents and how they affect the central nervous system (Durães et al. 2018). Antibiotics are known to induce significant gut dysbiosis. Wang et al. experiments demonstrated that treating animals with ampicillin for 1 month could lead to disruption of gut microbiota, to increased serum corticosterone, and to increased anxiety-like behavior and impairment of memory (Wang et al. 2015).

Furthermore, mice treated by chronic antibiotics had depleted and restructured microbiota components in adulthood. These modifications were associated with decreased anxiety and with reduced oxytocin, vasopressin, and BDNF levels in adult brain tissue (Desbonnet et al. 2015). Another study shows that in adult mice treated with antibiotics, hippocampal neurogenesis is decreased. The reconstitution with normal flora and the use of probiotics led to the reversal of this reduction. These two interventions are thought to be involved in the exhibition of a higher number of Ly6C(hi) monocytes in mice brain. This particular type of monocyte plays a partial role in the rescue of neurogenesis and memory retention (Mohler et al. 2016). The chronic use of antibiotics treatment gives rise to intestinal gut perturbation inducing changes in peripheral cytokines and chemokines. These changes are concurrent with the reduction of amyloids plaques formation in mice (Minter et al. 2016). All these findings make it plausible that neurodegenerative diseases might be treated in the future by modulating gut microbiota by antibiotics.

8.15 Fecal Microbiota Transplantation

Fecal microbiota transplantation (FMT) is an intervention that consists of replacing the gut microbiota of an unhealthy person (a receiver) with the microbiota of a healthy person (a donor) (Jiang et al. 2017). This technique has proven its effectiveness in treating patients who present recurrent *Clostridium difficile* infections which is a very delicate condition to treat (Tang et al. 2017). Guihua et al. some evidence showed that FMT was more efficient than the treatment with an antibiotic called vancomycin (van Nood et al. 2011). FMT helps also to improve the symptoms of some conditions that may be related to gut microbiota dysbiosis especially inflammatory bowel disease and functional gastrointestinal tract. FMT was also shown to have favorable outcomes in some extra-intestinal diseases including among other Parkinson's disease, chronic fatigue syndrome, multiple sclerosis, and metabolic syndrome (Choi and Cho 2016). According to Bruce-Keller et al., administrating microbiota from high-fat-fed mice into animals, engendered significant troubles of exploratory, cognitive, and memory functions. There is a study that was done in 2017 which reported fecal microbiota transplantation in the case of a 22-year-old girl diagnosed with epilepsy. After the administration, the girl was still free of epilepsy symptoms. The researchers suggested that FMT might help in treating certain neurological disorders (He et al. 2017). Despite the increasing data that support the application of fecal microbiota transplantation, randomized control trials are still necessary to prove its efficacy in neurological disorders treatment.

References

Agahi A, Hamidi GA, Daneshvar R, Hamdieh M, Soheili M, Alinaghipour A et al (2018) Does severity of Alzheimer's disease contribute to its responsiveness to modifying gut microbiota? A double blind clinical trial. Front Neurol 9(August):1–9

Akbari E, Asemi Z, Kakhaki RD, Bahmani F, Kouchaki E, Tamtaji OR, Hamidi GA, Salami M (2016) Effect of probiotic supplementation on cognitive function and metabolic status in Alzheimer's disease. a randomised double blind and controlled trial. Front Aging Neurosci 8:256

Asiimwe N, Yeo SG, Kim MS, Jung J, Jeong N (2016) Nitric oxide: exploring the contextual link with Alzheimer's disease. Oxidat Med cell Longev:1–10

Asti A, Gioglio L (2014) Can a bacterial endotoxin be a key factor in the kinetics of amyloid fibril formation? J Alzheimer's Dis 39(1):169–179

Aziz Q, Doré J, Emmanuel A (2013) Gut microbiota and gastrointestinal health: current concepts and future directions. Neurogastroenterol Motil 25:4–15

Bajaj JS, Ridlon JM, Hylemon PB, Thacker LR, Heuman DM, Smith S, Sikaroodi M, Gillevet PM (2011) Linkage of gut microbiome with cognition in hepatic encephalopathy. Am J Physiol Liver Physiol 302:G168–G175

Banack SA, Caller TA, Stommel EW (2010) The cyanobacteria derived toxin beta-N-mthylamino-L-alanine and amyotrophic lateral sclerosis. Toxins (Basel) 2(12):2837–2850

Banks WA (2008) The blood-brain barrier: connecting the gut and the brain. Regul Pept 149:11–14

Barberger-Gateau P, Raffaitin C, Letenneur L, Berr C, Tzourio C, Dartigues JF, Alperovitch A (2007) Dietary patterns and risk of dementia: the three-city cohort study. Neurology 69:1921–1930

Barrett E, Ross RP, O'Toole PW, Fitzgerald GF, Stanton C (2012) gamma-Aminobutyric acid production by culturable bacteria from the human intestine. J Appl Microbiol 113:411–417

Bekris LM, Yu CE, Bird TD, Tsuang DW (2011) Genetics of Alzheimer disease. J Geriatr Psychiatry Neurol 23(4):213–227

Biagi E, Candela M, Turroni S, Garagnani P, Franceschi C, Brigidi P (2013) Ageing and gut microbes: perspectives for health maintenance and longevity. Pharmacol Res 69:11–20

Bonfili L, Cecarini V, Berardi S, Scarpona S, Suchodolski JS, Nasuti C, Fiorini D, Grame MCB (2017) Microbiota modulation countacts Alzheimer's disease progression influencing neurnal proteolysis and gut hormones plasma levels. Sci Rep:7, 2426

Borre YE, O'Keeffe GW, Clarke G, Stanton C, Dinan TG, Cryan JF (2014) Microbiota and neurodevelopmental windows: implications for brain disorders. Trends Mol Med 20(9):509–518

Bradley WGMD (2009) Beyond Guam: the cyanobacteria/BMAA hypothesis of the cause of ALS and other neurodegenerative diseases. Amytroph Lateral Sclera 10:7–20

Braniste V, Al-Asmakh M, Kowal C, Anuar F, Abbaspour A, Toth M, Korecka A, Bakocevic N, Ng LG, Kundu P, Gulyas B, Halldin C, Hultenby K, Nilsson H, Hebert H, Volpe BT, Diamod B, Pettersson S (2014) The gut microbiota influences blood-brain barrier permeability in mice. Sci Transl Med 6:263ra158

Bravo JA, Julio-Pieper M, Forsythe P, Kunze W, Dinan TG, Bienenstock J et al (2012) Communication between gastrointestinal bacteria and the nervous system. Curr Opin Pharmacol 12(6):667–672

Brenner SR (2013) Blue-green algae or cyanobacteria in the intestinal micro-flora may produce neurotoxins such as Beta-N-Methylamino-l-Alanine (BMAA) which may be related to development of amyotrophic lateral sclerosis, Alzheimer's disease and Parkinson-Dementia-Complex. Med Hypotheses 80(1):103

Bruce-Keller AJ, Salbaum JM, Luo M, Blanchard ET, Taylor CM, Welsh DA, Berthoud H (2015) Obese-type gut microbiota induce neurobehavioral changes in the absence of obesity. Biol Psychiatry 77:607–615

Buford TW (2017) Trust your gut: the gut microbiome in age-related inflammation, health, and disease:1–11

Chen SG, Stribinskis V, Rane MJ, Demuth DR, Gozal E, Roberts AM et al (2016) Exposure to the functional bacterial amyloid protein curli enhances alpha-synuclein aggregation in aged fischer 344 RAts and caenorhabditis elegans. Sci Rep 6(September):1–10

Chen G, Xu T, Yan Y, Zhou Y, Jiang Y, Melcher K et al (2017) Amyloid beta: structure , biology and structure-based therapeutic development. Acta Pharmacologica Sinica 38(9):1205–1235

Choi HH, Cho YS (2016) Fecal microbiota transplantation: current applications, effectiveness, and future perspectives. Clin Endosc. Taylor & Francis 49:257–265

Cho I, Blaser MJ (2012) The human microbiome: at the interface of health and disease. Nat Rev Genet 13(4):260–70. Available from: http://www.ncbi.nlm.nih.gov/pubmed/22411464

Claesson MJ, Jeffery IB, Conde S, Power SE, O'Connor EM, Cusack S, Harris HMB, Coakley M, Lakshminarayanan B, O'Sullivan O, Fitzgerald GF, Deane J, O'Connor M, Harnedy N, O'Connor K, O'Mahony D, van Sinderen D, Wal P (2012) Gut microbiota composition correlates with diet and health in the elderly. Nature 488:178–184

Collins SM, Surette M, Bercik P (2012) The interplay between the intestinal microbiota and the brain. Nat Rev Microbiol 10:735–742

Cox PA, Davis DA, Mash DC, Metcalf JS, Banack S (2016) Dietary exposure to an environmental toxin triggers neurofibrillary tangles and amyloid deposits in the brain. Proc Biol Sci 283:20152397

Crane PK, Walker R, Hubbard RA, Li G, Nathan DM, Zheng H, Haneuse S, Craft S, Montine TJ, Kahn SE, McCormick W, McCurry SM, Bowen JD, Larson E (2013) Glucose levels and risk of dementia. N Engl J Med 369:540–548

Cryan JF, Dinan TG (2012) Mind-altering microorganisms: the impact of the gut microbiota on brain and behaviour. Nat Rev Neurosci 13(10):701–712

Cryan JF, O'Mahony SM (2011) The microbiome-gut-brain axis: from bowel to behavior. Neurogastroenterol Motil 23(3):187–192

Davari S, Talaei SA, Alaei HSM (2013) Probotics treatment improves diabetes-induced impairment of synaptic activity and cognitive function: behavioral and electrophysiological proofs for microbiome-gut-brain axis. Neuroscience 240:287–296

De Filippo C, Cavalieri D, Di Paola M, Ramzzotti M, Poullet JB, Massart S, Collini S, Pieraccini G, Lionetti P (2010) Impact of diet in shaping gut microbiota revealed by a comparative study in children from Europe and rural Africa. Proc Natl Acad Sci U S A 107:14691

Desbonnet L, Clarke G, Traplin A, O'Sullivan O, Crispie F, Moloney RD et al (2015) Gut microbiota depletion from early adolescence in mice: implications for brain and behaviour. Brain Behav Immun 48(April):165–173

Distrutti E, O'Reilly JA, McDonald C, Cipriani S, Renga B, Lynch MA et al (2014) Modulation of intestinal microbiota by the probiotic VSL#3 resets brain gene expression and ameliorates the age-related deficit in LTP. PLoS One 9(9):176–184

Dominicé PF, Lehto L (1991) Koulutuselämäkertojen laatiminen ryhmäreflektion välineenä. Amyloid Depos as Cent event Aetiol Alzheimer's Dis:214–232

Durães F, Pinto M, Sousa E (2018) Old drugs as new treatments for neurodegenerative diseases. Pharmaceuticals 11(2):1–21

Erny D, Hrabe de Angelis AL, Prinz M (2017) Communicating systems in the body: how microbiota and microglia cooperate. Immunology 150:7–15

Fassbender K, Walter S, Kühl S, Landmann R, Ishii K, Bertsch T et al (2004) The LPS receptor (CD14) links innate immunity with Alzheimer's disease. FASEB J 18(1):203–205

Foster JA, Neufeld KA (2018) Gut-brain axis: how the microbiome influences anxiety and depression. Trends Neurosci 36:305–312

Friedland RP (2015) Mechanisms of molecular mimicry involving the microbiota in neurodegeneration. J Alzheimer's Dis 45:349–362

Gareau MG (2014) Microbiota-gut-brain axis and cognitive function. Adv Exp Med Biol 817:357–372

Gareau MG, Wine E, Rodrigues DM, Cho JH, Whary MT, Philpott DJ et al (2011) Bacterial infection causes stress-induced memory dysfunction in mice. Gut 60(3):307–317

Gill SR et al (2006) Metagenomic analysis of the human distal gut microbiome. Science 312:1355–1359

Hauss-Wegrzyniak B, Vraniak PD, Wenk GL (2000) LPS-induced neuroinflammatory effects do not recover with time. Neuroreport 11(8):1759–1763

He Z, Cui B, Zhang T, Li P, Long C, Ji G et al (2017) Fecal microbiota transplantation cured epilepsy in a case with Crohn's disease: the first report. World J Gastroenterol 23(19):3565–3568

Heijitz RD, Wang S, Anuar F, Qian Y, Bjorkholm B, Samuelsson A et al (2011) Normal gut microbiota modulates brain development and behavior. Proc Natl Acad Sci USA 108:3047–3052

Heneka MT, Golenbock DT, Latz E (2015) Innate immunity in Alzheimer's disease. Nat Immunol 16:229–236

Hoban AE, Stilling RM, Ryan FJ, Dinan TG, Claesson MJ, Clarke G, Cryan J (2016) Regulation of prefrontal cortex myelination by the microbiota. Transl Psychiatry 6:e774

Hornig M (2013) The role of microbes and autoimmunity in the pathogenesis of neuropsychiatric illness. Curr Opin Rheumatol 25:488–795

Hu X, Wang T, Jin F (2016) Alzheimer's disease and gut microbiota. Sci China Life Sci 59:1006–1023

Hufeldt MR, Nielsen DS, Vogensen FK, Midtvedt T, Hansen A (2010) Variation in the gut microbiota of laboratory mice is related to both genetic and environmental factors. Comp Med 60:336–347

Hufnagel DA, Tükel Ç, Chapman MR (2013) Disease to dirt: the biology of microbial amyloids. PLoS Pathog 9(11):1–5

Jiang C, Li G, Huang P, Liu Z, Zhao B (2017) The gut microbiota and Alzheimer's disease. J Alzheimer's Dis 58(1):1–15

Kobayashi Y, Sugahara H, Shimada K, Mitsuyama E, Kuhara T, Yasuoka A et al (2017) Therapeutic potential of Bifidobacterium breve strain A1 for preventing cognitive impairment in Alzheimer's disease. Sci Rep 7(1):8614–8622

Kowalski K, Mulak A, Words K (2019) Brain-gut-microbiota axis in Alzheimer's disease. J Neurogastroenterol Motil 25(1):48–60

Kumar DKV, Choi HS, Washicosky KJ, Eimer WA, Tucker S, Ghofrani J et al (2016) Amyloid-β peptide protects against microbial infection in mouse and worm models of Alzheimer's disease. Sci Transl Med 8(340):340ra72

La Rosa F, Clerici M, Ratto D, Occhinegro A, Licito A, Romeo M, Lorio CD, Rossi P (2018) The gut-brain axis in Alzheimer's disease and omega-3. A critical overview of clinical trials. Nutrients 10:E1267

Ladner-Keay CL, LeVatte M, Wishart DS (2016) Role of polysaccharide and lipid in lipopolysaccharide induced prion protein conversion. Prion 10(6):466–483

Lakhan SE, Caro M, Hadzimichalis N (2013) NMDA receptor activity in neuropsychiatric disorders. Front Psychiatry 4(June):1–7

Larsen P, Nielson JL, Dueholm MS, Wetzel R, Otzen D, Nielsen P (2007) Biofilms, amyloid adhesins are abundant in natural. Environ Microbiol 9:3077–3090

Larsen P, Nielsen JL, Otzen D, Nielsen PH (2008) Amyloid-like adhesins produced by floc-forming and filamentous bacteria in activated sludge. Appl Environ Microbiol 74(5):1517–1526

Liu Y, Walter S, Stagi M, Cherny D, Letiembre M, Schulz-Schaeffer W et al (2005) LPS receptor (CD14): a receptor for phagocytosis of Alzheimer's amyloid peptide. Brain 128(8):1778–1789

Lu B, Nagappan G, Guan X, Nathan PJ, Wren P (2013) BDNF-based synaptic repair as a disease-modifying strategy for neurodegenerative diseases. Nat Rev Neurosci 14:401–416

Lukiw WJ (2016) Bacteroides fragilis lipopolysaccharide and inflammatory signaling in Alzheimer's disease. Front Microbiol 7(September):1–6

Meek PD, McKeithan K, Schumock GT (1998) Economic considerations in Alzheimer's disease. Pharmacotherapy 18:68–72

Mendelsohn D, Riedel WJ, Sambeth A (2009) Effects of acute tryptophan depletion on memory, attention and executive functions: a systematic review. Neurosci Biobehav Rev 33(6):926–952

Minter MR, Zhang C, Leone V, Ringus DL, Zhang X, Oyler-Castrillo P et al (2016) Antibiotic-induced perturbations in gut microbial diversity influences neuro-inflammation and amyloidosis in a murine model of Alzheimer's disease. Sci Rep 6(July):1–12

Mohler L, Mattei D, Heimesaat MM, Bereswill S, Fisher A, Alutis M, French T, Hambardzumyan D, Matzinger P, Dunay IR, Wolf S (2016) Ly6C(hi) monocytes provide a link between antibiotic induced changes in gut microbiota and adult hippocampal neurogenesis. Cell Rep 15:1945–1956

Nishida A, Inoue R, Inatomi O, Bamba S, Naito Y, Andoh A (2018) Gut microbiota in the pathogenesis of inflammatory bowel disease. Clin J Gastroenterol 11(1):1–10
Nunn PB, Ponnusamy M (2009) Beta-N-methylalanine (BMAA): metabolism and metabolic effects in model systems and in neural and other tissues of the rat in vitro. Toxicon 54:85–94
Ohara T (2016) Epidemiology of diabetes and risk of dementia. Brain Nerve 68:719–728
Ozawa M, Ninomiya T, Ohara T, Doi Y, Uchida K, Shirota T et al (2013) Dietary patterns and risk of dementia in an elderly Japanese. Am J Clin Nutr 97(5):1076–1082
Pérez Martinez G, Bauerl C, Collado M (2014) Understanding gut microbiota in elderly's health will enable intervention through probiotics. Benef Microbes 5:235–246
Peuhkuri K, Vapaatalo H, Korpela R (2016) Even low-grade inflammation impacts on small intestinal function. World J Gastroenterol 15(5):477–491
Pistollato F, Sumalla CS, Elio I, Masias VM, Giampieri F (2016) Role of gut microbiota and nutrients in amyloids formation and pathogenesis of Alzheimer disease. Nutr Rev 74:624–634
Priyamvada S, Gomes R, Gill RK, Saksena S, Alrefai WA, Dudeja PK (2016) Mechanisms underlying dysregulation of electrolyte absorption in IBD associated diarrhea. Inflamm Bowel Dis 21(12):2926–2935
Proctor C, Thiennimitr P, Chattipakorn N, Chattipakorn SC (2017) Diet, gut microbiota and cognition. Metab Brain Dis 32:1–17
Qin J, Li R, Raes J, Arumugam M, Burgdorf KS, Manichanh C et al (2010) A human gut microbial gene catalogue established by metagenomic sequencing. Nature 464(7285):59–65
Rampelli S, Candela M, Turroni S, Biagi E, Collino S, Toole PWO et al (2013) Functional metagenomic profiling of intestinal microbiome in extreme ageing. Aging 5(12):902–912
Reitz C, Brayne C, Mayeux R (2011) Epidemiology of Alzheimer disease. Nat Rev Neurol. 7(3):137–152
Rothhammer V, Borucki DM, Tjon EC, Takenaka MC, Chao CC, Ardura-Fabregat A et al (2018) Microglial control of astrocytes in response to microbial metabolites. Nature 557(7707):724–728
Salazar N, Valdés-Varela L, González S, Gueimonde M, de los Reyes-Gavilán CG (2017) Nutrition and the gut microbiome in the elderly. Gut Microbes 8(2):82–97
Saleem F, Bjorndahl TC, Ladner CL, Perez-Pineiro R, Ametaj BN, Wishart DS (2014) Lipopolysaccharide induced conversion of recombinant prion protein. Prion 8(2):1–12
Sampson TR, Mazmanian S (2015) Control of brain development, function, and behavior by the microbiome. Cell Host Microbe 17:565–576
Scott KP, Gratz SW, Sheridan PO, Flint HJ, Duncan SH (2013) The influence of diet on the gut microbiota. Pharmacol Res 69(1):52–60
Semrin G, Fishman DS, Athos Bousvaros AZ, Grand RJ, Weinstein DA (2016) Impaired intestinal iron absorption in Crohn's disease correlates with disease activity and markers of inflammation. Inflamm Bowel Dis 33(2):557–573
Smith QR, Nagura H, Takada Y, Duncan M (1992) Facilitated transport of the neurotoxin, beta-N-methylamino-L-alanine, across the blood-brain-barrier. J Neurochem 58(4):1330–1337
Solas M, Milagro FI, Ramírez MJ, Martínez JA (2017) Inflammation and gut-brain axis link obesity to cognitive dysfunction: plausible pharmacological interventions. Curr Opin Pharmacol 37:87–92
Strooper BDE (2010) Proteases and proteolysis in Alzheimer disease: a multifactorial view on the disease process. Physiol Rev 90:465–494
Swerdlow R, Burns J, Khan S (2010) The AD mitochondrial cascade hypothesis. J Alzheimers Dis 20(Suppl 2):265–279
Swerdlow RH, Khan SM (2004) A "mitochondrial cascade hypothesis" for sporadic Alzheimer's disease. Med Hypotheses 63(1):8–20
Taichi A, Suzuki MW (2014) Geographical variation of human gut microbial composition. Biol Lett 10:20131037
Tang G, Yin W, Liu W (2017) Is frozen fecal microbiota transplantation as effective as fresh fecal microbiota transplantation in patients with recurrent or refractory Clostridium difficile infection: a meta-analysis? Diagn Microbiol Infect Dis. Elsevier Inc. 88:322–329

Tiihonen K, Ouwehand AC, Rautonen N (2010) Human intestinal microbiota and healthy aging. Ageing Res Rev 9:107–116

Tran L, Greenwood-Van Meerveld B (2013) Age-associated remodeling of the intestinal epithelial barrier. J Gerontol 68(9):1045–1056

Tse JK (2017) Gut microbiota, nitric oxide and microglia as pre-requisites for neurodegenerative disorders. ACS Chem Neurosci 8:1438–1447

Tükel Ç, Nishimori JH, Wilson RP, Winter MG, Keestra AM, van Putten JPM, Baumler AJ (2010) Toll-like receptors 1 and 2 cooperatively mediate immune responses to curli, a common amyloid from enterobacterial biofilms. Cell Microbiol 12:1495–1505

United Nations (2017) World Population Prospects The 2017 Revision

van Nood E, Vrieze A, Nieuwdorp M, Fuentes S, Zoetendal EG, de Vos WM, Visser CE, Kuijper EJ, Bartelsman JFWM, Tijssen JGP, Speelman P, Dijkgraaf MGW, Keller JJ (2011) Duodenal infusion of donor feces for recurrent clostridium difficile. Univ Bucharest Rev Lit Cult Stud Ser 13(1):7–20

Walker LC, Schelle J, Jucker M (2016) The prion-like properties of amyloid-β assemblies: implications for Alzheimer's disease. Cold Spring Harb Perspect Med 6(7):1–14

Wang T, Hu X, Liang S, Li W, Wu X, Wang L, Jin F (2015) Lactobacillus fermentum NS9 restores the antibiotic induced physiological and psychological abnormalities in rats. Benef Microbes 6:707–717

Wang Y, Wang Z, Wang Y, Li F, Jia J, Song X, Qin S, Wang R, Jin F, Kitazato K, Wang Y (2018) The gut miroglia connection: implications for central nervous system diseases. Front Immunol 9:2325

Woodmansey EJ (2007) Intestinal bacteria and ageing. J Appl Microbiol 102:1178–1186

Xu R, Wang QQ (2016) Towards understanding brain-gut-microbiome connections in Alzheimer's disease. BMC Syst Biol 10(Suppl 3)

Zhao Y, Jaber V, Lukiw WJ (2017a) Secretory products of the human GI tract microbiome and their potential impact on Alzheimer's disease (AD): detection of Lipopolysaccharide (LPS) in AD hippocampus. Front Cell Infect Microbiol 7(July):1–9

Zhao Y, Cong L, Lukiw WJ (2017b) Lipopolysaccharide (LPS) accumulates in neocortical neurons of Alzheimer's disease (AD) brain and impairs transcription in human neuronal-glial primary co-cultures. Front Aging Neurosci 9(December):1–9

Zhao Y, Cong L, Jaber V, Lukiw WJ (2017c) Microbiome-derived lipopolysaccharide enriched in the perinuclear region of Alzheimer's disease brain. Front Immunol 8(September):1–6

Zhou Y, Smith D, Leong BJ, Brännström K, Almqvist F, Chapman MR (2012) Promiscuous cross-seeding between bacterial amyloids promotes interspecies biofilms. J Biol Chem 287(42):35092–35103

9 Stem Cell Therapy: A Great Leap Forward in Alzheimer's Treatment

Nazish Tabassum, Chandra Bhan Yadav, Anshuman Singh, and Vinod Verma

Abstract

Alzheimer's disease (AD) is a neurodegenerative disorder and one of the substantial socioeconomic and medical calamities of our time. It is characterized by progressive neurodegenerative disorder featuring dementia and cognitive impairment. It is caused by synaptic failure and the excessive aggregation of two types of misfolded proteins, namely, amyloid-β (Aβ) and tau protein. AD is a complex disease that affects neurons in different parts of the brain. The treatment of AD is difficult with currently available medications and treatment approaches due to inadequate knowledge of its etiology and versatile nature of its pathology. Stem cell holds a promising approach to regenerate the tissue systems and has been explored in various studies of neurodegenerative disorders and provides great research opportunity. Stem cell research-based therapies emit a new hope for AD treatment as a regenerative approach. This chapter focuses on recent advances in stem cell therapies according to cell types and pathophysiology of AD along with human clinical trials of stem cell therapies for the AD.

Keywords

Stem cell · Alzheimer's disease · Amyloid precursor protein

N. Tabassum · C. B. Yadav
Centre of Biotechnology, Nehru Science Complex, University of Allahabad, Allahabad, Uttar Pradesh, India

A. Singh · V. Verma (✉)
Stem Cell Research Centre (SCRC), Department of Hematology, Sanjay Gandhi Postgraduate Institute of Medical Sciences (SGPGIMS), Lucknow, Uttar Pradesh, India
e-mail: vverma@sgpgi.ac.in

G. M. Ashraf, A. Alexiou (eds.), *Biological, Diagnostic and Therapeutic Advances in Alzheimer's Disease*, https://doi.org/10.1007/978-981-13-9636-6_9

9.1 Introduction

In 1907, Alois Alzheimer first reported Alzheimer's disease (AD) (Choi et al. 2014a, b). However, 60–80% of dementia cases are caused by AD, characterized by memory loss and cognitive disorder which affects the quality of life. AD is a progressive neurodegenerative disorder, where the symptom of dementia is gradually elevated over a number of years. In the early stages of the AD, mild memory loss occurs, but in the later stage, the patients lose the ability to converse and the ability to respond to their environment. AD severely affects the human health, as it is the sixth leading cause of mortality in the United States. The average life duration of Alzheimer's patient is 8 years after the AD's symptoms become noticeable, but survival range can vary from 4 to 20 years, depending on the age, lifestyle, diet, and health condition. In 2010, the estimated economic burden of AD's treatment was $172 billion in the United States and $604 billion worldwide that will be tripled by 2050 (Wimo et al. 2010). In India, approximately 3.7 million people were suffering from AD, and this number is expected to double by the year 2030 (Alzheimer's and Related Disorders Society of India (ARDSI) 2010).

Two types of AD are reported: (i) familial and (ii) sporadic. Familial AD is caused by an autosomal genetic mutation in the genes responsible for Aβ plaques. This genetic mutation is related to the amyloid precursor protein (APP), presenilin-1 (PSEN-1), and presenilin-2 (PSEN-2). However, familial AD is rare in prevalence and less than 5% of familial AD cases are reported (Selkoe 2001; Prasher et al. 1998; Rosen et al. 2010; Marchetti and Marie 2011; Genin et al. 2011). Sporadic AD is ubiquitous in nature and caused by the interaction between genetic profile and environmental factors (Duncan and Valenzuela 2017; Persson et al. 2014). The cardinal pathologic features of AD include the aggregation of two types of misfolded proteins (amyloid beta and tau) (Allen et al. 2011; Eckman and Eckman 2007). Amyloid beta (Aβ) protein is a pathological cleavage product of the APP. Aβ protein accumulates into plaques and minor oligomers. Mutations in APP genes or in APP processing pathway genes are linked to the inherited familial AD (Huang and Mucke 2012). Tau is a microtubule-associated protein that accumulates intracellularly as neurofibrillary tangles (NFTs) which is a pathological feature closely linked with cognitive decline in the AD. However, mutations in tau protein lead to cause frontotemporal dementia, not AD (Huang and Mucke 2012).

A rising accord inside the field is that treatment of AD patients with currently available medicines comes late, which is the result of vital neuronal cell loss within the brain. To combat these problems, human embryonic stem cell (hESC)/induced pluripotent stem cell (iPSC)/mesenchymal stem cell (MSC)-derived neural cells have been suggested as powerful replacement therapy for AD (Fig. 9.1). In this chapter, the current state of research in the etiology of AD, probable challenges, and techniques for using stem cell-based treatment will be discussed briefly. Recent studies that have developed promising cell types and clinical investigations that could be used to combat this detrimental disease in the future will also be highlighted.

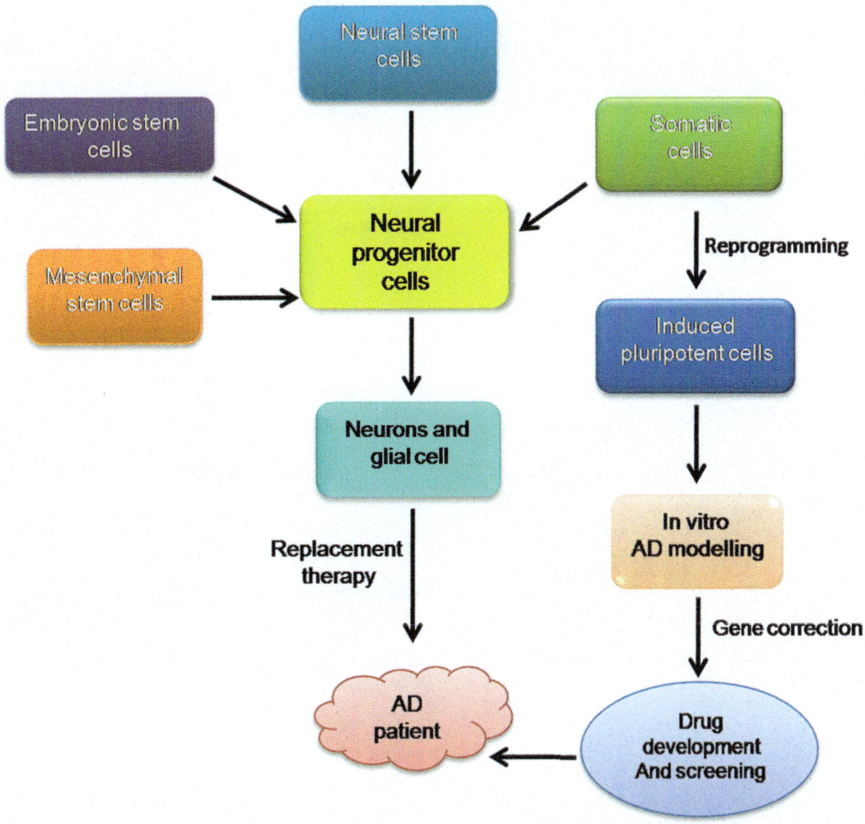

Fig. 9.1 Stem cell therapy in AD

9.2 Pathophysiology of Alzheimer's Disease

AD is distinguished by extracellular amyloid plaques and intracellular NFT features. Amyloid-β (Aβ) protein is the major constituent of plaques associated with AD (Fig. 9.2). The pathophysiology of AD involves several neurotransmitters system and processes (Lin et al. 2001). Three hallmarks of the AD are β-amyloid plaques, neurofibrillary tangles, and neuronal cell death.

Recently, recognized characteristics of AD include degeneration of synapses, aneuploidy, neuronal loss, granulovacuolar degeneration, and amyloid plaques. Three types of amyloid plaques are known in the brain of AD patients:

1. *Diffuse plaques*: contain no amyloid core
2. *Neuritic plaques*: consist of a central amyloid core surrounded by neurites
3. *Burnt-out plaques*: consist of an isolated amyloid core

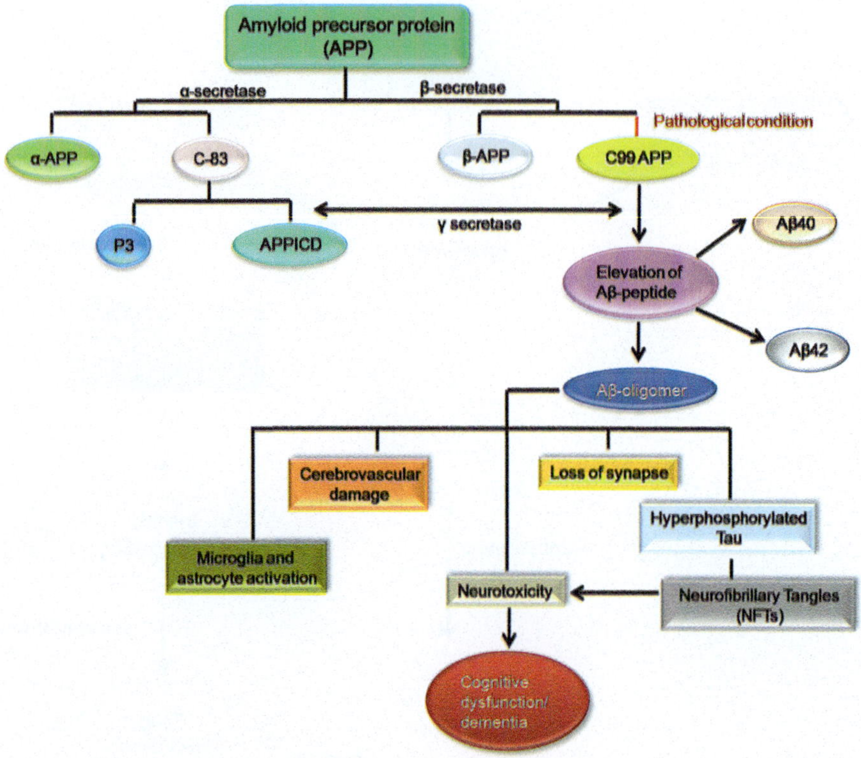

Fig. 9.2 Pathogenesis of AD represented by interacting damage pathways lead by soluble oligomers of the amyloid beta peptide

Apart from the amyloid plaques and tangles, globular and non-fibrillar proteins are continuously released in the AD patient's brain. Cellular changes include short-term and rapid degeneration of neurons which leads to neuronal death when Aβ proteins remain globular.

A few theories related to AD such as the cholinergic, Aβ, tau, and inflammation hypothesis have been explained. Some of them are listed below to understand the mechanism of this disorder:

1. *Changes in brain structure*: The characteristic of the AD on a macro level is the progressive loss of brain tissue. The cortex atrophies are responsible for memory formation in the brain.
2. *Degenerative processes in AD*: AD is characterized on a micro level by three neuropathologic hallmarks: extracellular β-amyloid plaques, intracellular NFTs, and neuronal degeneration. β-Amyloid plaques play an important role in AD pathogenesis which is known as "amyloid cascade" (Swerdlow 2007).

9.3 β-Amyloid Hypothesis

β-Amyloid plaques are aggregates of insoluble peptides formed after the cleavage of APP. Three enzymes, namely γ-secretase, β-secretase, and α-secretase, participate in the APP cleavage. However, APP cleavage by β-secretase followed by γ-secretase produces a soluble 40-amino acid peptide. In addition, γ-secretase cleaves APP that forms nonsoluble 42-amino acid peptide Aβ42 or Aβ which aggregates as β-amyloid plaques. There are three genes involved in the formation of Aβ: APP, PS1, and PS2. PS1 and PS2 genes code for presenilin which is a subunit of γ-secretase. Tau protein hyperphosphorylation occurs after plaque formation in the brain (Selkoe 2002). Neurofibrillary tangles (NFTs) result from damage of neuronal microtubules caused by tau protein modification (Imbimbo et al. 2005). Tau protein disrupts the collapse structure of microtubules and destroys the neuron's transport and communication system. Modifications in tau lead to its oligomerization and NFT production (Maccioni et al. 2010).

9.4 Cholinergic Hypothesis

Loss of cholinergic neurons is one of the pathologies of AD. In that case, more than 75% of cholinergic neurons are reduced in AD patient's brain (Perry et al. 1978). However, acetylcholine is involved in memory; thus, loss of cholinergic activity relates with impairment of memory. Acetylcholine attaches to the post-synaptic receptors: muscarinic and nicotinic. Pre-synaptic nicotinic receptors influence the release of acetylcholine, serotonin, norepinephrine, and glutamate which have a role in AD pathophysiology.

9.5 Glutamatergic Hypothesis

Glutamatergic neurons form the projections which influence the cognition in the brain. AD pathology is linked to only one type of receptor, that is, NMDA receptor which then undergoes low-level activation in AD patient's brain. However, dysregulation of the glutamate NMDA receptor is responsible for neuronal damage which interferes with normal signal transduction (Danysz et al. 2000). It can lead to the production of APP which is related to plaque development and tau hyperphosphorylation.

9.6 Oxidative Stress Hypothesis

Aβ generates the reactive oxygen and nitrogen species which have an unpaired extra electron and also induces lipid peroxidation. The free radicals cause cellular and molecular damage in neuronal cells. The brain can be damaged from oxidative stress because of high oxygen utilization rate and antioxidant enzymes as compared with the other organs. Upregulation of cytokines and DNA damage in neurons have an essential role in AD progression.

9.7 Chronic Inflammation Hypothesis

β-Amyloid deposition in neurons and NFTs causes inflammation in response to cellular damage. Inflammation leads to the increased number of prostaglandins, produced by COX-1 and COX-2, localized in distinct areas of the brain. Inflammation occurs within or adjacent to the neuritic plaque. Antichymotrypsin, macroglobulin in neuritic plaques, and activated microglia codes for interleukin-1 and interleukin-6 also are detectable in case of the inflammation-related AD.

9.8 Cholesterol and Other Factors

Cholesterol is also implicated in AD pathogenesis. Elevated cholesterol levels raise Aβ production, and thus, the risk of AD progression increases (Reiss 2005). During the AD progression, brain regions become altered, and reduced serotonin levels play an important role in depression and anxiety which are common in an AD patient (Mössner et al. 2000; Lai et al. 2005).

9.9 Stem Cells Used in Alzheimer's Treatment

Stem cells are undifferentiated cells that possess self-renewal and differentiation property. Self-renewal is described as the ability to undergo numerous cell cycle divisions, resulting in identical daughter cells, and differentiation capability is the development of specialized cells from the undifferentiated stem cells (Tabassum et al. 2017). On the virtue of origin, stem cells can be categorized into embryonic stem cells (ESCs) and adult stem cells, and based on potency, these cells are categorized into totipotent, multipotent, pluripotent, and unipotent. Due to the differentiation properties of stem cells into neuronal-like cells, they can be used for the treatment of Alzheimer's disease. The human body generates four types of stem cells: neural stem cells (NSCs), MSCs, ESCs, and iPSCs. These cells have unique properties; thus, they are the most suitable candidates for stem cell therapy.

Embryonic Stem Cells (ESCs) ESCs are pluripotent stem cells which are obtained from the inner cell mass of the blastocyst that gives rise to all cell types except placenta. Researchers successfully differentiated the ESCs into several specific neural cell types including dopaminergic neurons in vitro (Krencik et al. 2011; Malmersjo et al. 2009). The direct transplantation of ESCs showed high risks of teratoma formation due to their potent differentiation ability (Kooreman and Wu 2010). Moreover, various rodent studies demonstrated that the transplantation of ESC-derived NSCs shows no tumorigenesis, but to confirm these results, further research is needed

(Araki et al. 2013; Tang et al. 2008). Along with tumorigenesis, rejection of transplanted ESC-derived tissues by the immune system occurred (Pearl et al. 2012).

Induced Pluripotent Stem Cells (iPSCs) iPSCs are pluripotent stem cells which are reprogrammed from adult fibroblasts by using four transcription factors including Oct3/4, Sox2, Klf4, and c-Myc that are pretty much similar to the ESCs (Takahashi and Yamanaka 2006). These cells are reprogrammed into pluripotency state, having the capability to differentiate into different types of cells including neurons (Cooper et al. 2010) and neurospheres (Nori et al. 2011). Researchers used the iPSC-derived glia cells regarding the inflammatory response in Alzheimer's disease (Holtman et al. 2015). In 2014, Takamatsu revealed that iPSC-derived macrophages express neprilysin and β-amyloid-degrading protease (Takamatsu et al. 2014). However, certain unsolved problems are still present regarding the clinical usage of iPSCs such as tumor formation, immunogenicity, long-time safety, genetic defects, and optimal reprogramming (Tolosa et al. 2016; Araki et al. 2013). Therefore, iPSC-based treatment for AD has been more focused on the establishment of cell-based disease models as compared to treatments (Choi et al. 2014a, b; Yagi et al. 2012; Sproul et al. 2014). Israel and coworkers highlighted the cholinergic neurons of the basal forebrain because of their dysfunction in the early stage of AD (Israel et al. 2012). We know that there is a widespread degeneration in the later stage of the AD, so the protocol using iPSCs should be more elaborated (Pen and Jensen 2017).

Neuronal Stem Cells (NSCs) NSCs are found within the brain. In the past few decades, it was thought that the process of neurogenesis takes place in the fetus; however, the recent studies demonstrated that neurogenesis also occurs in an adult's brain. NSCs were found in the sub-granular zone and sub-ventricular zone of the brain (Taupin 2006; Mu and Gage 2011). These cells are differentiated into neurons, astrocytes, and oligodendrocytes (Taupin 2006). Due to the differentiation capability, NSCs are considered as the best choice for the replacement of injured neurons. In 2001, for the first time, Qu and coworkers proved the replacement of injured neuron by implanting human NSCs into the mature rat's brain (Qu et al. 2001). The results showed that NSCs survived and differentiated into neurons and astrocytes in rat's brain. Moreover, memory impairment was also observed in mature rats after the transplantation when evaluated with the control (Qu et al. 2001). However, NSC isolation from the adult's brain is complicated, so current studies mainly use fetal NSCs, which could also raise ethical problems. To combat these problems, researchers focused on the MSCs, and it was found that bone marrow MSCs (BM-MSCs), adipose tissue (AT-MSCs), and umbilical cord blood MSCs (UC-MSCs) could be trans-differentiated into neuronal cells (Brazelton et al. 2000; Mezey et al. 2000; Kim et al. 2012a, b).

Mesenchymal Stem Cells (MSCs) MSC-based therapy has an advantage over other cell-based therapy because it can be given intravenously and has blood barrier penetration and low tumorigenicity (Oh et al. 2015; Ra et al. 2011). The in vitro transplantation of MSCs in AD cell model augmented the metabolic activity and survival which help to rescue the patients with AD. Co-culturing of human MSCs and mouse microglia cells increased the expression of neprilysin (Aβ-degrading enzyme) (Kim et al. 2012a, b). BM-MSCs show the immunomodulatory capability by releasing the soluble factors including TGF-β, IL-6, IL-10, and PGE2 (Ramasamy et al. 2007; Aggarwal and Pittenger 2005). These factors inhibit the functioning of monocyte-derived dendritic cells and modify the phenotype of the natural killer cell (Sotiropoulou et al. 2006). In 2012, Chen and coworkers demonstrated that AT-MSCs can be differentiated into astrocytes and neuronal-like cells (Chen et al. 2012). The transcriptional profile of AT-MSCs showed some similarity with BM-MSCs (Peroni et al. 2008). AT-MSCs also secrete various neurotrophic factors (Gutiérrez-Fernández et al. 2013; Yang et al. 2012). UC-MSCs can be also differentiated into neuron-like cells. Researchers studied these cells in mouse model having Alzheimer's disease and clinically (Kang et al. 2016). Table 9.1 summarizes the studies of stem cell therapy on AD-diseased animal models.

9.10 Some Clinical Trials of Stem Cell Therapies for Alzheimer's Disease

Since 2011, animal model evidence supported the approval of MSC-based therapies in clinical trials for patients with Alzheimer's disease. UC-MSCs were preferred, and the route of administration of stem cell is intravenous (Table 9.2).

9.11 Conclusion and Future Prospects

Stem cell therapy exhibits therapeutic benefits in several neurodegenerative disorders. Stem cell transplantation increases the expression of synaptic protein markers in AD animal models. The transplantation of MSCs elevated the level of Aβ-degrading enzyme and reduced the level of Aβ due to microglial expression. With the ongoing development of reprogramming technology, there is an immense potential in the utilization of iPSCs in the treatment of AD. For reprogramming, somatic cells from patients could be used to generate iPSCs. After that, it can be differentiated into neural precursor cells for transplantation. This means that tissue rejections will never again an issue and there will be negligible ethical problems. Also, it can ameliorate the modeling of neurodegenerative diseases like AD, because iPSCs could differentiate into neurons, having the inimitable genetic phenotype of the patient. Thus, stem cell-derived neuronal cells create a cellular model which offers the closest relation to the sporadic form of the AD disease and expectantly translated into human studies to find a cure for the AD.

Table 9.1 Outline of studies of stem cell therapy on Alzheimer's-diseased animal models

S. no.	Models for study	Type of stem cell transplanted	Site of administration of stem cell	Outcome of the study	References
1.	DS model mice (Ts65Dn)	Murine NSCs	Hippocampus	Reduction of tau-positive clusters in trisomic and disomic mice	Kern et al. (2011)
2.	Ibotenic acid-induced NBM lesion mice	Murine ESCs and ESC-derived NSCs	Frontal association cortex and barrel field of S1 cortex	NPC restored memory, ES lowers working memory and induced massive teratoma formation	Wang et al. (2006)
3.	Acute Aβ-induced model mice	Murine BM-MSCs	Hippocampus (dentate gyrus)	BM-MSCs enhanced microglial activation. Lowers Aβ deposits of acutely induced AD mice	Lee et al. (2009)
4.	APP/PS1 transgenic mice	BM-MSC	Intra-hippocampus	Reduction of senile plaques.Significant increased DeltaNp73 protein expression.	Wen et al. (2011)
5.	Triple transgenic AD model mice (3 X Tg-AD)	Murine NSCs	Hippocampus	NSC transplantation rescued learning and memory loss. No change in Aβ, tau pathology but elevated synaptic density in mice's hippocampus	Blurton-Jones et al. (2009)
6.	Transgenic AD model mice (Tg2576)	Human ASCs	1. Intravenous 2. Hippocampus (bilateral dentate gyrus)	Rescued memory impairment and recover spatial learning; diminish amyloid plaque formation, upregulated interleukin-10 and neurotrophic factors in the brain of Tg2576 mice	Kim et al. (2012a, b)
7.	Rat Fimbria-Fornix transaction	Murine NSCs and NSC-derived glial cells	Basal forebrain	Recovered memory and learning; number of p75NGFR-positive neurons increased	Xuan et al. (2009)
8.	1. Aged rats (30 months) 2. Ibotenic acid-induced NBM lesion rats	Murine BM-MSC	Hippocampus (CA-1 region)	1. Aged rats learn quickly 2. Ibo-induced memory impairment	Babaei et al. (2012)

(continued)

Table 9.1 (continued)

S. no.	Models for study	Type of stem cell transplanted	Site of administration of stem cell	Outcome of the study	References
9.	1. Matured rats (6 months) 2. Aged rats (24 months) – memory impaired and unimpaired	Human NSCs	Right lateral ventricle	Cognitive function gets better in immature and aged memory-impaired groups. Morphologically functional hNSC-derived cells were found in the hippocampus and cortex	Qu et al. (2001)
10.	Rat hippocampus l Aβ injection	EPI-NSCs	Hippocampus	Improvement in cognitive tasks increased neuron number and differentiation into a different cell type	Esmaeilzade et al. (2012)
11.	APP and presenilin-1 (PSEN-1) double-transgenic mice	Human UC-MSCs	Hippocampus	Spatial learning and memory improved. Reduction of Aβ load and tau hyperphosphorylation, proinflammatory cytokine release from microglia inhibited	Lee et al. (2012)
12.	AF64A cholinotoxin injection in rats impaired	Human NSCs	Right lateral ventricle	Rats receiving NSCs over-expressing ChAT showed a full resurgence in learning and memory functions, whereas those receiving NSCs only remained a memory	Park et al. (2012)

Table 9.2 List of some main clinical trials of stem cell therapies for Alzheimer's disease

Trial number	Title	Type of stem cells and administration	Status (start-end date)	Sponsor	Location
NCT02833792	A Phase IIa Study of Allogeneic Human Mesenchymal Stem Cells in Subjects with Mild to Moderate Dementia Due to Alzheimer's Disease	Human adult ischemia-tolerant MSCs via intravenous administration	June 2016–June 2020	Stemedica Cell Technologies, Inc., United States	United States
NCT01297218	Open-Label, Single-Center, Phase 1 Clinical Trial to Evaluate the Safety and the Efficacy of NEUROTSTEM®-AD in Patients with Dementia of the Alzheimer's Type	Human umbilical cord blood-MSC via intravenous administration	February 2011–April 2012	Medipost Co. Ltd.	Korea
NCT02054208	A Phase I, Prospective, Randomized, Double-Blinded, Placebo-Controlled Trial to Evaluate the Safety and Potential Efficacy of Longeveron Allogeneic Human Mesenchymal Stem Cell (LMSCs) Infusion Versus Placebo in Patients With Alzheimer's Disease	Longeveron MSCs (high dose or low dose) via peripheral intravenous infusion	August 2016–August 2018	Longeveron LLC, United States	United States
NCT01547689	Open-Label, Single-Center, Self-Control, Phase I/II Clinical Trial to Evaluate the Safety and the Efficacy of UC-MSC in Patients with Alzheimer's Disease	Human umbilical cord blood-MSC via intravenous administration	March 2012–December 2016	Affiliated Hospital to Academy of Military Medical Sciences Peking University Third Hospital	China
NCT02899091	A Randomized, Double-Blind, Placebo-Controlled, Phase I/IIa Clinical Trial for Evaluation of Safety and Potential Therapeutic Effect After Transplantation of CB-AC-02 in Patients with Alzheimer's Disease	CB-AC-02 (placenta-derived mesenchymal stem cells) via injection	September 2016 (not yet recruiting)	CHABiotech CO., Ltd	Korea

(continued)

Table 9.2 (continued)

Trial number	Title	Type of stem cells and administration	Status (start-end date)	Sponsor	Location
NCT01617577	Efficacy and Safety of Filgrastim as a Pro-cognitive Agent in Alzheimer's Disease	Subcutaneous filgrastim (G-CSF)	June 2012–December 2012	University of South Florida	United States, Florida
NCT02912169	An Open-Label, Non-randomized, Multi-center Study to Assess the Safety and Effects of Autologous Adipose-Derived Stromal Vascular Fraction (AD-SVF) Cells Delivered Intravenous (IV) and Intranasal in Patients with Alzheimer's Disease	Autologous adipose-derived stromal vascular fraction (AD-SVF) cells via intravenous (IV) and intranasal administration	September 2016–November 2017	Ageless Regenerative Institute, United States	United States, Florida
NCT02600130	A Phase I Prospective, Randomized, Double-Blinded, Placebo-Controlled Trial to Evaluate the Safety and Potential Efficacy of Longeveron Allogeneic Human Mesenchymal Stem Cell (LMSCs) Infusion Versus Placebo in Patients With Alzheimer's Disease	Longeveron MSC via peripheral intravenous	August 2016–October 2019	Longeveron LLC	United States, Florida
NCT03297177	Use of Autologous Stem Cell Use in Neurological Non-neoplastic Disorders and Disease	Autologous stem/stromal cells derived from subdermal fat deposit via the intravenous parenteral route	January 2020–January 2023	Healeon Medical Inc., United States	United States, Massachusetts

Acknowledgment This work is supported by SERB (EEQ/2018/000114) grant.

Disclosure of Potential Conflicts of Interest The author(s) declare no potential conflicts of interest with respect to the research, authorship, and/or publication of this article.

References

Aggarwal S, Pittenger MF (2005) Human mesenchymal stem cells modulate allogeneic immune cell responses. Blood 105(4):1815–1822. https://doi.org/10.1182/blood-2004-04-1559

Allen S, Watson J, Dawbarn D et al (2011) The neurotrophins and their role in Alzheimer's disease. Curr Neuropharmacol 9(4):559–573. https://doi.org/10.2174/157015911798376190

Araki R, Uda M, Hoki Y et al (2013) Negligible immunogenicity of terminally differentiated cells derived from induced pluripotent or embryonic stem cells. Nature 494(7435):100. https://doi.org/10.1371/journal.pone.0069617

ARSI. Alzheimer's and Related Disorders Society of India (ARDSI) (2010)

Babaei P, Soltani Tehrani B, Alizadeh A (2012) Transplanted bone marrow mesenchymal stem cells improve memory in rat models of Alzheimer's disease. Stem Cells Int 2012:369417. https://doi.org/10.1155/2012/369417

Blurton-Jones M, Kitazawa M, Martinez-Coria H et al (2009) Neural stem cells improve cognition via BDNF in a transgenic model of Alzheimer disease. Proc Natl Acad Sci 106(32):13594–13599. https://doi.org/10.1073/pnas.0901402106

Brazelton TR, Rossi FM, Keshet GI et al (2000) From marrow to brain: expression of neuronal phenotypes in adult mice. Science 290(5497):1775–1779. https://doi.org/10.1126/science.290.5497.1775

Chen J, Tang YX, Liu YM et al (2012) Transplantation of adipose-derived stem cells is associated with neural differentiation and functional improvement in a rat model of intracerebral hemorrhage. CNS Neurosci Ther 18(10):847–854. https://doi.org/10.1111/j.1755-5949.2012.00382.x

Choi SH, Kim YH, Hebisch M et al (2014a) A three-dimensional human neural cell culture model of Alzheimer's disease. Nature 515(7526):274

Choi SS, Lee SR, Kim SU et al (2014b) Alzheimer's disease and stem cell therapy. Exp Neurobiol 23(1):45–52. https://doi.org/10.5607/en.2014.23.1.45

Cooper O, Hargus G, Deleidi M et al (2010) Differentiation of human ES and Parkinson's disease iPS cells into ventral midbrain dopaminergic neurons requires a high activity form of SHH, FGF8a and specific regionalization by retinoic acid. Mol Cell Neurosci 45(3):258–266. https://doi.org/10.1016/j.mcn.2010.06.017

Danysz W, Parsons CG, MÖbius HJ et al (2000) Neuroprotective and symptomatological action of memantine relevant for Alzheimer's disease – a unified glutamatergic hypothesis on the mechanism of action. Neurotox Res 2(2–3):85–97

Duncan T, Valenzuela M (2017) Alzheimer's disease, dementia, and stem cell therapy. Stem Cell Res Ther 8(1):111. https://doi.org/10.1186/s13287-017-0567-5

Eckman CB, Eckman EA (2007) An update on the amyloid hypothesis. Neurol Clin 25(3):669–682. https://doi.org/10.1016/j.ncl.2007.03.007

Esmaeilzade B, Nobakht M, Joghataei MT et al (2012) Delivery of epidermal neural crest stem cells (EPI-NCSC) to hippocamp in Alzheimer's disease rat model. Iran Biomed J 16(1):1. https://doi.org/10.6091/IBJ.1029.2012

Genin E, Hannequin D, Wallon D et al (2011) APOE and Alzheimer's disease: a major gene with semi-dominant inheritance. Mol Psychiatry 16(9):903. https://doi.org/10.1038/mp.2011.52

Gutiérrez-Fernández M, Rodríguez-Frutos B, Ramos-Cejudo J et al (2013) Effects of intravenous administration of allogenic bone marrow-and adipose tissue-derived mesenchymal stem cells on functional recovery and brain repair markers in experimental ischemic stroke. Stem Cell Res Ther 4(1):11. https://doi.org/10.1186/scrt159

Holtman IR, Raj DD, Miller JA et al (2015) Induction of a common microglia gene expression signature by aging and neurodegenerative conditions: a co-expression meta-analysis. Acta Neuropathol Commun 3(1):31. https://doi.org/10.1186/s40478-015-0203-5

Huang Y, Mucke L (2012) Alzheimer mechanisms and therapeutic strategies. Cell 148(6):1204–1222. https://doi.org/10.1016/j.cell.2012.02.040

Imbimbo BP, Lombard J, Pomara N et al (2005) Pathophysiology of Alzheimer's disease. Neuroimag Clin 15(4):727–753. https://doi.org/10.1016/j.nic.2005.09.009

Israel MA, Yuan SH, Bardy C et al (2012) Probing sporadic and familial Alzheimer's disease using induced pluripotent stem cells. Nature 482(7384):216

Kang JM, Yeon BK, Cho SJ et al (2016) Stem cell therapy for Alzheimer's disease: A review of recent clinical trials. J Alzheimers Dis 54(3):879–889. https://doi.org/10.3233/JAD-160406

Kern DS, Maclean KN, Jiang H et al (2011) Neural stem cells reduce hippocampal tau and reelin accumulation in aged Ts65Dn Down syndrome mice. Cell Transplant 20(3):371–379

Kim JY, Kim DH, Kim JH et al (2012a) Soluble intracellular adhesion molecule-1 secreted by human umbilical cord blood-derived mesenchymal stem cell reduces amyloid-β plaques. Cell Death Differ 19(4):680. https://doi.org/10.1038/cdd.2011.140

Kim S, Chang KA, Park HG et al (2012b) The preventive and therapeutic effects of intravenous human adipose-derived stem cells in Alzheimer's disease mice. PLoS One 7(9):e45757

Kooreman NG, Wu JC (2010) Tumorigenicity of pluripotent stem cells: biological insights from molecular imaging. J R Soc Interface 7(Suppl 6):S753–S763. https://doi.org/10.1098/rsif.2010.0353.focus

Krencik R, Weick JP, Liu Y et al (2011) Specification of transplantable astroglial subtypes from human pluripotent stem cells. Nat Biotechnol 29(6):528. https://doi.org/10.1038/nbt.1877

Lai MK, Tsang SW, Alder JT et al (2005) Loss of serotonin 5-HT 2A receptors in the postmortem temporal cortex correlates with rate of cognitive decline in Alzheimer's disease. Psychopharmacology 179(3):673–677. https://doi.org/10.1007/s00213-004-2077-2

Lee HJ, Lee JK, Lee H et al (2012) Human umbilical cord blood-derived mesenchymal stem cells improve neuropathology and cognitive impairment in an Alzheimer's disease mouse model through modulation of neuroinflammation. Neurobiol Aging 33(3):588–602. https://doi.org/10.1016/j.neurobiolaging.2010.03.024

Lee JK, Jin HK, Bae JS (2009) Bone marrow-derived mesenchymal stem cells reduce brain amyloid-β deposition and accelerate the activation of microglia in an acutely induced Alzheimer's disease mouse model. Neurosci Lett 450(2):136–141. https://doi.org/10.1016/j.neulet.2008.11.059

Lin HAI, Bhatia R, Lal R et al (2001) Amyloid β protein forms ion channels: implications for Alzheimer's disease pathophysiology. FASEB J 15(13):2433–2444. https://doi.org/10.1096/fj.01-0377com

Maccioni RB, Farías G, Morales I, Navarrete L et al (2010) The revitalized tau hypothesis on Alzheimer's disease. Arch Med Res 41(3):226–231. https://doi.org/10.1016/j.arcmed.2010.03.007

Malmersjö S, Liste I, Dyachok O et al (2009) Ca2+ and cAMP signaling in human embryonic stem cell–derived dopamine neurons. Stem Cells Dev 19(9):1355–1364. https://doi.org/10.1089/scd.2009.0436

Marchetti C, Marie H (2011) Hippocampal synaptic plasticity in Alzheimer's disease: what have we learned so far from transgenic models? Rev Neurosci 22(4):373–402. https://doi.org/10.1515/RNS.2011.035

Mezey E, Chandross KJ, Harta G et al (2000) Turning blood into brain: cells bearing neuronal antigens generated in vivo from bone marrow. Science 290(5497):1779–1782. https://doi.org/10.1126/science.290.5497.1779

Mössner R, Schmitt A, Syagailo Y et al (2000) The serotonin transporter in Alzheimer's and Parkinson's disease. In: Advances in research on neurodegeneration. Springer, Vienna, pp 345–350

Mu Y, Gage FH (2011) Adult hippocampal neurogenesis and its role in Alzheimer's disease. Mol Neurodegener 6(1):85. https://doi.org/10.1186/1750-1326-6-85

Nori S, Okada Y, Yasuda A et al (2011) Grafted human-induced pluripotent stem-cell–derived neurospheres promote motor functional recovery after spinal cord injury in mice. Proc Natl Acad Sci 108(40):16825–16830. https://doi.org/10.1073/pnas.1108077108

Oh SH, Kim HN, Park HJ et al (2015) Mesenchymal stem cells increase hippocampal neurogenesis and neuronal differentiation by enhancing the Wnt signaling pathway in an Alzheimer's disease model. Cell Transplant 24(6):1097–1109

Park D, Joo SS, Kim TK et al (2012) Human neural stem cells overexpressing choline acetyltransferase restore cognitive function of kainic acid-induced learning and memory deficit animals. https://doi.org/10.3727/096368911X586765

Pearl JI, Kean LS, Davis MM et al (2012) Pluripotent stem cells: immune to the immune system? Sci Transl Med 4(164):164ps25. https://doi.org/10.1126/scitranslmed.3005090

Pen AE, Jensen UB (2017) Current status of treating neurodegenerative disease with induced pluripotent stem cells. Acta Neurol Scand 135(1):57–72. https://doi.org/10.1111/ane.12545

Peroni D, Scambi I, Pasini A et al (2008) Stem molecular signature of adipose-derived stromal cells. Exp Cell Res 314(3):603–615. https://doi.org/10.1016/j.yexcr.2007.10.007

Perry EK, Tomlinson BE, Blessed G et al (1978) Correlation of cholinergic abnormalities with senile plaques and mental test scores in senile dementia. Br Med J 2(6150):1457–1459

Persson T, Popescu BO, Cedazo-Minguez A et al (2014) Oxidative stress in Alzheimer's disease: why did antioxidant therapy fail? Oxidative Med Cell Longev. https://doi.org/10.1155/2014/427318

Prasher VP, Farrer MJ, Kessling AM et al (1998) Molecular mapping of Alzheimer-type dementia in Down's syndrome. Ann Neurol 43(3):380–383. https://doi.org/10.1002/ana.410430316

Qu T, Brannen CL, Kim HM et al (2001) Human neural stem cells improve cognitive function of aged brain. Neuroreport 12(6):1127–1132

Ra JC, Shin IS, Kim SH et al (2011) Safety of intravenous infusion of human adipose tissue-derived mesenchymal stem cells in animals and humans. Stem Cells Dev 20(8):1297–1308. https://doi.org/10.1089/scd.2010.0466

Ramasamy R, Fazekasova H, Lam EWF et al (2007) Mesenchymal stem cells inhibit dendritic cell differentiation and function by preventing entry into the cell cycle. Transplantation 83(1):71–76. https://doi.org/10.1097/01.tp.0000244572.24780.54

Reiss AB (2005) Cholesterol and apolipoprotein E in Alzheimer's disease. Am J Alzheimers Dis Other Dement 20(2):91–96. https://doi.org/10.1177/153331750502000208

Rosen KM, Moussa CEH, Lee HK et al (2010) Parkin reverses intracellular β-amyloid accumulation and its negative effects on proteasome function. J Neurosci Res 88(1):167–178. https://doi.org/10.1002/jnr.22178

Selkoe DJ (2001) Alzheimer's disease: genes, proteins, and therapy. Physiol Rev 81(2):741–766. https://doi.org/10.1152/physrev.2001.81.2.741

Selkoe DJ (2002) Alzheimer's disease is a synaptic failure. Science 298(5594):789–791. https://doi.org/10.1126/science.1074069

Sotiropoulou PA, Perez SA, Gritzapis AD et al (2006) Interactions between human mesenchymal stem cells and natural killer cells. Stem Cells 24(1):74–85. https://doi.org/10.1634/stemcells.2004-0359

Sproul AA, Jacob S, Pre D et al (2014) Characterization and molecular profiling of PSEN1 familial Alzheimer's disease iPSC-derived neural progenitors. PLoS One 9(1):e84547. https://doi.org/10.1371/journal.pone.0084547

Swerdlow RH (2007) Pathogenesis of Alzheimer's disease. Clin Interv Aging 2(3):347

Tabassum N, Verma V, Yadav CB et al (2017) Tissue engineering in regenerative medicine. In: Stem cells from culture dish to clinic. Nova Science Publishers Inc, New York, pp 17–32

Takahashi K, Yamanaka S (2006) Induction of pluripotent stem cells from mouse embryonic and adult fibroblast cultures by defined factors. Cell 126(4):663–676. https://doi.org/10.1016/j.cell.2006.07.024

Takamatsu K, Ikeda T, Haruta M et al (2014) Degradation of amyloid beta by human induced pluripotent stem cell-derived macrophages expressing Neprilysin-2. Stem Cell Res 13(3):442–453. https://doi.org/10.1016/j.scr.2014.10.001

Tang J, Xu H, Fan X et al (2008) Embryonic stem cell-derived neural precursor cells improve memory dysfunction in Aβ (1–40) injured rats. Neurosci Res 62(2):86–96. https://doi.org/10.1016/j.neures.2008.06.005

Taupin P (2006) Adult neural stem cells, neurogenic niches, and cellular therapy. Stem Cell Rev 2(3):213–219

Tolosa L, Pareja E, Gómez-Lechón MJ (2016) Clinical application of pluripotent stem cells: an alternative cell-based therapy for treating liver diseases? Transplantation 100(12):2548–2557. https://doi.org/10.1097/TP.0000000000001426

Wang Q, Matsumoto Y, Shindo T et al (2006) Neural stem cells transplantation in cortex in a mouse model of Alzheimer's disease. J Med Investig 53(1–2):61–69. https://doi.org/10.2152/jmi.53.61

Wen SR, Qi HP, Ren YJ et al (2011) Expression of deltaNp73 in hippocampus of APP/PS1 transgenic mice following GFP-BMSCs transplantation. Neurol Res 33(10):1109–1114. https://doi.org/10.1179/1743132811Y.0000000051

Wimo A, Winblad B, Jönsson L (2010) The worldwide societal costs of dementia: estimates for 2009. Alzheimers Dement 6(2):98–103. https://doi.org/10.1016/j.jalz.2010.01.010

Xuan AG, Luo M, Ji WD et al (2009) Effects of engrafted neural stem cells in Alzheimer's disease rats. Neurosci Lett 450(2):167–171. https://doi.org/10.1016/j.neulet.2008.12.001

Yagi T, Kosakai A, Ito D et al (2012) Establishment of induced pluripotent stem cells from centenarians for neurodegenerative disease research. PLoS One 7(7):e41572. https://doi.org/10.1371/journal.pone.0041572

Yang KL, Lee JT, Pang CY et al (2012) Human adipose-derived stem cells for the treatment of intracerebral hemorrhage in rats via femoral intravenous injection. Cell Mol Biol Lett 17(3):376. https://doi.org/10.2478/s11658-012-0016-5

Diet and Nutrition in Alzheimer's Disease and Healthy Aging

10

Poonam Sharma, Vivek Kumar Gaur, and Janmejai Kumar Srivastava

Abstract

Aging is a universally natural phenomenon which is associated with cognitive decline and several neurological disorders such as Alzheimer disease. As this phenomenon is inevitable, many factors affect the progression and development of age-associated cognitive decline. Among these factors, lifestyle pattern such as exercise and diet plays a major role in contributing to neurological fitness. In this chapter, we discuss the relationship and effects of food and nutrition over aging and Alzheimer disease. Mediterranean diet rich in antioxidants and bioactive compounds is most efficient in delaying the onset and progression of age-related neurological disorders. Flavonoids and polyphenols are the major antiaging food component which also serves as antioxidants. These effectively reduce the generation of stress-induced reactive oxygen species. Also, omega-3 fatty acid such as docosahexaenoic acid is an essential fatty acid whose supplementation in the diet improves mental health.

P. Sharma
Department of Bioengineering, Integral University, Lucknow, Uttar Pradesh, India

V. K. Gaur
Environmental Biotechnology Laboratory, Environmental Toxicology Group, CSIR-Indian Institute of Toxicology Research, Lucknow, Uttar Pradesh, India

Amity Institute of Biotechnology, Amity University, Lucknow, Uttar Pradesh, India

J. K. Srivastava (✉)
Amity Institute of Biotechnology, Amity University, Lucknow, Uttar Pradesh, India
e-mail: jksrivastava@lko.amity.edu

G. M. Ashraf, A. Alexiou (eds.), *Biological, Diagnostic and Therapeutic Advances in Alzheimer's Disease*, https://doi.org/10.1007/978-981-13-9636-6_10

10.1 Introduction

Aging is universally natural phenomenon that occurs in all living beings. However, successful aging includes low possibilities of diseases, high physical functioning, and cognitive capacity (Rowe and Kahn 1997). Aging brain provides a favorable microenvironment for the occurrence of neurodegenerative diseases and age-related disorders (Joseph et al. 2005). Since aging is the main risk factor, the increase in life expectancy is associated with an increased rate of neurodegenerative diseases like Parkinson's and Alzheimer's (Mandel et al. 2012). As evident from several studies, the diet and nutrition play a significant role in the development and progression of age-related disorders more importantly AD (Parkinson and Cicerale 2016; Taylor et al. 2017; Weiser et al. 2016; Zeng et al. 2017). Diets supplemented with phenolic-rich vegetables and fruits are reported to be effective in reversing the detrimental effect of aging caused on behavior and neuronal functioning (Lau et al. 2007). Phenolic compounds present in the foods confer anti-inflammatory and antioxidant property, thereby exhibiting their protective role (Rice-Evans and Miller 1996). Reactive oxygen species are a product of normal aerobic metabolism (Beckman and Ames 1998), and as reported, 2–5% of the consumed oxygen is reduced to generate free radical (Floyd and Hensley 2002). Some of these radicals escape elimination which gradually accumulates causing oxidative damage leading to increased oxidative stress in aged individuals (Berger 2005; Sohal and Weindruch 1996). This suggests the warranted role of antioxidants as a supplement in food.

In the year 2016, 44 million people were reported to be suffering from AD worldwide (Yusufov et al. 2017). As reported by Alzheimer's Association in the year 2015, the annual cost of AD globally was $605 billion, whereas approximately $226 billion is estimated in the United States (Yusufov et al. 2017). Family history, presence of apolipoprotein (ApoE4), and history of head injuries are the major risk factors of AD along with a dietary pattern (Yusufov et al. 2017), thus suggesting many claims including Mediterranean diet (Gardener et al. 2012), alcohol consumption (Piazza-Gardner et al. 2013), increased vitamins intake (Engelhart et al. 2002), and fat and carbohydrate consumption (Morris et al. 2003; Seneff et al. 2011).

10.2 Aging and Alzheimer

Aging is a process accompanied by diminished cognitive and motor performance. The psychological and cognitive health of older adults is a major concern for researchers and healthcare services. Cognitive decline in older adults is a major cause of AD in western countries (Blennow et al. 2006). AD is among the most prevalent neurodegenerative diseases whose pervasiveness increase exponentially after an age of 60 and a 47% prevalence rate had been observed for people above 80 years of age, with a doubling rate of 5 years (Preston 1986; Evans et al. 1989; Brayne et al. 1995). A pathophysiological and histological change that occurs due to aging of central nervous system also leads to compromised cognitive state (Farkas and Luiten, 2001). This cognitive failure may be due to cerebrovascular dearth or deranged microvascular integrity in the brain. This is marked by a reduced supply of blood to the brain, vasoconstriction, and a decrease in the perfusion rate of the brain, thereby reducing the cerebral glucose utilization (Farkas and Luiten 2001).

To combat this disease efficiently, it is necessary to distinguish between AD and normal aging, which may require an optimized neuropsychological assessment. AD is a disorder that can be treated, whereas no medical treatment benefits age-associated cognitive decline (Vandenberghe and Tournoy 2005; Kane et al. 2017). The transformation from normal aging to AD, however, passes through a transition phase referred to as mild cognitive impairment (MCI) (Petersen et al. 1999). Few aged people experience successful aging by maintaining their cognitive status at the time of senescence (Vandenberghe and Tournoy 2005). The episodic memory of most of the healthy persons above 50 years of age begins to decline (Small et al. 1999) affecting the familiarity judgment (Vandenberghe and Tournoy 2005). No verifiable effective treatment exists for MCI; therefore, the patients must undergo regular neuropsychological or clinical follow-up to ensure the timely diagnosis for the transition from MCI to AD (Petersen et al. 2001).

The keystone for the diagnosis of AD is episodic memory impairment. Additionally, to meet the criteria for probable AD, one other cognitive region must be distressed, such as semantic memory, word finding (Locascio et al. 1995), or executive functions (Baddeley et al. 2001), that will significantly affect auxiliary activities of daily living (Vandenberghe and Tournoy 2005). These can be diagnosed based on the scores obtained from classical tests, such as Mini-Mental State Examination (MMSE), animal verbal fluency (AVF) test, and clock drawing test (Petersen et al. 2001; Vandenberghe and Tournoy 2005). Studies performed on rat's prefrontal cortex and hippocampus do not provide any clear evidence of age-related neuronal damage (Rapp and Gallagher 1996), whereas, in humans, a reduction in synaptic density, dendritic spin, and dendritic segments of dorsolateral prefrontal cortex was observed with aging (Uylings and Brabander 2002). Aging is the prominent risk factor for AD, as over our life span, pernicious events accumulate which may gradually affect the brain, thereby increasing the possibility of neurodegenerative disorders. Also, aging process is itself coupled with detrimental changes that may increase the risk of AD (Vandenberghe and Tournoy 2005).

10.3 Diet and Aging

Diet plays a vital role in easing the complications of elderly age. During aging, volume of brain in terms of gray matter and white matter reduces and development of amyloid plaques, Lewy bodies, neurofibrillary tangles, synaptic dystrophy, and neuron loss takes places; hence, the burden of diseases increases (Dianne et al. 2011; Moore et al. 2018).

There are various diet patterns recommended for different populations, such as Dietary Approaches to Stop Hypertension Diet (DASH), The Dietary Guidelines for Americans, The Healthy Eating Plate, Mediterranean diet, ketogenic diet, etc.

10.3.1 Mediterranean Diet

In today's world, diet plays a major role in the well-being of an individual. It plays a significant role in reducing the risk factors such as neuronal disabilities and premature death. Sticking on recommended diet style can make positive changes in human health. During the 1960s, Mediterranean diet was a popular dietary pattern for population neighboring the Mediterranean Sea, involved in olive oil cultivation, so a generic term "Mediterranean diet" is used for it (Aridi et al. 2017; Willett et al. 1995). The existence of MedDiet is noticed because, during the 1960s, the rate of occurrence of certain cancer, coronary heart diseases, and certain diet-related chronic diseases was less reported in this region. This diet was first discovered by Ancel Keys in the 1960s as low saturated fat diet, enriched with vegetable oil (Davis et al. 2015). The following dietary pattern had been recommended:

(a) Olive oil as the main dietary fat source.
(b) Low or moderate amount of animal foods (dairy products, poultry and eggs, fishes, and red meat).
(c) Higher plant food intake (fruits, vegetables, whole cereals, legumes, nuts, and seeds).
(d) Moderate wine intake along with meals (Salas-Salvadó et al. 2011; Bonaccio et al. 2013).

These are the vital source of vitamins, polyphenols, minerals, and unsaturated fatty acids (Miranda et al. 2017). It is a dietary pattern in the research spotlight due to its multiple health benefits during last decade. MedDiet strengthens focus on unprocessed foods which fulfils the calorific and nutritional requirements for the well-being of an individual. In MedDiet, carbohydrate need is fulfilled primarily by beans/legumes, whole grains, fruits, and vegetables (excluding potatoes); it provides energy to body by metabolism. Unsaturated fat is major source of energy which comprises olive oil, avocados and nuts accounting for reduction of inflammatory responses by reducing cytokines production (Miranda et al. 2017; Estruch et al. 2013); omega-3 fatty acid is provided by flax seeds, nuts, and cold-water fishes. Animal products like dairy products (cheese, yoghurt), meat, poultry, and egg are rich sources of amino acids. MedDiet is also rich in various antioxidants, minerals, polyphenols, dietary fiber contents, bioactive elements, and added sugar accounting for 5–10% of daily calorific intake which makes it low glycemic index diet (Martinez-Gonzalez and Martin-Calvo 2016; Gardener and Caunca 2018). This dietary pattern is reported with strong evidence to promote beneficiary function in reduction and prevention of a number of lifestyle-related diseases such as cancer, type 2 diabetes mellitus, cardiovascular disease, AD, and obesity (Fig. 10.1).

10.3.1.1 Benefits of Mediterranean Diet

MedDiet comprises an array of nutritional components which a normal diet does not confer; it provides ample quantity of antioxidants and bioactive compounds with anti-inflammatory effects and low glycemic index, so it is recommended to

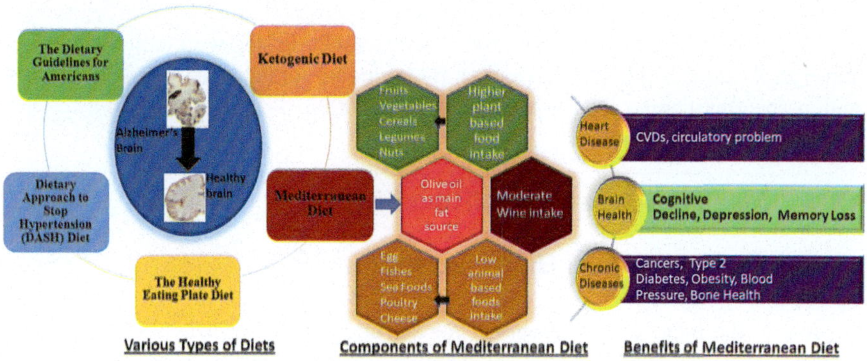

Fig. 10.1 Components of Mediterranean diet and its beneficial effects

people suffering from health ailments like obesity and reduces the risk of chronic diseases like CVD, type 2 diabetes, cognitive decline, and certain cancers. It is also reported to be helpful in body weight management and increases longevity (Estruch et al. 2013).

MedDiet contains extra virgin olive oil which is loaded with polyphenols like oleuropein aglycone and oleocanthal and has high oleic acid content. Oleocanthal is reported to attenuate amyloid-beta oligomer-mediated astrocytes inflammation and synaptic proteins while oleic acid poses antiatherogenic properties. Unsaturated fatty acids present in olive oil reduce inflammatory responses by inhibiting or lowering the synthesis of inflammatory cytokines (Miranda et al. 2017). Red wine consists of resveratrol, a polyphenol known for its antioxidants, cardioprotective effects, and antiapoptotic activity. Nuts like almonds, hazelnuts, and walnut are rich in polyphenols as well as essential vitamins, minerals, fibers, and MUFAs and PUFAs. Walnut consists of ellagitannins, a polyphenol reported to be antioxidant in nature. It regulates body lipid profile and shows antiatherogenic activities and reduces obesity (Castro-Barquero et al. 2018; Finicelli et al. 2018). Researchers have revealed that adherence to MedDiet can improve neurodegenerative disease condition or mental health of an individual. It can improve the condition of dementia, mild cognitive decline, Alzheimer's disease, etc. (Gardener and Caunca 2018; Aridi et al. 2017).

10.3.1.2 Combined Effect with Nutrients

Mediterranean diet is full of nutrients but still lacks few essential nutritional components such as calcium; enrichment of Mediterranean diet along with dairy products results in improvement in brain processing speed, tension, confusion, anger, total mood disturbance, and depression (Wade et al. 2018). Calcium is a vital component for brain functioning as learning and memory process of neurons is dependent on calcium signaling (Brini et al. 2014). This suggests that dairy product supplementation with MedDiet is beneficial for cognitive function and psychological well-being in the elderly population at risk of dementia, as it supplies a range of peptides; micronutrients and minerals like calcium, iodine, magnesium, potassium, zinc, and phosphorus; and an array of essential vitamins comprising riboflavin, vitamin A, B6, B12, etc. (Wade et al. 2017). Supplementing diet with antioxidants like resveratrol reduces plaque formation by reducing glutathione production in a region-specific manner (Karuppagounder et al. 2009).

10.4 Diet Constituents and Their Effects

10.4.1 Carbohydrate or High Calorific Intake

Glucose is the main source of energy for the brain, and it utilizes 25% of glucose consumed in the body (Sokoloff 1999). Consumption of glucose transiently increases the function of prefrontal cortex (Kumar et al. 2016). Also, an improvement in cognitive behavioral tasks was observed in aged rodent models after glucose injection (Gold, 2005). Stefanidis and Watt (2012) suggested that glucose consumption in chronic excess can cause high inflammation and reduced synaptic plasticity leading to cognitive deficiency. It was reported that with age, glucose consumption in neurons decreases which is one of the causes of cognitive fatigue commonly seen in aged individuals (Galeffi et al. 2015). This may be attributed to reduced insulin sensitivity of brain (Ryu et al. 2014).

It was reported that peripheral hyperglycemia and defective glucose metabolism increase the risk of AD. There is an increased risk of MCI progression to AD in an individual having high blood glucose (Crane et al. 2013) or who are suffering from type 2 diabetes (Huang et al. 2014; Matsuzaki et al. 2010). Taylor et al. (2017) concluded that a high glycemic diet can increase the risk of AD by increasing cerebral amyloid burden. Alternatively, caloric restriction (CR) causes a reduction in body fat and was reported to increase the life span (Heilbronn and Ravussin 2003). Several in-vivo studies showed that during aging, the CR reduced the formation of reactive oxygen species and activation of glial and thus decreases the oxidative stress and neuroinflammation (Lee et al. 2000; Morgan et al. 1999; Olgun et al. 2002).

10.4.2 Proteins

Proteinopathy causes AD and is characterized by aggregation of proteins, namely, β-amyloid and hyperphosphorylated Tau. Diet containing high amount of animal-derived proteins is correlated with increased oxidative stress, thereby increasing the risk of AD (Pistollato et al. 2016). A meta-analysis performed by Nakagawa et al. (2012) on 15 animals also showed that protein restriction is the most important caloric restriction that is responsible for extension of life span. NHANES data showed that diets containing >20% proteins are associated with increased risk of cancer and mortality in subjects aged <65 years (Levine et al. 2014; Le Couteur et al. 2016). Proteins majorly drive the intake of food, and diets having low protein content usually are associated with an increase in food intake and vice versa. This is also termed as protein leverage. Furthermore, the health outcomes of a diet having low protein content also depend on whether it had been replaced by carbohydrate or fats (Simpson and Raubenheimer 2005).

10.4.3 Lipids and Fatty Acids

Several nutritional factors are involved in the pathogenesis of AD, among which lipid plays a major role. The production of Aβ peptide can be prevented by maintaining the membrane lipid content, thus preventing the deleterious effect that can be caused by interaction of Aβ with synaptic membrane, thereby preventing neurodegeneration (Florent-Bechard et al. 2009). An ApoE protein encoded by apolipoprotein 34 allele (ApoE4) (its inheritance majorly contributes to sporadic AD) belongs to plasma lipid-binding proteins family (Corder et al. 1993; Deane et al. 2008). This protein is involved in cholesterol and triglycerides delivery and transport (Deane et al. 2008). During the early stages of AD, inflammation takes place which is coupled with increased levels of free fatty acids. This deleterious cascade is associated with phospholipases which in turn is involved in the synthesis of secondary messengers (Sastre et al. 2006). It was reported that sphingolipid metabolism is altered in AD brain. It was observed having elevated levels of sphingosine and ceramide, whereas sphingomyelin levels decrease (He et al. 2010). Neuronal death is also contributed by proapoptotic environment formed in AD brain that may be caused due to decreased sphingosine-1-phosphate and increased sphingosine and ceramides concentrations (Florent-Bechard et al. 2009). Docosahexaenoic acid (DHA) deficiency is often observed in elderly people; thus, it is an essential fatty acid that must be supplemented in diet. Several studies report that omega-3 fatty acids play an essential role in the management and pathology of AD; for details, refer to Sect. 10.4.3.3.

10.4.3.1 Virgin Olive Oil

Virgin olive oil (VOO) is the purer form of olive oil having less than 2% acidity, whereas extra virgin olive oil (EVOO) has not more than 0.8% acidity. VOO is rich in molecules exhibiting anti-inflammatory property namely polyphenols (Parkinson and Cicerale 2016). There are approximately 36 phenolic compounds present in VOO which contributes to its health-promoting qualities (Estruch et al. 2013, Parkinson and Cicerale 2016). They are present in the form of flavonoids, phenolic alcohols, phenolic acids, secoiridoids, lignans, and hydroxy-isocromans. Of these, secoiridoids constitute the major fraction and the smallest are phenolic acids (Parkinson and Cicerale 2016). It is evident from several studies that tyrosol and hydroxytyrosol, two most abundant phenolics present in VOO, absorb significantly in humans in a dose-dependent manner (Rodríguez-Morato et al. 2016, Parkinson and Cicerale 2016). VOO itself provides protection to breakdown of phenolics in the gastrointestinal tract before they are being absorbed (Tuck and Hayball 2002).

EVOO contains a variety of polyphenols essentially oleuropein aglycone (OLE) (Valls-Pedret et al. 2012; Pitozzi et al. 2010) exhibiting neuroprotective effect by reducing cognitive decline in elderly people (Berr et al. 2009). As evident by in vivo and in vitro studies, the EVOO decreases cognitive deterioration and advancement of Tau and Aβ pathology (Abuznait et al. 2013; Farr et al. 2012; Qosa et al. 2015); thus, consumption of EVOO over a long time promotes mental health and reduces the risk of AD (Berr et al. 2009; Qosa et al. 2015).

10.4.3.2 Coconut

Coconut, *Cocos nucifera*, belongs to Arecaceae family (Lopes and Larkins 1993) and is known for its medicinal and nutritional values and provides a large number of valuable products including vitamins, dietary fiber minerals, hormones, and phenolic compounds (Fernando et al. 2015). Coconut oil is reported to exhibit antioxidant and antiaging properties (Marina et al. 2009; Hanne et al. 2003). Cytokinins and phenolics present in coconut prevent the aggregation of Aβ peptide inhibiting the risk of AD pathogenesis (Fernando et al. 2015). Unlike other dietary fats, coconut oil is rich in saturated fats containing medium-chain fatty acids (MCFA) which are easily metabolized and utilized by the liver (Krishna et al. 2010; Bach and Babayan 1982; Chandrashekar et al. 2010). MCFA can be metabolized to ketones and induce metabolic ketosis which serves as a therapy for brain disorders such as neurodegeneration and epilepsy (Fernando et al. 2015); ketones, monocarboxylic acids, and lactate serve as alternate fuel to brain maintaining energy homeostasis (Hasselbalch et al. 1994; Page et al. 2009). Of these, ketones are largely used during glucose deficiency termed as ketosis (Morris 2005; Sumithran et al. 2013). Ketone bodies once absorbed can be converted to acetyl-CoA entering Krebs cycle and generating energy in mitochondria in form of ATP. Alternatively, in neurons, they can be used as precursors of acetylcholine (Hasselbalch et al. 1994).

10.4.3.3 Fishes and Omega-3 Fatty Acid

Among the essential components of nutrition, fish is considered an important nutritional constituent as it is rich in omega-3 fatty acids such as DHA (Weiser et al. 2016; Zeng et al. 2017). Approximately 60% of omega-3 polyunsaturated fatty acid (PUFA) in human brain is DHA. Since de novo synthesis of DHA in humans is not efficient, the majority of it is provided through dietary intake, predominantly from fish fats (Burdge 2004; Ouellet et al. 2009). DHA is a constituent of brain tissue membranes, and diets supplemented with high DHA were reported to efficiently reduce cognitive decline in aged rats (Weiser et al. 2016; Lim and Suzuki 2000; Calon and Cole 2007).

Protective effect of eicosapentaenoic acid (EPA) for the nervous system of humans was also reported (Wei et al. 2008). DHA is able to cross blood–brain barrier by diffusion and gets incorporated in the phospholipids of the membranes (Ouellet et al. 2009). In addition to cognitive decline, a corresponding decrease in the levels of blood and brain's DHA characterizes the neurodegenerative diseases and normal aging (McGahon et al. 1999; Afshordel et al. 2015; Belkouch et al. 2016).

10.4.4 Vitamins

Vitamins are essential micronutrients required for proper growth and development. Several studies have reported the neuroprotective role of specific vitamins. All known vitamins serve as cofactor or are involved in more than one key pathway associated with AD or MCI. Vitamin B serves as a cofactor in all the major pathways, and together with vitamin E, they exhibit protection against dementia. Douaud et al. (2013) reported that vitamin B complex supplementation slows the progression and onset of AD. Vitamin E is divided into two major groups, i.e., tocotrienols and tocopherols. In vitro studies suggested the effectiveness of tocotrienols in reducing oxidative stress thereby safeguarding neuronal cells from glutamate-induced cell death (Selvaraju et al. 2014) and cytotoxicity that leads to AD (Chin and Tay 2018). α-Tocopherol, a cytosolic antioxidant, inhibits changes in the brain of mouse models preventing AD (Nishida et al., 2006). The benefits and adverse association of vitamin C with cognitive function and AD had been suggested by several studies (Bowman 2012; Swaminathan and Jicha 2014). It protects the neurons from oxidative stress, reduces inflammation, modulates epigenetic function, and regulates neurotransmission. Vitamin C supplementation exhibits preventive effect against AD in addition to major depressive disorder, anxiety, and schizophrenia, whereas its deficiency may cause cognitive impairment (Han et al. 2018).

10.4.5 Minerals

Minerals are an essential component of diet which plays an important role in various physiological and developmental processes. A 17-year study performed on 1000 older peoples revealed the association of minerals with stroke, hypertension (Cherbuin et al. 2014), and dementia (Ozawa et al. 2012). They reported that the probability of developing dementia decreases with increased intake of potassium, magnesium, and calcium. Plasma magnesium level was reported to be lesser in peoples suffering from AD (Barbagallo et al. 2011). Positive correlation of plasma iron level was also reported with cognitive function in aged individual (Yavuz et al. 2012). Iron co-localizes with amyloid plaques leading to pathological process of their aggregation (Roberts et al. 2012; Cherbuin et al. 2014).

Zinc (Zn) serves as an essential component for cell survival participating in various molecular and biochemical processes. Its concentration in brain varies and is found responsible for memory, mood, learning, and cognition (Opoka et al. 2011; Pavlica and Gebhardt 2010). In the presence of increased concentration of Cu++ ions, the oxidized metal ions were reported to interact with amyloid-beta (Aβ) protein, leading to neuronal loss and amyloid neurotoxicity (Brewer 2012), thus contributing in Alzheimer's pathophysiology. A decrease in copper concentration was reported in brain of AD patients, whereas no increase in concentration of Zn was observed (Schrag et al. 2011). Furthermore, Alzheimer's patient was found to have low Zn and high Cu concentration in plasma and serum (Schrag et al. 2011; Bucossi et al. 2011). Contrastingly, Rodrigues et al. (2017) reported that in aged individuals, the erythrocyte and plasma concentrations of Cu, Zn, and Cu/Zn ratio are not related to AD, whereas a correlation of dementia severity is observed with the antioxidant activity of erythrocyte superoxide dismutase enzyme (Perrin et al. 1990).

10.5 Antiaging Components of Food: Polyphenols and Flavonoids

Polyphenols, secondary metabolites of plants, are the most inexhaustible source of antioxidants in diet. These are used as nutraceuticals that are efficient to counteract inflammation and stress-induced free radicals (Pandey and Rizvi 2009; Mandel et al. 2012). Epidemiological research and meta-analysis suggest that intake of diets rich in polyphenols serves several health benefits as antioxidants and imparts protection against cardiovascular diseases, viz., cancers, osteoporosis, neurodegenerative diseases, and diabetes (Pandey and Rizvi 2009). Flavonoids are polyphenolic compounds with phenolic structures which are found in grains, vegetables, flowers, bark, wine, tea, and fruits (Panche et al. 2016). Catechins are black and green flavonoids accounting for 30–40% of the dry weight of green tea leaves (Wang et al. 1994). They are reported to protect the aging brain, thereby reducing the risk of dementia and AD (Mandel et al. 2012).

Nonheme iron complexed with ferritin accumulates in particular region of the brain, such as the prefrontal cortex, motor cortex, sensory cortex, thalamus, and putamen during first 30 years of life (Zecca et al. 2004; Hallgren and Sourander 1958). During aging of brain, it is converted to hemosiderin, containing iron at higher reactivity thus causing an increased risk of oxidative stress (Crichton et al. 2002). Aging brain has increased monoamine oxidase activity generating more hydrogen peroxide (Fowler et al. 1980; Cohen 2000; Reinikainen et al. 1988), and this in turn reacts with nonheme iron generating Fe^{2+} viz. Fenton reaction. Fe^{2+} increases the aggregation of proteins such as beta-amyloid (Aβ) peptide and alpha-synuclein, and these proteins majorly contribute to AD (Mandel et al. 2012).

A number of antioxidants/iron chelators had been reported to possess neuroprotective activity (Mandel et al. 2012). Plant flavonoids possess strong transitional metal chelating property due to the presence of one or more hydroxyl (OH) group at different positions (Nanjo et al. 1996; Mandel et al. 2012; Perron et al. 2008). Epigallocatechin-3-gallate (EGCG), a green tea polyphenol, inhibits above 90% of the DNA damage caused by iron. It was reported that catechins were able to scavenge the hydrogen-donating radical of reactive oxygen species (ROS) with the ability to chelate metal ions, thus inhibiting the genesis of iron free radicals such as Fe^{2+} in in vivo and in vitro studies (Salah et al. 1995; Nanjo et al. 1996). Polyphenols were also reported to inhibit lipid peroxidation induced by iron; of these polyphenols, most effective is EGCG because of the presence of trihydroxyl group and gallate moiety at 3′ position in B and C ring, respectively (Nanjo et al. 1996).

Several studies had reported that polyphenols not only affect any single specific pathway say for inhibiting Aβ aggregation, but it modulates different tissue and cellular processes, which together contributes to their neuroprotective action exhibiting a reduction in cognitive impairment (Yang et al. 2010). The prime effect of polyphenols is to reduce the ROS produced by neuroinflammation and mitochondria, also inhibiting the toxic aggregates of Aβ protein. These positive effects of polyphenols are attributed to their property to induce autophagy (Stefani and Rigacci 2014). Alterations in autophagy are considered crucial for the pathology of AD and had been reported as a promising target for treatment and pharmacology (Cordero et al. 2018). Polyphenols can inflect autophagy by multiple pathways (Hasima and Ozpolat 2014), including Beclin-1-independent (noncanonical) and Beclin-1-dependent (canonical) signaling pathways. A polyphenol, rottlerin isolated from Asian *Mallotus philippensis* tree induces autophagy via its antioxidant property, blocking the production of free radicals and triggering NF-κB inhibition and AMPK activation. This leads to reduced ATP concentration, thus activating regulatory mechanisms, i.e., autophagy (Maioli et al. 2010). Polyphenols such as resveratrol, quercetin, and genistein exhibit their effect by activating sirtuins (Cordero et al. 2018). These in turn activate autophagy either indirectly by regulating a transcription factor FOXO3a or directly by causing deacetylation of autophagic gene products such as Atg 5, 7, and 8 (Hanna et al. 2016).

10.6 Fruits and Vegetables as Antiaging Agents

Aging is a natural phenomenon cum combined effect of various oxidative stress and neuroinflammation. Upon aging, potential of vital organs (including neuronal cells) diminishes due to apoptosis (Winter et al. 2018). The incidences of neurodegenerative diseases are increasing rapidly as a report of the World Health Organization (WHO) states that by 2040 neurodegenerative diseases would be the second-leading cause of death after cardiovascular diseases defeating cancer (Tejada et al. 2017). Consumption of antioxidant-rich fruits and vegetables lowers the incidences of chronic diseases (Dastmalchi et al. 2011). Aging can also be delayed by supplementing fruits and vegetables in diet. They act like reservoir of bioactive agents, antioxidants, organic acids, and phenolic acids, thereby exhibiting antiaging food properties. Different types of foods contain different active ingredients which exhibit various health benefits (Table 10.1).

10.6.1 Berries

Fruits are categorized into pome, drupe, berries, and aggregate fruit. Berries exhibit health-promoting properties due to containment of nutritive, non-nutritive, and bioactive compounds, where nutritive compounds include sugars, vitamins, carotenoids, minerals, and essential oils, while bioactive compounds comprise flavonoids, anthocyanins, stilbenes, tannins, phenolics, and phenolic acids (Szajdek and Borowska 2008). These bioactive compounds have potent anticancer, antimicrobial, antioxidants, anti-inflammatory, anti-neurodegenerative, and immune-booster properties (Sara et al. 2018; Kelly et al. 2017; Nile and Park 2014; Costa et al. 2013);

Table 10.1 Active ingredients present in natural food and their health benefits

Natural foods	Active ingredients	Health benefits	References
Cranberries	Anthocyanins, flavonols, flavan-3-ols, antioxidants, proanthocyanidins, and the phenolic acid derivatives	CVDs, various infections – urinary tract, dental health, and *Helicobacter pylori*-induced stomach ulcers, and cancers	Lorenzo et al. (2018) and Cote et al. 2010
Blueberries	Antioxidants, anthocyanins, the flavonols, the flavan-3-ols, proanthocyanidins, phenolic acid derivatives, and cinnamic acid	Anticancerous, prevents CVDs, antimutagenic, antitumor, immunomodulatory, neuroprotective, anti-inflammatory, anti-obesity	Sara et al. (2018)
Strawberries	Anthocyanins, polyphenols, antioxidants, organic acids	Anti-inflammatory, anti-obesity, antioxidant	Enomoto et al. (2018) and Gasparrini et al. (2018)
Grapes and grape products	Anthocyanins, flavan-3-ol, proanthocyanidins, phenolic acids, gallic acid, flavonols hydroxycinnamates, resveratrol	Prevents chronic–degenerative diseases – CVDs, cancer, neurodegenerative disease, cognitive decline, Alzheimer's disease, diabetes	Vislocky and Fernandez (2010)
Pomegranate	Antioxidants, anthocyanins – cyanidin, delphinidin, pelargonidin, flavonols – punicalagins, catechins, epicatechin, gallocatechin, epigallocatechin, tannins – ellagitannins and ellagic acid	Antioxidative, antiaging, anti-inflammatory, reduces oxidative stress and blood pressure, antiatherosclerotic, CVDs, vasculoprotective	Wang et al. (2018), Essa et al. (2015), and Stowe (2011)
Apple	Antioxidants, polyphenols, phenolic acids – phloretin and phloridzin	Cognitive performance, oxidative damage, neuroprotective, improves synaptic signaling, Alzheimer's disease, chemopreventive/ chemotherapeutic agent	Tu et al. (2017) and Essa et al. (2012)
Fig	Polyphenols, minerals, vitamins, dietary fibers, antioxidants	CVDs, antioxidative, anticancer, antituberculosis, antidiabetic, immunomodulatory	Essa et al. (2015)

(continued)

Table 10.1 (continued)

Natural foods	Active ingredients	Health benefits	References
Date palm	Antioxidants – anthocyanins, ferulic acid, protocatechuic acid, caffeic acid, flavonoids, phenolic acids, polyphenols, sterols, carotenoids	Neuroprotective, antioxidative, anti-inflammatory, anticancer, immunomodulatory, reduces oxidative stress, antidiabetic	El-Far et al. (2019), Subash et al. (2015), and Essa et al. (2015)
Avocados	MUFA, PUFA, dietary fiber, folate, potassium, essential micronutrients, bioactive phytochemicals	Antioxidants, anti-obesity, reduces CVDs, LDL cholesterols	Mahmassani et al. (2018) and Park et al. (2018)
Drum sticks	Phenolics, flavonoids – quercetin, kaempferol, apigenin	Anticancer, anti-inflammatory, immunomodulatory, hypoglycemic, antioxidative	Saucedo-Pompa et al. (2018)
Turmeric	Polyphenols – curcumin, antioxidants – bisdemethoxycurcumin, demethoxycurcumin	Antioxidant and anti-inflammatory, antihyperlipidemia, cancer, Alzheimer's disease, antiaging	Hewlings and Kalman (2017) and Sikora et al. (2010)
Spinach	Antioxidants, carotenoids, flavonoids, phenolic compounds, vitamins, glycolipids, thylakoids	Alzheimer's disease, cognitive decline, brain healing, memory restoring, reduces CVDs	Jiraungkoorskul (2016) and Volpe (2013)
Tomato	Carotenoids:alpha, β-carotene, lutein, lycopene, vitamins A, B, C, phytosterols	Antioxidative, cancers, diabetes, CVDs	Bhowmik et al. (2012)
Ginger	Gingerol, antioxidants, tannins, vitamins A, E, ascorbic acid, carotenoids – β-carotene lycopene, lutein, bioflavonoids – genistein and quercetin, tannins – catechins	Anticancerous, oxidative stress, diabetes, anti-inflammatory, antihyperlipidemia, reduces platelet aggregation, hypertension	De Lima et al. (2018) and Singletary (2010)
Garlic	Allicin, thiosulfinates, thiacremonon	Antioxidant, immunomodulatory, antidiabetic, antithrombotic, anti-inflammatory, anticancer, CVDs	Tsai et al. (2012)
Green tea	Polyphenols-methylated catechin, epigallocatechin gallate	Cognitive impairment, CVDs preventions, anti-allergy, anti-obesity	Zhang et al. (2018)
Ginkgo biloba	Ginkogalic acid, terpene trilactones – bilobalide, ginkgolide, flavonoids	Cognitive improvement, vasodilation, cerebral vascularization	Olivera-Pueyo and Pelegrín-Valero (2017) and Ude et al. (2013)

inclusion of berries or berry product in diet may reduce chronic inflammation, improve plasma lipid profile, and thereby support cardiovascular health. Berry seed oil is proven to reduce prevalence of atopic death in early stage of life (Yang and Kortesniemi 2015).

10.6.1.1 Strawberry

In MedDiet, strawberry (*Arbutus unedo*) is a common and important fruit. It is a rich source of antioxidant polyphenols, such as ellagitannins, anthocyanins, catechin, quercetin and kaempferol, ellagic acid, gallic acid, tannins, vitamin C, vitamin E, and carotenoids (Essa et al. 2012; Giampieri et al. 2012). It bears potential to combat inflammation disorders and oxidative stress, reduce obesity-related disorders and heart disease risk, and protect against various types of cancer (Afrin et al. 2016).

Ellagitannins are phenolic phytochemical exhibiting neuroprotective properties which protect against neurodegeneration. It is commonly present in many fruits like pomegranate and shows anti-inflammatory, antioxidants, and antiapoptotic abilities to provide therapeutic benefits for treatment of various neuronal disabilities (DaSilva et al. 2017; Tejada et al. 2017; González-Trujano et al. 2015). Anthocyanins are flavonoid with antioxidant properties capable of scavenging hydroxyl and peroxyl radicals responsible for aging and onset of lifestyle-related diseases. Strawberries improve neuronal signaling by reducing oxidative stress and inflammation (Poulose et al. 2014)

10.6.1.2 Blueberries

Blueberry (*Vaccinium angustifolium*) or whortleberry is recognized as "Superfood" due to the presence of bioactive constituents like chlorogenic acid, flavonoids, pterostilbene, resveratrol, alpha-linolenic acid, vitamins, and cytoprotective antioxidant polyphenols, i.e., anthocyanins. Anthocyanins are glycosidic and acyl glycosidic forms of anthocyanidins, which are polyhydroxy and polymethoxy derivatives of 2-phenylbenzopyrilium (flavilium salts) conjugated with sugars (Kelly et al. 2017; Routray and Orsat 2011; Chen et al. 2010; Goswami and Das 2009). These anthocyanins possess potential to restore memory, attenuate age-related brain aging, and impair cognitive deficits in the brain in order to combat reactive oxygen species (Bowtell et al. 2017; Dastmalchi et al. 2011; Kalt et al. 2007). Anthocyanins are proven to be anticancerous in breast cancer, have anti-invasive and chemo-inhibitor properties, as well as inhibit proteasome activities, thus furnishing health benefits (Routray and Orsat 2011; Faria et al. 2010). Anthocyanins can improve ocular health in terms of visual function by increasing rhodopsin regeneration and prevent age-related chronic diseases, such as hypertension, cancer, diabetes, urinary tract infection, obesity, hyperlipidemia, neurodegeneration, and osteoporosis through its apoptosis, antiangiogenesis, anti-inflammation, and antioxidant effects (Chen et al. 2010).

Blueberries were reported to possess phytochemicals with the ability to reduce age-associated oxidative stress and inflammation by altering neuronal signaling involved in communication, calcium buffering, neuroprotective stress shock proteins, plasticity, stress signaling pathways, and obesity (Lewis et al. 2018; McGill et al. 2018; Shukitt-Hale et al. 2008). Blueberries enhance memory and motor performance and reverse deleterious effect of aging on motor behavior (Poulose et al. 2014; Joseph et al. 2003; Tan et al. 2017; Essa et al. 2012). In aging population, blueberries can improve cognition by mediating hippocampal plasticity (Tan et al. 2017).

10.6.2 Spinach

Spinach, *Spinacia oleracea* (SO), is alluded as "power food" due to its 20 diverse nutritional constituents, which include bioactive phytochemicals (flavonoids comprising glycosides, glycolipids, luteolin, kaempferol, apigenin, quercetin, and coumaric acid derivatives) and essential nutrients such as proteins, vitamins (vitamin A, ascorbic acid, β-carotene, carotenoids), and minerals (calcium, magnesium, potassium, folate, folic acid, nitrate, oxalate) that promote health beyond basic nutrition (Jiraungkoorskul 2016; Bondonno et al. 2014; Volpe 2013; Koh et al. 2012). Spinach is used as antioxidant, anti-anemia, antidiabetic, anthelmintic, antiulcer, antibacterial, anticonvulsant, antihyperlipidemic, anti-inflammatory, and antiviral agent. It was reported to aid in recovery from nervous, hepatic, and respiratory diseases (Jiraungkoorskul 2016; Heo et al. 2010). Spinach phytochemicals are capable of scavenging reactive oxygen species produced by oxidative stress; it can modulate metabolism of body by altering gene expression, cell proliferation, inflammation, and defense mechanism of human body, and it supports weight loss by inducing secretion of satiety hormones and curbs food intake. It possesses insulin-sensitizing activity. Nonessential phytochemicals and bioactives, such as glycolipids and thylakoids, favor health benefits (Fiorito et al. 2019; Roberts and Moreau 2016). Flavonoid component of spinach exhibits protection from AD by significantly reducing β-amyloid-induced neuronal cell death. These in turn activate autophagy either indirectly by regulating a transcription factor FOXO3a or directly by causing deacetylation of autophagic gene products such as Atg 5, 7, and 8 (Hanna et al. 2016).

Spinach is rich in vitamin B, which plays a significant role in brain healing and memory restoring. Vitamins B2, B6, and B12 are necessary for monoamine neurotransmitter production (Kim et al. 2014). Level of magnesium is crucial for prevention of a number of chronic diseases, including hypertension, diabetes, cardiovascular diseases, migraine, and AD. It plays a major role in nerve transmission, glucose and insulin metabolism, neuromuscular conduction, vasomotor tone, blood pressure, cardiac excitability, and muscular conduction (Volpe 2013).

10.7 Role of Hydration in Alzheimer's disease

Water is the most important element on earth. Its importance relies on the fact that every biological organism requires it for maintaining the normal body homeostasis. The normal human body comprises about 60% water, the brain is made up of 75%, and our blood contains approximately 83% of water (Parretti et al. 2015). The regulation of fluid balance is critical to several important components which include normal cell metabolism and its physiochemical properties. Body fluid balance is primarily determined by cells osmolality; osmotic gradient across the cell membrane causes change in the osmolality which either moves to cell swelling or its shrinkage (Biller et al. 2015). It was well reported that alteration in the cell osmolality is directly or indirectly involved in the formation of multitude of diseases which also include neuronal dysfunction, brain atrophy, chronic cerebrovasculopathy, and AD (Cowen et al. 2013).

Even very small variation in the osmotic balance is detected by hypothalamus osmoreceptor (Cowen et al. 2013). Vasopressin, a peptide hormone, regulates these diseases by stimulating the vasopressinergic nerve ending present in the neurohypophysis region. Binding of vasopressin to the receptor present in the kidney decreases the excretion of water, which subsequently decreases the filtered water that returned to the blood thus lowering the plasma osmolality. Failure of this mechanism results in a variety of water balance disorders primarily neuronal dysfunction and Alzheimer's disease. Hypohydration can cause chronic hypovolemia which is hypothesized to be the primary mechanism for the development of hypertension, obesity, and AD (Thornton 2014). A considerable serum hyperosmolality risk was observed in patients with MCI and AD patients. The central nervous system is a lipophilic organ, whereas it contains 80% of water which is majorly stored in astrocytes, and dehydration can cause upregulation of their AQP-4 proteins (Duning et al. 2005).

10.8 Conclusion

Diet is suggested to be a major companion in the management of mild cognitive impairment, memory loss, and Alzheimer's. The components of diet such as polyphenols possess potential to combat progression of memory loss due to anti-inflammatory response. Antioxidants present in various fruits and vegetables scavenge ROS, thereby altering the process of aging in natural way. Mediterranean diet recommended for Alzheimer's patients is an optimized combination of natural products such as fruits vegetables, nuts, milk and milk products, meat, seafoods, and olive oil. Each one of them is loaded with different active ingredients like omega-3 fatty acid, DHA, essential amino acids, vitamins, polyphenols, and flavonoids, etc., and they, therefore, are useful in age-related discomforts. By combining Mediterranean diet with different added nutrients and a mild exercise, the situation of Alzheimer can be reversed to noticeable levels.

References

Abuznait AH, Qosa H, Busnena BA, El Sayed KA, Kaddoumi A (2013) Olive-oil-derived oleocanthal enhances β-amyloid clearance as a potential neuroprotective mechanism against Alzheimer's disease: in vitro and in vivo studies. ACS Chem Neurosci 4(6):973–982

Afrin S, Gasparrini M, Forbes-Hernandez TY, Reboredo-Rodriguez P, Mezzetti B, Varela-López A, Giampieri F, Battino M (2016) Promising health benefits of the strawberry: a focus on clinical studies. J Agric Food Chem 64(22):4435–4449

Afshordel S, Hagl S, Werner D, Röhner N, Kögel D, Bazan NG, Eckert GP (2015) Omega-3 polyunsaturated fatty acids improve mitochondrial dysfunction in brain aging–impact of Bcl-2 and NPD-1 like metabolites. Prostaglandins, Leukot Essent Fatty Acids (PLEFA) 92:23–31

Aridi Y, Walker J, Wright O (2017) The association between the Mediterranean dietary pattern and cognitive health: a systematic review. Nutrients 9(7):674

Bach AC, Babayan VK (1982) Medium-chain triglycerides: an update. Am J Clin Nutr 36(5):950–962

Baddeley AD, Baddeley HA, Bucks RS, Wilcock GK (2001) Attentional control in Alzheimer's disease. Brain 124(8):1492–1508

Barbagallo M, Belvedere M, Di Bella G, Dominguez LJ (2011) Altered ionized magnesium levels in mild-to-moderate Alzheimer's disease. Magnes Res 24(3):115–121

Beckman KB, Ames BN (1998) The free radical theory of aging matures. Physiol Rev 78(2):547–581

Belkouch M, Hachem M, Elgot A, Van AL, Picq M, Guichardant M, Lagarde M, Bernoud-Hubac N (2016) The pleiotropic effects of omega-3 docosahexaenoic acid on the hallmarks of Alzheimer's disease. J Nutr Biochem 38:1–1

Berger MM (2005) Can oxidative damage be treated nutritionally? Clin Nutr 24(2):172–183

Berr C, Portet F, Carriere I, Akbaraly TN, Feart C, Gourlet V, Combe N, Barberger-Gateau P, Ritchie K (2009) Olive oil and cognition: results from the three-city study. Dement Geriatr Cogn Disord 28(4):357–364

Bhowmik D, Kumar KS, Paswan S, Srivastava S (2012) Tomato-a natural medicine and its health benefits. J Pharmacogn Phytochem 1(1):33–43

Biller A, Reuter M, Patenaude B, Homola GA, Breuer F, Bendszus M, Bartsch AJ (2015) Responses of the human brain to mild dehydration and rehydration explored in vivo by 1H-MR imaging and spectroscopy. Am J Neuroradiol 36:2277–2284

Blennow K, de Leon MJ, Zetterberg H (2006) Alzheimer's disease. Lancet 368:387–403

Bonaccio M, Di Castelnuovo A, Bonanni A, Costanzo S, De Lucia F, Pounis G, Zito F, Donati MB, De Gaetano G, Iacoviello L (2013) Adherence to a Mediterranean diet is associated with a better health-related quality of life: a possible role of high dietary antioxidant content. BMJ Open 3(8):e003003

Bondonno CP, Downey LA, Croft KD, Scholey A, Stough C, Yang X, Considine MJ, Ward NC, Puddey IB, Swinny E, Mubarak A (2014) The acute effect of flavonoid-rich apples and nitrate-rich spinach on cognitive performance and mood in healthy men and women. Food Funct 5(5):849–858

Bowman GL (2012) Ascorbic acid, cognitive function, and Alzheimer's disease: a current review and future direction. Biofactors 38(2):114–122

Bowtell JL, Aboo-Bakkar Z, Conway ME, Adlam AL, Fulford J (2017) Enhanced task-related brain activation and resting perfusion in healthy older adults after chronic blueberry supplementation. Appl Physiol Nutr Metab 42(7):773–779

Brayne C, Gill C, Huppert FA, Barkley C, Gehlhaar E, Girling DM, O'connor DW, Paykel ES (1995) Incidence of clinically diagnosed subtypes of dementia in an elderly population: Cambridge Project for Later Life. Br J Psychiatry 167:255–262

Brewer GJ (2012) Copper excess, zinc deficiency, and cognition loss in Alzheimer's disease. Biofactors 38(2):107–113

Brini M, Calì T, Ottolini D, Carafoli E (2014) Neuronal calcium signaling: function and dysfunction. Cell Mol Life Sci 71(15):2787–2814

Bucossi S, Ventriglia M, Panetta V, Salustri C, Pasqualetti P, Mariani S, Siotto M, Rossini PM, Squitti R (2011) Copper in Alzheimer's disease: a meta-analysis of serum, plasma, and cerebrospinal fluid studies. J Alzheimers Dis 24(1):175–185

Burdge G (2004) α-Linolenic acid metabolism in men and women: nutritional and biological implications. Curr Opin Clin Nutr Metab Care 7:137–144

Calon F, Cole G (2007) Neuroprotective action of omega-3 polyunsaturated fatty acids against neurodegenerative diseases: evidence from animal studies. Prostaglandins Leukot Essent Fat Acids 77(5–6):287–293

Castro-Barquero S, Lamuela-Raventós R, Doménech M, Estruch R (2018) Relationship between Mediterranean dietary polyphenol intake and obesity. Nutrients 10(10):1523

Chandrashekar P, Lokesh BR, Krishna AG (2010) Hypolipidemic effect of blends of coconut oil with soybean oil or sunflower oil in experimental rats. Food Chem 123(3):728–733

Chen C, Li Y, Xu Z (2010) Chemical principles and bioactivities of blueberry. Yao xue xue bao=Acta pharmaceutica Sinica 45(4):422–429

Cherbuin N, Kumar R, Sachdev P, Anstey KJ (2014) Dietary mineral intake and risk of mild cognitive impairment: the PATH through life project. Front Aging Neurosci 6:4

Chin KY, Tay S (2018) A review on the relationship between tocotrienol and Alzheimer disease. Nutrients 10(7):881

Cohen G (2000) Oxidative stress, mitochondrial respiration, and Parkinson's disease. Ann N Y Acad Sci 899(1):112–120

Corder EH, Saunders AM, Strittmatter WJ, Schmechel DE, Gaskell PC, Small G, Roses AD, Haines JL, Pericak-Vance MA (1993) Gene dose of apolipoprotein E type 4 allele and the risk of Alzheimer's disease in late onset families. Science 261(5123):921–923

Cordero JG, García-Escudero R, Avila J, Gargini R, García-Escudero V (2018) Benefit of oleuropein aglycone for Alzheimer's Disease by promoting autophagy. Oxid Med Cell Longev 2018

Costa AG, Garcia-Diaz DF, Jimenez P, Silva PI (2013) Bioactive compounds and health benefits of exotic tropical red–black berries. J Funct Foods 5(2):539–549

Côté J, Caillet S, Doyon G, Sylvain JF, Lacroix M (2010) Bioactive compounds in cranberries and their biological properties. Crit Rev Food Sci Nutr 50(7):666–679

Cowen LE, Hodak SP, Verbalis JG (2013) Age-associated abnormalities of water homeostasis. Endocrinol Metab Clin 42(2):349–370

Crane PK, Walker R, Hubbard RA, Li G, Nathan DM, Zheng H, Haneuse S, Craft S, Montine TJ, Kahn SE, McCormick W (2013) Glucose levels and risk of dementia. N Engl J Med 369(6):540–548

Crichton RR, Wilmet S, Legssyer R, Ward RJ (2002) Molecular and cellular mechanisms of iron homeostasis and toxicity in mammalian cells. J Inorg Biochem 91(1):9–18

DaSilva NA, Nahar PP, Ma H, Eid A, Wei Z, Meschwitz S, Zawia NH, Slitt AL, Seeram NP (2017) Pomegranate ellagitannin-gut microbial-derived metabolites, urolithins, inhibit neuroinflammation in vitro. Nutr Neurosci 8:1–1

Dastmalchi K, Flores G, Petrova V, Pedraza-Penalosa P, Kennelly EJ (2011) Edible neotropical blueberries: antioxidant and compositional fingerprint analysis. J Agric Food Chem 59(7):3020–3026

Davis C, Bryan J, Hodgson J, Murphy K (2015) Definition of the Mediterranean diet; a literature review. Nutrients 7(11):9139–9153

de Lima RM, dos Reis AC, de Menezes AA, Santos JV, Filho JW, Ferreira JR, de Alencar MV, da Mata AM, Khan IN, Islam A, Uddin SJ (2018) Protective and therapeutic potential of ginger (Zingiber officinale) extract and [6]-gingerol in cancer: a comprehensive review. Phytother Res 32(10):1885–1907

Deane R, Sagare A, Hamm K, Parisi M, Lane S, Finn MB, Holtzman DM, Zlokovic BV (2008) apoE isoform–specific disruption of amyloid β peptide clearance from mouse brain. J Clin Invest 118(12):4002–4013

Dianne F, Ions LJ, Alatawi F, Wakeling LA (2011) The potential role of epigenetic responses to diet in ageing. Proc Nutr Soc 70(3):374–384

Douaud G, Refsum H, de Jager CA, Jacoby R, Nichols TE, Smith SM, Smith AD (2013) Preventing Alzheimer's disease-related gray matter atrophy by B-vitamin treatment. Proc Natl Acad Sci 110(23):9523–9528

Duning T, Kloska S, Steinsträter O, Kugel H, Heindel W, Knecht S (2005) Dehydration confounds the assessment of brain atrophy. Neurology 64(3):548–550

El-Far AH, Oyinloye BE, Sepehrimanesh M, Gab Allah MA, Abu-Reidah I, Shaheen HM, Razeghian-Jahromi I, AA Alsenosy, Noreldin AE, Al Jaouni SK, Mousa SA (2019) Date palm (Phoenix dactylifera): novel findings and future directions for food and drug discovery. Curr Drug Dis Technol 16:2. https://doi.org/10.2174/1570163815666180320111937

Engelhart MJ, Geerlings MI, Ruitenberg A, van Swieten JC, Hofman A, Witteman JC, Breteler MM (2002) Dietary intake of antioxidants and risk of Alzheimer disease. JAMA 287(24):3223–3229

Enomoto H, Sato K, Miyamoto K, Ohtsuka A, Yamane H (2018) Distribution analysis of anthocyanins, sugars, and organic acids in strawberry fruits using matrix-assisted laser desorption/ionization-imaging mass spectrometry. J Agric Food Chem 66(19):4958–4965

Essa MM, Vijayan RK, Castellano-Gonzalez G, Memon MA, Braidy N, Guillemin GJ (2012) Neuroprotective effect of natural products against Alzheimer's disease. Neurochem Res 37(9):1829–1842

Essa MM, Subash S, Akbar M, Al-Adawi S, Guillemin GJ (2015) Long-term dietary supplementation of pomegranates, figs and dates alleviate neuroinflammation in a transgenic mouse model of Alzheimer's disease. PLoS One 10(3):e0120964

Estruch R, Ros E, Salas-Salvadó J, Covas MI, Corella D, Arós F, Gómez-Gracia E, Ruiz-Gutiérrez V, Fiol M, Lapetra J, Lamuela-Raventos RM (2013) Primary prevention of cardiovascular disease with a Mediterranean diet. N Engl J Med 368(14):1279–1290

Evans DA, Funkenstein HH, Albert MS, Scherr PA, Cook NR, Chown MJ, Hebert LE, Hennekens CH, Taylor JO (1989) Prevalence of Alzheimer's disease in a community population of older persons: higher than previously reported. JAMA 262(18):2551–2556

Faria A, Pestana D, Teixeira D, De Freitas V, Mateus N, Calhau C (2010) Blueberry anthocyanins and pyruvic acid adducts: anticancer properties in breast cancer cell lines. Phytother Res 24(12):1862–1869

Farkas E, Luiten PG (2001) Cerebral microvascular pathology in aging and Alzheimer's disease. Prog Neurobiol 64(6):575–611

Farr SA, Price TO, Dominguez LJ, Motisi A, Saiano F, Niehoff ML, Morley JE, Banks WA, Ercal N, Barbagallo M (2012) Extra virgin olive oil improves learning and memory in SAMP8 mice. J Alzheimers Dis 28(1):81–92

Fernando WM, Martins IJ, Goozee KG, Brennan CS, Jayasena V, Martins RN (2015) The role of dietary coconut for the prevention and treatment of Alzheimer's disease: potential mechanisms of action. Br J Nutr 114(1):1–4

Finicelli M, Squillaro T, Di Cristo F, Di Salle A, Melone MA, Galderisi U, Peluso G (2018) Metabolic syndrome, Mediterranean diet, and polyphenols: evidence and perspectives. J Cell Physiol 14

Fiorito S, Preziuso F, Epifano F, Scotti L, Bucciarelli T, Taddeo VA, Genovese S (2019) Novel biologically active principles from spinach, goji and quinoa. Food Chem 276:262–265

Florent-Bechard S, Desbene C, Garcia P, Allouche A, Youssef I, Escanye MC, Koziel V, Hanse M, Malaplate-Armand C, Stenger C, Kriem B (2009) The essential role of lipids in Alzheimer's disease. Biochimie 91(6):804–809

Floyd RA, Hensley K (2002) Oxidative stress in brain aging: implications for therapeutics of neurodegenerative diseases. Neurobiol Aging 23(5):795–807

Fowler CJ, Wiberg Å, Oreland L, Marcusson J, Winblad B (1980) The effect of age on the activity and molecular properties of human brain monoamine oxidase. J Neural Transm 49(1–2):1–20

Galeffi F, Shetty PK, Sadgrove MP, Turner DA (2015) Age-related metabolic fatigue during low glucose conditions in rat hippocampus. Neurobiol Aging 36(2):982–992

Gardener H, Caunca MR (2018) Mediterranean diet in preventing neurodegenerative diseases. Curr Nutr Rep 7(1):10–20

Gardener S, Gu Y, Rainey-Smith SR, Keogh JB, Clifton PM, Mathieson SL, Taddei K, Mondal A, Ward VK, Scarmeas N, Barnes M (2012) Adherence to a Mediterranean diet and Alzheimer's disease risk in an Australian population. Transl Psychiatry 2(10):e164

Gasparrini M, Giampieri F, Forbes-Hernandez TY, Afrin S, Cianciosi D, Reboredo-Rodriguez P, Varela-Lopez A, Zhang J, Quiles JL, Mezzetti B, Bompadre S (2018) Strawberry extracts efficiently counteract inflammatory stress induced by the endotoxin lipopolysaccharide in Human Dermal Fibroblast. Food Chem Toxicol 114:128–140

Giampieri F, Tulipani S, Alvarez-Suarez JM, Quiles JL, Mezzetti B, Battino M (2012) The strawberry: composition, nutritional quality, and impact on human health. Nutrition 28(1):9–19

Gold PE (2005) Glucose and age-related changes in memory. Neurobiol Aging 26(1):60–64

González-Trujano ME, Pellicer F, Mena P, Moreno DA, García-Viguera C (2015) Antinociceptive and anti-inflammatory activities of a pomegranate (Punica granatum L.) extract rich in ellagitannins. Int J Food Sci Nutr 66(4):395–399

Goswami SK, Das DK (2009) Resveratrol and chemoprevention. Cancer Lett 284(1):1–6

Hallgren B, Sourander P (1958) The effect of age on the non-haemin iron in the human brain. J Neurochem 3(1):41–51

Han QQ, Shen TT, Wang F, Wu PF, Chen JG (2018) Preventive and therapeutic potential of vitamin C in mental disorders. Curr Med Sci 38(1):1–10

Hanna L, Monika K, Włodzimierz L, Stępkowski TM, Kamil B (2016) The role of natural polyphenols in cell signaling and cytoprotection against cancer development. J Nutr Biochem 32:1–19

Hanne M, Lindman AS, Blomfeldt A, Seljeflot I, Pedersen JI (2003) A diet rich in coconut oil reduces diurnal postprandial variations in circulating tissue plasminogen activator antigen and fasting lipoprotein (a) compared with a diet rich in unsaturated fat in women. J Nutr 133(11):3422–3427

Hasima N, Ozpolat B (2014 Nov) Regulation of autophagy by polyphenolic compounds as a potential therapeutic strategy for cancer. Cell Death Dis 5(11):e1509

Hasselbalch SG, Knudsen GM, Jakobsen J, Hageman LP, Holm S, Paulson OB (1994) Brain metabolism during short-term starvation in humans. J Cereb Blood Flow Metab 14(1):125–131

He X, Huang Y, Li B, Gong CX, Schuchman EH (2010) Deregulation of sphingolipid metabolism in Alzheimer's disease. Neurobiol Aging 31(3):398–408

Heilbronn LK, Ravussin E (2003) Calorie restriction and aging: review of the literature and implications for studies in humans. Am J Clin Nutr 78(3):361–369

Heo JC, Park CH, Lee HJ, Kim SO, Kim TH, Lee SH (2010) Amelioration of asthmatic inflammation by an aqueous extract of Spinacia oleracea Linn. Int J Mol Med 25(3):409–414

Hewlings S, Kalman D (2017) Curcumin: a review of its' effects on human health. Foods 6(10):92

Huang CC, Chung CM, Leu HB, Lin LY, Chiu CC, Hsu CY, Chiang CH, Huang PH, Chen TJ, Lin SJ, Chen JW (2014) Diabetes mellitus and the risk of Alzheimer's disease: a nationwide population-based study. PLoS One 9(1):e87095

Jiraungkoorskul W (2016) Review of neuro-nutrition used as anti-Alzheimer plant, spinach, spinacia oleracea. Pharmacogn Rev 10(20):105

Joseph JA, Arendash G, Gordon M, Diamond D, Shukitt-Hale B, Morgan D, Denisova NA (2003) Blueberry supplementation enhances signaling and prevents behavioral deficits in an Alzheimer disease model. Nutr Neurosci 6(3):153–162

Joseph JA, Shukitt-Hale B, Casadesus G (2005) Reversing the deleterious effects of aging on neuronal communication and behavior: beneficial properties of fruit polyphenolic compounds. Am J Clin Nutr 81(1):313S–316S

Kalt W, Joseph JA, Shukitt-Hale B (2007) Blueberries and human health: a review of current reseach. J Am Pomol Soc 61(3):151

Kane RL, Butler M, Fink HA, Brasure M, Davila H, Desai P, Jutkowitz E, McCreedy E, Nelson VA, McCarten JR, Calvert C (2017) Interventions to prevent age-related cognitive decline, mild cognitive impairment, and clinical Alzheimer's-type dementia. Agency for Healthcare Research and Quality, Rockville

Karuppagounder SS, Pinto JT, Xu H, Chen HL, Beal MF, Gibson GE (2009) Dietary supplementation with resveratrol reduces plaque pathology in a transgenic model of Alzheimer's disease. Neurochem Int 54(2):111–118

Kelly E, Vyas P, Weber J (2017) Biochemical properties and neuroprotective effects of compounds in various species of berries. Molecules 23(1):26

Kim H, Kim G, Jang W, Kim SY, Chang N (2014) Association between intake of B vitamins and cognitive function in elderly Koreans with cognitive impairment. Nutr J 13(1):118

Koh E, Charoenprasert S, Mitchell AE (2012) Effect of organic and conventional cropping systems on ascorbic acid, vitamin C, flavonoids, nitrate, and oxalate in 27 varieties of spinach (Spinacia oleracea L.). J Agric Food Chem 60(12):3144–3150

Krishna AG, Gaurav R, Singh BA, Kumar PP, Preeti C (2010) Coconut oil: chemistry, production and its applications-a review. Indian Coconut J 53(3):15–27

Kumar N, Wheaton LA, Snow TK, Millard-Stafford M (2016) Carbohydrate ingestion but not mouth rinse maintains sustained attention when fasted. Physiol Behav 153:33–39

Lau FC, Shukitt-Hale B, Joseph JA (2007) Nutritional intervention in brain aging. In: Inflammation in the pathogenesis of chronic diseases. Springer, Dordrecht, pp 299–318

Le Couteur DG, Solon-Biet S, Wahl D, Cogger VC, Willcox BJ, Willcox DC, Raubenheimer D, Simpson SJ (2016) New horizons: dietary protein, ageing and the Okinawan ratio. Age Ageing 45(4):443–447

Lee CK, Weindruch R, Prolla TA (2000) Gene-expression profile of the ageing brain in mice. Nat Genet 25(3):294

Levine ME, Suarez JA, Brandhorst S, Balasubramanian P, Cheng CW, Madia F, Fontana L, Mirisola MG, Guevara-Aguirre J, Wan J, Passarino G (2014) Low protein intake is associated with a major reduction in IGF-1, cancer, and overall mortality in the 65 and younger but not older population. Cell Metab 19(3):407–417

Lewis ED, Ren Z, DeFuria J, Obin MS, Meydani SN, Wu D (2018) Dietary supplementation with blueberry partially restores T-cell-mediated function in high-fat-diet-induced obese mice. Br J Nutr 119(12):1393–1399

Lim SY, Suzuki H (2000) Intakes of dietary docosahexaenoic acid ethyl ester and egg phosphatidylcholine improve maze-learning ability in young and old mice. J Nutr 130(6):1629–1632

Locascio JJ, Growdon JH, Corkin S (1995) Cognitive test performance in detecting, staging, and tracking Alzheimer's disease. Arch Neurol 52(11):1087–1099

Lopes MA, Larkins BA (1993) Endosperm origin, development, and function. Plant Cell 5(10):1383

Lorenzo JM, Pateiro M, Domínguez R, Barba FJ, Putnik P, Kovačević DB, Shpigelman A, Granato D, Franco D (2018) Berries extracts as natural antioxidants in meat products: a review. Food Res Int 106:1095–1104

Mahmassani HA, Avendano EE, Raman G, Johnson EJ (2018) Avocado consumption and risk factors for heart disease: a systematic review and meta-analysis. Am J Clin Nutr 107(4):523–536

Maioli E, Greci L, Soucek K, Hyzdalova M, Pecorelli A, Fortino V, Valacchi G (2010) Rottlerin inhibits ROS formation and prevents NF κ B activation in MCF-7 and HT-29 cells. Biomed Res Int 3:2009

Mandel SA, Weinreb O, Amit T, Youdim MB (2012) Molecular mechanisms of the neuroprotective/neurorescue action of multi-target green tea polyphenols. Front Biosci (Schol Ed) 4:581–598

Marina AM, Che Man YB, Nazimah SA, Amin I (2009) Antioxidant capacity and phenolic acids of virgin coconut oil. Int J Food Sci Nutr 60(sup2):114–123

Martinez-Gonzalez MA, Martin-Calvo N (2016) Mediterranean diet and life expectancy; beyond olive oil, fruits and vegetables. Curr Opin Clin Nutr Metab Care 19(6):401

Matsuzaki T, Sasaki K, Tanizaki Y, Hata J, Fujimi K, Matsui Y, Sekita A, Suzuki SO, Kanba S, Kiyohara Y, Iwaki T (2010) Insulin resistance is associated with the pathology of Alzheimer disease The Hisayama Study. Neurology 75(9):764–770

McGahon BM, Martin DS, Horrobin DF, Lynch MA (1999) Age-related changes in synaptic function: analysis of the effect of dietary supplementation with ω-3 fatty acids. Neuroscience 94(1):305–314

McGill CM, Brown TJ, Cheng YY, Fisher LN, Shanmugavelandy SS, Gustafson SJ, Dunlap KL, Lila MA, Kester M, Toran PT, Claxton DF (2018) Therapeutic effect of blueberry extracts for acute myeloid leukemia. Int J Biopharm Sci 1(1)

Miranda A, Gomez-Gaete C, Mennickent S (2017) Role of Mediterranean diet on the prevention of Alzheimer disease. Revista medica de Chile 145(4):501–507

Moore K, Hughes CF, Ward M, Hoey L, McNulty H (2018) Diet, nutrition and the ageing brain: current evidence and new directions. Proc Nutr Soc 77(2):152–163

Morgan TE, Xie Z, Goldsmith S, Yoshida TS, Lanzrein AS, Stone D, Rozovsky I, Perry G, Smith MA, Finch CE (1999) The mosaic of brain glial hyperactivity during normal ageing and its attenuation by food restriction. Neuroscience 89(3):687–699

Morris AA (2005) Cerebral ketone body metabolism. J Inherit Metab Dis 28(2):109–121

Morris MC, Evans DA, Bienias JL, Tangney CC, Bennett DA, Aggarwal N, Schneider J, Wilson RS (2003) Dietary fats and the risk of incident Alzheimer disease. Arch Neurol 60(2):194–200

Nakagawa S, Lagisz M, Hector KL, Spencer HG (2012) Comparative and meta-analytic insights into life extension via dietary restriction. Aging Cell 11(3):401–409

Nanjo F, Goto K, Seto R, Suzuki M, Sakai M, Hara Y (1996) Scavenging effects of tea catechins and their derivatives on 1, 1-diphenyl-2-picrylhydrazyl radical. Free Radic Biol Med 21(6):895–902

Nile SH, Park SW (2014) Edible berries: bioactive components and their effect on human health. Nutrition 30(2):134–144

Nishida Y, Yokota T, Takahashi T, Uchihara T, Jishage KI, Mizusawa H (2006) Deletion of vitamin E enhances phenotype of Alzheimer disease model mouse. Biochem Biophys Res Commun 350(3):530–536

Olgun A, Akman S, Serdar MA, Kutluay T (2002) Oxidative phosphorylation enzyme complexes in caloric restriction. Exp Gerontol 37(5):639–645

Olivera-Pueyo J, Pelegrín-Valero C (2017 Sep 2) Dietary supplements for cognitive impairment. Actas espanolas de psiquiatria, 45

Opoka W, Jakubowska M, Baś B, Sowa-Kućma M (2011) Development and validation of an anodic stripping voltammetric method for determination of Zn2+ ions in brain microdialysate samples. Biol Trace Elem Res 142(3):671–682

Ouellet M, Emond V, Chen CT, Julien C, Bourasset F, Oddo S, LaFerla F, Bazinet RP, Calon F (2009) Diffusion of docosahexaenoic and eicosapentaenoic acids through the blood–brain barrier: an in situ cerebral perfusion study. Neurochem Int 55(7):476–482

Ozawa M, Ninomiya T, Ohara T, Hirakawa Y, Doi Y, Hata J, Uchida K, Shirota T, Kitazono T, Kiyohara Y (2012) Self-reported dietary intake of potassium, calcium, and magnesium and risk of dementia in the Japanese: the Hisayama Study. J Am Geriatr Soc 60(8):1515–1520

Page KA, Williamson A, Yu N, McNay EC, Dzuira J, McCrimmon RJ, Sherwin RS (2009) Medium-chain fatty acids improve cognitive function in intensively treated type 1 diabetic patients and support in vitro synaptic transmission during acute hypoglycemia. Diabetes 58(5):1237–1244

Panche AN, Diwan AD, Chandra SR (2016) Flavonoids: an overview. J Nutr Sci 5:e47

Pandey KB, Rizvi SI (2009) Plant polyphenols as dietary antioxidants in human health and disease. Oxidative Med Cell Longev 2(5):270–278

Park E, Edirisinghe I, Burton-Freeman B (2018) Avocado fruit on postprandial markers of cardio-metabolic risk: a randomized controlled dose response Trial in overweight and obese men and women. Nutrients 10(9):1287

Parkinson L, Cicerale S (2016) The health benefiting mechanisms of virgin olive oil phenolic compounds. Molecules 21(12):1734

Parretti HM, Aveyard P, Blannin A, Clifford SJ, Coleman SJ, Roalfe A, Daley AJ (2015) Efficacy of water preloading before main meals as a strategy for weight loss in primary care patients with obesity: RCT. Obesity 23(9):1785–1791

Pavlica S, Gebhardt R (2010) Comparison of uptake and neuroprotective potential of seven zinc-salts. Neurochem Int 56(1):84–93

Perrin R, Briançon S, Jeandel C, Artur Y, Minn A, Penin F, Siest G (1990) Blood activity of Cu/Zn superoxide dismutase, glutathione peroxidase and catalase in Alzheimer's disease: a case-control study. Gerontology 36(5–6):306–313

Perron NR, Hodges JN, Jenkins M, Brumaghim JL (2008) Predicting how polyphenol antioxidants prevent DNA damage by binding to iron. Inorg Chem 47(14):6153–6161

Petersen RC, Smith GE, Waring SC, Ivnik RJ, Tangalos EG, Kokmen E (1999) Mild cognitive impairment: clinical characterization and outcome. Arch Neurol 56(3):303–308

Petersen RC, Stevens JC, Ganguli M, Tangalos EG, Cummings JL, DeKosky ST (2001) Practice parameter: early detection of dementia: mild cognitive impairment (an evidence-based review): report of the Quality Standards Subcommittee of the American Academy of Neurology. Neurology 56(9):1133–1142

Piazza-Gardner AK, Gaffud TJ, Barry AE (2013) The impact of alcohol on Alzheimer's disease: a systematic review. Aging Ment Health 17(2):133–146

Pistollato F, Sumalla Cano S, Elio I, Masias Vergara M, Giampieri F, Battino M (2016) Associations between sleep, cortisol regulation, and diet: possible implications for the risk of Alzheimer disease. Adv Nutr 7(4):679–689

Pitozzi V, Jacomelli M, Zaid M, Luceri C, Bigagli E, Lodovici M, Ghelardini C, Vivoli E, Norcini M, Gianfriddo M, Esposto S (2010) Effects of dietary extra-virgin olive oil on behaviour and brain biochemical parameters in ageing rats. Br J Nutr 103(11):1674–1683

Poulose SM, Bielinski DF, Carrihill-Knoll KL, Rabin BM, Shukitt-Hale B (2014) Protective effects of blueberry-and strawberry diets on neuronal stress following exposure to 56Fe particles. Brain Res 1593:9–18

Preston GA (1986) Dementia in elderly adults: prevalence and institutionalization. J Gerontol 41(2):261–267

Qosa H, Mohamed LA, Batarseh YS, Alqahtani S, Ibrahim B, LeVine H III, Keller JN, Kaddoumi A (2015) Extra-virgin olive oil attenuates amyloid-β and tau pathologies in the brains of TgSwDI mice. J Nutr Biochem 26(12):1479–1490

Rapp PR, Gallagher M (1996) Preserved neuron number in the hippocampus of aged rats with spatial learning deficits. Proc Natl Acad Sci 93(18):9926–9930

Reinikainen KJ, Paljarvi L, Halonen T, Malminen O, Kosma VM, Laakso M, Riekkinen PJ (1988) Dopaminergic system and monoamine oxidase-B activity in Alzheimer's disease. Neurobiol Aging 9:245–252

Rice-Evans CA, Miller NJ (1996) Antioxidant activities of flavonoids as bioactive components of food. Biochem Soc Trans 24(3):790–795

Roberts JL, Moreau R (2016) Functional properties of spinach (Spinacia oleracea L.) phytochemicals and bioactives. Food Funct 7(8):3337–3353

Roberts BR, Ryan TM, Bush AI, Masters CL, Duce JA (2012) The role of metallobiology and amyloid-β peptides in Alzheimer's disease. J Neurochem 120:149–166

Rodrigues GP, Cozzolino SM, do Nascimento Marreiro D, Caldas DR, da Silva KG, de Sousa Almondes KG, Neto JM, Pimentel JA, de Carvalho CM, do Nascimento Nogueira N (2017) Mineral status and superoxide dismutase enzyme activity in Alzheimer's disease. J Trace Elem Med Biol 44:83–87

Rodríguez-Morató J, Boronat A, Kotronoulas A, Pujadas M, Pastor A, Olesti E, Perez-Mana C, Khymenets O, Fitó M, Farré M, de la Torre R (2016) Metabolic disposition and biological significance of simple phenols of dietary origin: hydroxytyrosol and tyrosol. Drug Metab Rev 48(2):218–236

Routray W, Orsat V (2011) Blueberries and their anthocyanins: factors affecting biosynthesis and properties. Compr Rev Food Sci Food Saf 10(6):303–320

Rowe JW, Kahn RL (1997) Successful aging. The Gerontologist 37(4):433–440

Ryu SY, Coutu JP, Rosas HD, Salat DH (2014) Effects of insulin resistance on white matter microstructure in middle-aged and older adults. Neurology 82(21):1862–1870

Salah N, Miller NJ, Paganga G, Tijburg L, Bolwell GP, Riceevans C (1995) Polyphenolic flavanols as scavengers of aqueous phase radicals and as chain-breaking antioxidants. Arch Biochem Biophys 322(2):339–346

Salas-Salvadó J, Bulló M, Babio N, Martínez-González MÁ, Ibarrola-Jurado N, Basora J, Estruch R, Covas MI, Corella D, Arós F, Ruiz-Gutiérrez V (2011) Reduction in the incidence of type 2 diabetes with the Mediterranean diet: results of the PREDIMED-Reus nutrition intervention randomized trial. Diabetes Care 34(1):14–19

Sara S, Costa EM, Mariana V, Morais RM, Conceição C, Manuela P (2018) Health promoting properties of blueberries: a review. Crit Rev Food Sci Nutr 66:1–20

Sastre M, Klockgether T, Heneka MT (2006) Contribution of inflammatory processes to Alzheimer's disease: molecular mechanisms. Int J Dev Neurosci 24(2-3):167–176

Saucedo-Pompa S, Torres-Castillo JA, Castro-López C, Rojas R, Sánchez-Alejo EJ, Ngangyo-Heya M, Martínez-Ávila GC (2018) Moringa plants: bioactive compounds and promising applications in food products. Food Res Int 26

Schrag M, Mueller C, Oyoyo U, Smith MA, Kirsch WM (2011) Iron, zinc and copper in the Alzheimer's disease brain: a quantitative meta-analysis. Some insight on the influence of citation bias on scientific opinion. Prog Neurobiol 94(3):296–306

Selvaraju TR, Khaza'ai H, Vidyadaran S, Mutalib MS, Vasudevan R (2014) The neuroprotective effects of tocotrienol rich fraction and alpha tocopherol against glutamate injury in astrocytes. Bosn J Basic Med Sci 14(4):195

Seneff S, Wainwright G, Mascitelli L (2011) Nutrition and Alzheimer's disease: the detrimental role of a high carbohydrate diet. Eur J Intern Med 22(2):134–140

Shukitt-Hale B, Lau FC, Joseph JA (2008) Berry fruit supplementation and the aging brain. J Agric Food Chem 56(3):636–641

Sikora E, Scapagnini G, Barbagallo M (2010) Curcumin, inflammation, ageing and age-related diseases. Immun Ageing 7(1):1

Simpson SJ, Raubenheimer D (2005) Obesity: the protein leverage hypothesis. Obes Rev 6(2):133–142

Singletary K (2010) Ginger: an overview of health benefits. Nutr Today 45(4):171–183

Small SA, Stern Y, Tang M, Mayeux R (1999) Selective decline in memory function among healthy elderly. Neurology 52(7):1392

Sohal RS, Weindruch R (1996) Oxidative stress, caloric restriction, and aging. Science 273(5271):59–63

Sokoloff L (1999) Energetics of functional activation in neural tissues. Neurochem Res 24(2):321–329

Stefani M, Rigacci S (2014) Beneficial properties of natural phenols: highlight on protection against pathological conditions associated with amyloid aggregation. Biofactors 40(5):482–493

Stefanidis A, Watt MJ (2012) Does too much sugar make for lost memories? J Physiol 590(16):3633–3634

Stowe CB (2011) The effects of pomegranate juice consumption on blood pressure and cardiovascular health. Complement Ther Clin Pract 17(2):113–115

Subash S, Essa MM, Braidy N, Awlad-Thani K, Vaishnav R, Al-Adawi S, Al-Asmi A, Guillemin GJ (2015) Diet rich in date palm fruits improves memory, learning and reduces beta amyloid in transgenic mouse model of Alzheimer's disease. J Ayurv Integr Med 6(2):111

Sumithran P, Prendergast LA, Delbridge E, Purcell K, Shulkes A, Kriketos A, Proietto J (2013) Ketosis and appetite-mediating nutrients and hormones after weight loss. Eur J Clin Nutr 67(7):759

Swaminathan A, Jicha GA (2014) Nutrition and prevention of Alzheimer's dementia. Front Aging Neurosci 6:282

Szajdek A, Borowska EJ (2008) Bioactive compounds and health-promoting properties of berry fruits: a review. Plant Foods Hum Nutr 63(4):147–156

Tan L, Yang H, Pang W, Li H, Liu W, Sun S, Song N, Zhang W, Jiang Y (2017) Investigation on the role of BDNF in the benefits of blueberry extracts for the improvement of learning and memory in Alzheimer's disease mouse model. J Alzheimers Dis 56(2):629–640

Taylor MK, Sullivan DK, Swerdlow RH, Vidoni ED, Morris JK, Mahnken JD, Burns JM (2017) A high-glycemic diet is associated with cerebral amyloid burden in cognitively normal older adults. Am J Clin Nutr 106(6):1463–1470

Tejada S, Setzer W, Daglia M, Fazel Nabavi S, Sureda A, Braidy N, Gortzi O, Mohammad Nabavi S (2017) Neuroprotective effects of Ellagitannins: a brief review. Curr Drug Targets 18(13):1518–1528

Thornton SN (2014) Diabetes and hypertension, as well as obesity and Alzheimer's disease, are linked to hypohydration-induced lower brain volume. Front Aging Neurosci 6:279

Tsai CW, Chen HW, Sheen LY, Lii CK (2012) Garlic: Health benefits and actions. Biomedicine 2(1):17–29

Tu SH, Chen LC, Ho YS (2017) An apple a day to prevent cancer formation: reducing cancer risk with flavonoids. J Food Drug Anal 25(1):119–124

Tuck KL, Hayball PJ (2002) Major phenolic compounds in olive oil: metabolism and health effects. J Nutr Biochem 13(11):636–644

Ude C, Schubert-Zsilavecz M, Wurglics M (2013) Ginkgo biloba extracts: a review of the pharmacokinetics of the active ingredients. Clin Pharmacokinet 52(9):727–749

Uylings HB, De Brabander JM (2002) Neuronal changes in normal human aging and Alzheimer's disease. Brain Cogn 49(3):268–276

Valls-Pedret C, Lamuela-Raventós RM, Medina-Remón A, Quintana M, Corella D, Pintó X, Martínez-González MÁ, Estruch R, Ros E (2012) Polyphenol-rich foods in the Mediterranean diet are associated with better cognitive function in elderly subjects at high cardiovascular risk. J Alzheimers Dis 29(4):773–782

Vandenberghe R, Tournoy J (2005) Cognitive aging and Alzheimer's disease. Postgrad Med J 81(956):343–352

Vislocky LM, Fernandez ML (2010) Biomedical effects of grape products. Nutr Rev 68(11):656–670

Volpe SL (2013) Magnesium in disease prevention and overall health. Adv Nutr 4(3):378S–383S

Wade AT, Davis CR, Dyer KA, Hodgson JM, Woodman RJ, Keage HA, Murphy KJ (2017) A Mediterranean diet to improve cardiovascular and cognitive health: protocol for a randomised controlled intervention study. Nutrients 9(2):145

Wade AT, Davis CR, Dyer KA, Hodgson JM, Woodman RJ, Keage HA, Murphy KJ (2018) A Mediterranean diet supplemented with dairy foods improves mood and processing speed in an Australian sample: results from the MedDairy randomized controlled trial. Nutr Neurosci 9:1–3

Wang ZY, Huang MT, Lou YR, Xie JG, Reuhl KR, Newmark HL, Ho CT, Yang CS, Conney AH (1994) Inhibitory effects of black tea, green tea, decaffeinated black tea, and decaffeinated green tea on ultraviolet B light-induced skin carcinogenesis in 7, 12-dimethylbenz [a] anthracene-initiated SKH-1 mice. Cancer Res 54(13):3428–3435

Wang D, Özen C, Abu-Reidah IM, Chigurupati S, Patra JK, Horbanczuk JO, Jóźwik A, Tzvetkov NT, Uhrin P, Atanasov AG (2018) Vasculoprotective effects of pomegranate (Punica granatum L.). Front Pharmacol 9

Wei K, Luchtman D, Song C (2008) Eicosapentaenoic acid (EPA) increases cell viability and expression of neurotrophin receptors in retinoic acid and brain-derived neurotrophic factor differentiated SH-SY5Y cells. Eur J Nutr 47(2):104–113

Weiser M, Butt C, Mohajeri M (2016) Docosahexaenoic acid and cognition throughout the lifespan. Nutrients 8(2):99

Willett WC, Sacks F, Trichopoulou A, Drescher G, Ferro-Luzzi A, Helsing E, Trichopoulos D (1995) Mediterranean diet pyramid: a cultural model for healthy eating. Am J Clin Nutr 61(6):1402S–1406S

Winter AN, Ross EK, Wilkins HM, Stankiewicz TR, Wallace T, Miller K, Linseman DA (2018) An anthocyanin-enriched extract from strawberries delays disease onset and extends survival in the hSOD1G93A mouse model of amyotrophic lateral sclerosis. Nutr Neurosci 21(6):414–426

Yang B, Kortesniemi M (2015) Clinical evidence on potential health benefits of berries. Curr Opin Food Sci 2:36–42

Yang DS, Stavrides P, Mohan PS, Kaushik S, Kumar A, Ohno M, Schmidt SD, Wesson D, Bandyopadhyay U, Jiang Y, Pawlik M (2010) Reversal of autophagy dysfunction in the TgCRND8 mouse model of Alzheimer's disease ameliorates amyloid pathologies and memory deficits. Brain 134(1):258–277

Yavuz BB, Cankurtaran M, Haznedaroglu IC, Halil M, Ulger Z, Altun B, Ariogul S (2012) Iron deficiency can cause cognitive impairment in geriatric patients. J Nutr Health Aging 16(3):220–224

Yusufov M, Weyandt LL, Piryatinsky I (2017) Alzheimer's disease and diet: a systematic review. Int J Neurosci 127(2):161–175

Zecca L, Youdim MB, Riederer P, Connor JR, Crichton RR (2004) Iron, brain ageing and neurodegenerative disorders. Nat Rev Neurosci 5(11):863

Zeng LF, Cao Y, Liang WX, Bao WH, Pan JK, Wang Q, Liu J, Liang HD, Xie H, Chai YT, Guan ZT (2017) An exploration of the role of a fish-oriented diet in cognitive decline: a systematic review of the literature. Oncotarget 8(24):39877

Zhang M, Zhang X, Ho CT, Huang Q (2018) Chemistry and healthy effect of tea polyphenol,(-)-epigallocatechin 3-O-(3-O-methyl) gallate. J Agric Food Chem 67:5374–5378

Genetics, Neuronal Pathways, and Electrophysiology of Alzheimer's Disease

11

Mohammad Zubair

11.1 Introduction to Alzheimer

Alzheimer disease is characterized as a brain disorder that impairs an individual's ability of memorizing, thinking, and behavior. The condition is considered as the most common form of dementia, and it is likely to worsen over time; however, it is not a normal part of aging (Abraha et al. 2017). Dementia is associated with loss of memory and other cognitive capabilities of individuals. The individuals suffering from dementia have a comorbid health condition (Bunn et al. 2014). Forgetfulness is the major problem faced by the individuals suffering from Alzheimer, which affects their functionality at home and office. Dementia has been chosen as the most accurate term to describe Alzheimer's disease (Mace and Rabins 2011). The individual remains confused, misplaces things, gets lost in familiar places, and is also troubled while communicating with peers. The onset of Alzheimer results in the prevention of brain cells to perform well. This condition may lead to disorientation, confusion, intellectual impairment, and memory loss (Bennett et al. 2005).

Alzheimer's disease (AD) is an age-related, irreversible, progressive brain disorder that attacks the brain and results in increasingly impaired memory, thinking, reasoning, and behavior (National Institute on Aging 2016). Alzheimer's disease is the most common form of dementia worldwide. Around 50 million people have dementia, and there are nearly 10 million new cases every year (World Health Organization 2017). The family caregivers need to be capable enough to detect any sort of cognitive impairment and let the patient express his/her concerns. They also need to encourage them for following the physician's recommendations to obtain accurate diagnostic evaluations (Maslow and Fortinsky 2018; Austrom et al. 2018).

M. Zubair (✉)
Department of Medical Microbiology, Faculty of Medicine, University of Tabuk, Tabuk, Kingdom of Saudi Arabia
e-mail: zmohammad@ut.edu.sa

G. M. Ashraf, A. Alexiou (eds.), *Biological, Diagnostic and Therapeutic Advances in Alzheimer's Disease*, https://doi.org/10.1007/978-981-13-9636-6_11

Providing care for a person with the AD is very difficult and may have a negative effect on the health, employment, income, and financial security of the caregivers. This is because people living with dementia have complex care needs that have to be considered by the family caregivers and providers (Fazio et al. 2018). Family and other unpaid caregivers of people with Alzheimer's disease help the patients in assisting all of the daily living activities. More than one-half of the caregivers of people with Alzheimer's disease and other dementias reported providing help with getting in and out of bed. Potentially unsafe behaviors are likely to be reported by the older adults with probable dementia (Amjad et al. 2016).

11.1.1 Plaques and Tangles

There is abundance of plaques and tangles in the brains of individuals suffering from Alzheimer. The two cellular hallmarks characterizing the neuropathology of AD include accumulation of intracellular hyperphosphorylated tau in neurofibrillary tangles and extracellular plaques composed on amyloid (Gentier et al. 2015). The macroscopic plaques forming one of the hallmarks of AD pathology aggregate to form oligomers (Bertram et al. 2010; Hardy et al. 2014; Heppner et al. 2015). Plaques are defined as deposition of protein fragments known as beta-amyloid. These depositions are built within the spaces present between the nerve cells. However, twisted fibers of protein tau are known as tangles that are built inside the nerve cells. Although these plaques and tangles develop in every individual as they age, there is much difference in their development pattern within brain of an Alzheimer's patient.

The plaques and tangles in brains of an Alzheimer's patient start to appear in the areas that are important for memorizing. Majority of the scientists believe that plaques and tangles play an important role in blocking communication and disrupting processes between the nerve cells, although the exact role played by them is not yet clear (Alzheimer's Association 2005).

The symptoms of Alzheimer's disease appear when the nerve cells are destructed causing change in personality, problems in conducting daily routine activities, and memory failure. The process of spread of Alzheimer's in the brain has been illustrated in Fig. 11.1.

11.1.2 Risk Factors/Causes of Alzheimer's

The risk factors/causes increasing the likelihood of developing cancer are as follows:

- Genetics – Risk genes and deterministic genes are considered as the two major genes categories that affect the likelihood of developing Alzheimer's. The deterministic genes guarantee that inheritance of this gene would be the direct cause of developing Alzheimer's.

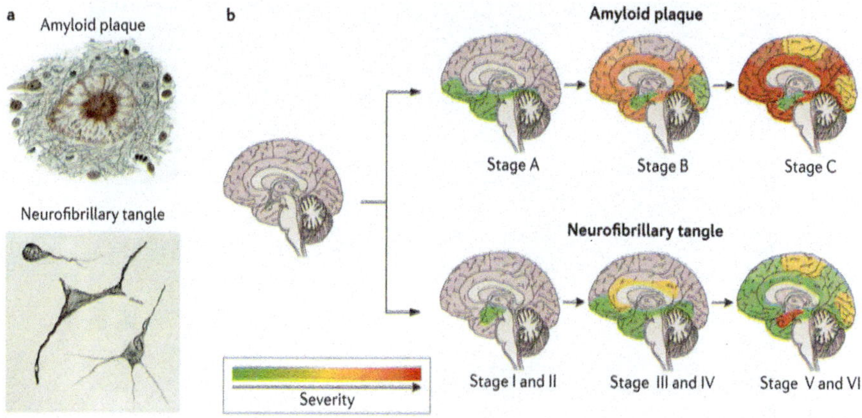

Fig. 11.1 The pathological evolution of Alzheimer's disease. (Adapted with permission from Masters et al. 2015)

- Family history – The risk of developing Alzheimer's increases among an individual having a family relative (parent, brother, sister) with Alzheimer's. Moreover, the risk is likely to increase among the individual with more than one family member with this illness.
- Age – It is the most common risk factor of developing Alzheimer's as majority of the patients are older than 65 years.

11.1.3 Association Between Hyperactivity of Hippocampus and Cortical Thinning

There are certain functional and structural alteration associated with AD within the distributed network of brain regions that support the memory and cognitive domains (Du et al. 2007; Putcha et al. 2011). There is significant association of hyperactivity of hippocampus with thinning of the cortex in precuneus and lateral temporoparietal cortices. This type of change is observed among presymptomatic amyloid-positive and dementia patients (Fig. 11.2). An important role is played by this network in the formation of episodic formation of memory. Moreover, this network is also implicated to target by the neurodegenerative process of Alzheimer's disease.

11.2 Genetics of Alzheimer's Disease

Genes, made from DNA, are characterized as basic unit of inheritance. They are packed in paired structure, known as chromosomes, and are present in all the body cells. Each individual carries two copies of each gene, which is inherited from each parent. Each of the gene variant plays a role in determining individual characteristics. However, combined effects of many variants acting together are reflected

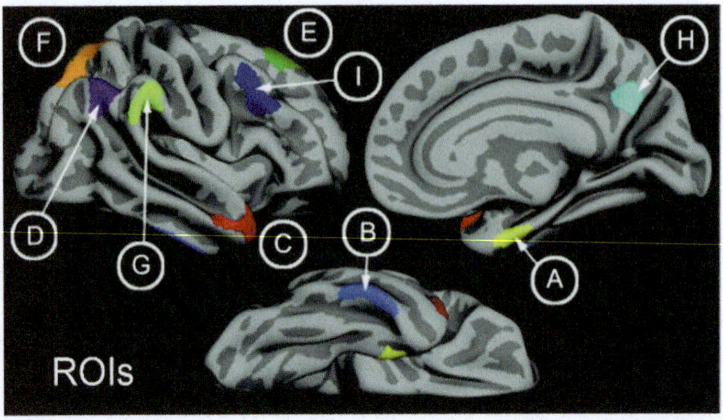

Fig. 11.2 Hyperactivity of hippocampus and cortical thinning (AD signature of cortical thinning. A, Medial temporal lobe; B, inferior temporal gyrus; C, temporal pole; D, angular gyrus; E, superior frontal gyrus; F, superior parietal lobule; G, supramarginal gyrus; H, precuneus; I, inferior frontal sulcus). (Adapted with permission from Masters et al. 2015)

through individual qualities. The inheritance follows a complex pattern; therefore, the inheritance of characteristic affected by a gene variant is not simple. However, mutations result in faulty gene that tends to be more harmful.

The nuclear bodies acting as interchromosomal hubs are demonstrated by the reconciling of the nuclear structure that help in shaping the overall 3-dimensional packaging of the genomic DNA within the nucleus (Quinodoz et al. 2017). Alzheimer's disease may follow either complex inheritance pattern due to multi-gene variants or simple inheritance pattern due to single-gene mutations (Loy et al. 2014).

11.2.1 Familial Alzheimer's Disease

The fact that mutation within single gene causing Alzheimer is suggested by the pattern of familial clustering, in which the mutations are passed down across several generations. Individuals in their 30s, 40s, and 50s with extremely rare mutations are likely to develop Alzheimer. This type of dementia is also known as early-onset or young-onset dementia. There is a total of three amyloid precursor protein (APP) genes and two presenilin genes (PSEN-1 and PSEN-2), resulting in Alzheimer's disease (Loy et al. 2014). Figure 11.3 presents family tree that has shown strong inheritance of mutation across three generations through the darkened images. The figure has clearly demonstrated that the grandfather had three children, among which two had inherited the mutation. Among the two children carrying mutation, the daughter gave birth to one affected male child, while her two daughters were normal. It has also been shown that the children of unaffected son had not inherited any mutation.

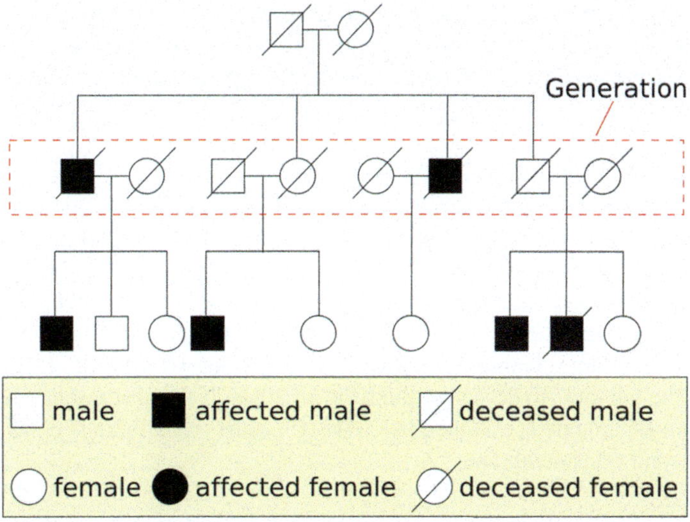

Fig. 11.3 Family tree showing strong inheritance of mutation. (Adapted with permission from Loy et al. 2014)

11.2.2 Genes Associated with Alzheimer's Disease

11.2.2.1 AD1: Amyloid Precursor Protein (APP)

The amyloid precursor protein (APP) is present on chromosome 21q that is similar to the observation depicted among the Down syndrome patients. Approximately, 10–15% of the early-onset Alzheimer's disease is associated with mutations in APP gene. The mutation on this gene is usually observed in or adjacent to the Aβ peptide sequence that is considered as an important factor of amyloid plaques. APP gene is spliced into different products and named on basis of the length of amino acids. The most relevant isoforms of APP leading to development of Alzheimer's are as follows (Bekris et al. 2010):

- APP695 that is restricted to the central nervous system
- APP751 and APP77 that are present in the tissues of central as well as peripheral nervous system

11.2.2.2 AD3: Presenilin 1 (PSEN1)

PSEN 1 has been identified on AD3 locus on chromosome number 14 that is most frequently mutated with around 215 mutations in 475 probands (Cacace et al. 2016). The mutation in this gene is identified through positional cloning that results in the encoding of a polytopic membrane protein (Genetics, provided, 93). PSEN1 being polytopic membrane protein plays an important role in the formation of catalytic core of the γ-secretase complex. Mutation taking place in this gene is the most common cause of Alzheimer's. There is decrease in activity of γ-secretase, as PSEN1 mutations increase the ratio Aβ42 to Aβ40 (Bekris et al. 2010). In PSEN1 mutation carriers, the deposition of Aβ42 is considered as an early preclinical event. The clinical diagnosis of Alzheimer's is confirmed by measuring the amyloid plaque. The presence of specific PSEN1 mutations helps in detecting the neurodegenerative changes.

11.2.2.3 AD4: Presenilin 2 (PSEN2)

PSEN2 is found on chromosome 1 AD4 that possess high homology with the AD3 locus, where PSEN1 is found. Unlike PSEN1, the missense mutations taking place in PSEN2 rarely result in Alzheimer's. The mutations taking place in PSEN2 possess lower penetrance; therefore, they are higher risk of being modified on the basis of environmental influences. PSEN2 displaying tissue-specific alternative splicing is likely to contain nine transmembrane domains along with a large loop structure between the 6th and 7th domains (Bekris et al. 2010). The substitution of an isoleucine for an asparagine at residues 141 (N141I) occurs when there is point mutation on the second transmembrane.

11.3 Neuronal Pathways of Alzheimer's Disease

The neurons tend to build a nerve impulse of electric charge after receiving messages from its surrounding cells. The release of chemical messengers known as neurotransmitters is triggered, when the electrical charge from the neuron travels down the axon till the end. The neurotransmitters move to the dendrites or cell body of neighboring neurons from the axons. The channels from cell membrane to the nerve cells are opened, when the receptor is activated.

The appearance of pathological emblems results in the establishment of Alzheimer's, which involve the neurofibrillary tangles and senile plaques. The extracellular deposits of amyloid beta and intracellular self-gathered clumps of tau proteins comprise these lesions that may cause neuronal death and synaptic failure. There is a central role of production and clearance of Aβ in describing the pathogenesis of Alzheimer's disease. The hypothesis known as amyloid cascade hypothesis (ACH) suggested that Aβ and processing of APP play an important role in the process of neurodegeneration. The aggregation of Aβ is considered as the initiation toward the formation of senile plaques (Ashraf et al. 2018). APP holds significance in neuropathogenesis due to its rapid production and metabolism features. The non-amyloidogenic and amyloidogenic pathways depict the proteolysis of APP.

Enzyme alpha (α)-secretase conducts the first step of non-amyloidogenic pathway that is responsible for breaking down APP into soluble amyloid precursor protein alpha (sAPPα) and alpha C-terminal fragment (αCTF/CTF83). The role of limiting protein in Aβ generation is played by BACE that is responsible for breaking down APP into soluble amyloid precursor protein beta (sAPPβ) and beta C-terminal fragments (βCTF/CTF99) (Ashraf et al. 2018).

CAST is likely to play an important role in explicating as a suicide substrate for calpain, and its ratio plays an important role too in controlling the activation level of calpain within a cell. The first interaction of CAST with calpain takes place at the

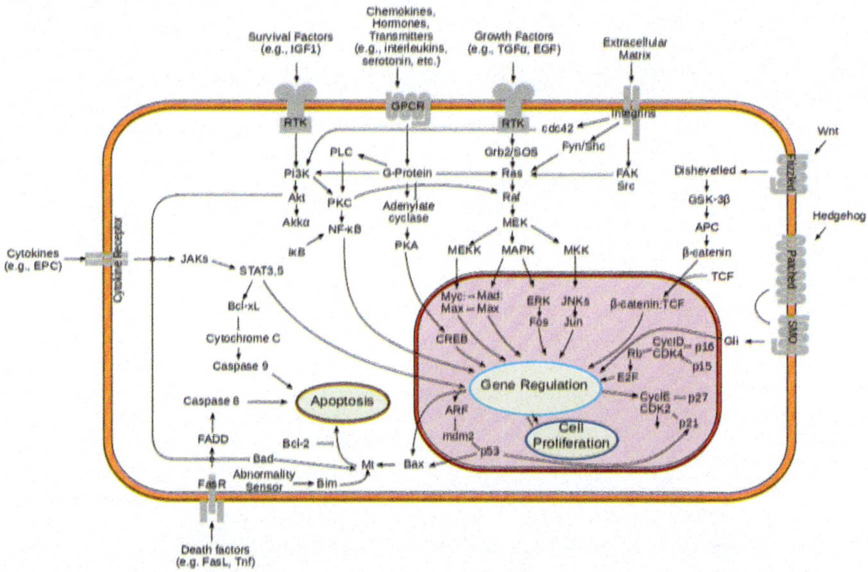

Fig. 11.4 Neuronal pathway. (ElSharkawy 2014)

membrane with a pro-calpain attached to it to interact with active calpain inside cytosol. The influx of calcium results in breaking of the reversible complex for releasing calpain. Calpain is likely to undergo autolysis inside the cytosol to get active conformation. The persistent activity of calpain is resisted through the rejoining of active calpain and CASE in a reversible complex. Figure 11.4 illustrates the slow digestion of CAST that is modulated by active calpain into small inactive fragments.

11.4 Electrophysiology of Alzheimer's

The cascade initiated by the deposition of Aβ presents a neuro-centric and linear model that may lead to synaptic dysfunction, neuronal loss, dementia, inflammation, and tau pathology. There is much controversy in the linearity of this cascade. Figure 11.5 depicts the association between Aβ and neurotoxicity that involves various molecular receptors and mechanisms. There is overexpression of APP, when there is increased concentration of Aβ or transgenic components. The neurons are likely to suffer oxidative stress, when there is interaction between Aβ and receptor at advanced glycation stage. This process is responsible for enhancing the inflammatory responses within the microglia. Moreover, across the blood-brain barrier, there is reversed transport of Aβ within the endothelial cells. The electrophysiology of Alzheimer's is depicted in as neuro-centric view after considering the contribution to different types of cells, their evolution, and interaction with each other (Fig. 11.5) (De Strooper and Karran 2016).

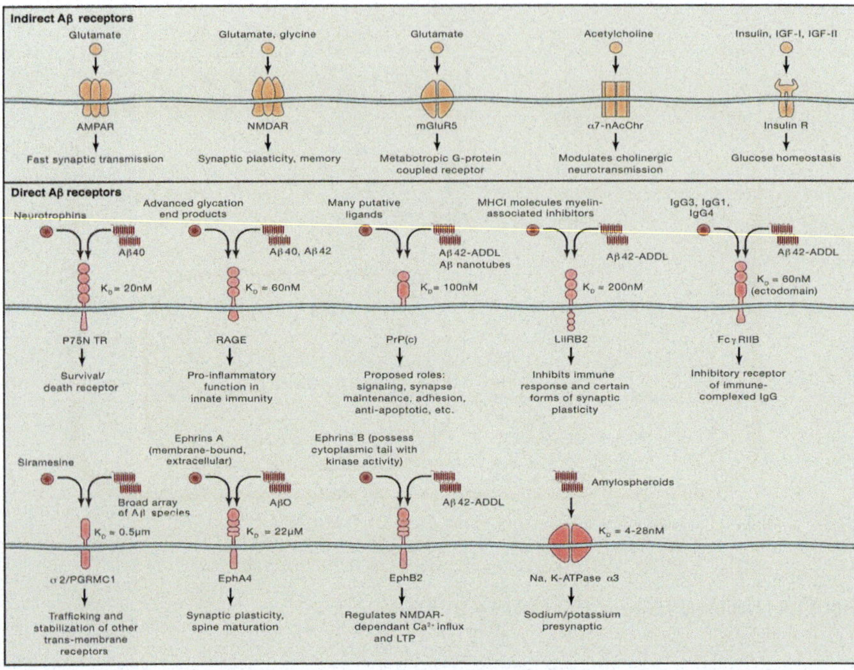

Fig. 11.5 Electrophysiology of Aβ receptors in the development of Alzheimer's disease. (Adapted with permission from De Strooper and Karran 2016)

The change in amplitude considering large-scale oscillations results due to alteration in the synchronization within a neural ensemble, which is also known as local synchronization. Oscillatory activity can be generated through the neural assembles via conducting local interactions between excitatory and inhibitory neurons.

Among Alzheimer's patients, the sources of alpha and beta activity are likely to shift toward the anterior side, with respect to the topography in anteroposterior direction. The best discriminating variables among Alzheimer's patient and normal individual are the combined alpha and theta GFP. The individual markers are helpful in clarifying the fact that combined outcomes are better than the classifications that are listed as follows (Poil et al. 2013):

- Peak width of dominant beta peak
- Bandwidth of subject-specific beta frequency
- Amplitude correlations in beta
- Range of amplitude in beta
- Power of alpha relative
- Ratio between theta and alpha power

11.4.1 Alzheimer's Disease at the Mild Cognitive Impairment (MCI) Stage

The difference between AD converters and mild cognitive impairment (MCI)-stable is easily identified at early stages through retrospective testing on data using classifier model that gives 92% sensitivity, 85% specificity, and 79% positive predictive value (Poil et al. 2013). At MCI stage, the integration of multiple biomarkers is responsible for predicting Alzheimer's disease using the logistic regression. The optimal set of biomarkers are derived from the beta frequency band that ranges between 13 and 30 Hz reflecting the early changes associated with Alzheimer's detected through EEG. Alzheimer's disease is associated with changes in the beta band; for instance, it is related to more of anterior distribution (Huang et al. 2000).

In Alzheimer's, individuals are presented with less efficient working memory as increased width of beta peak and bandwidth are associated with less stable frequency of beta (Koppel et al. 2011). Hyperexcitability suggested by increased beta frequency has also been observed among the cases of Alzheimer's. There is significant association between hippocampal hyperactivity and thin cortex as observed in precuneus and lateral temporoparietal cortices. The patients with presymptomatic amyloid-positive and suffering dementia are observed with thinning of their cortical pattern. The functional and anatomical connectivity network plays an important role in the formation of episodic memory. It has also been implicated by the neurodegenerative process of Alzheimer's disease.

The fluctuations in neuronal activity are demonstrated through the mechanisms tied to the amyloid pathology of Alzheimer's disease that are responsible for regulating levels of A-beta in the interstitial fluid (Castello and Soriano 2014). The progression of disease is likely to be modified on the basis of regulation of neural activity. The balance between pattern separation and pattern completion is assessed on the basis of lure items concerned with designing of the memory task. The entire process is mediated by DG/CA3. It has been shown that CA3 pyramidal neurons activate representations tied to prior experiences, which results failure to encode distinctive representations for new information. The greater pattern completion and diminished pattern separation indicate significant shift in the network function (Schaefer et al. 2006).

At the formative stages of Alzheimer's disease, the resulting disinhibition destabilizes network oscillatory activity due to the early failure of active inhibitory mechanisms. A new level of circuit-based pathophysiology is implicated for Alzheimer's as a result of hypersynchronous circuit activity, subclinical temporal lobe "silent" seizures, extensive rewiring of hippocampal networks, and cellular hyperexcitability. These consequences lead to the appearance of epilepsy that is responsible for aggravating memory loss among the affected individuals. The increase in interhemispheric coherence observed on frontal and temporal regions is associated with hippocampal atrophy among the individuals suffering from MCI. The increased temporal functional coupling is explained through an increase of neuronal excitability. The increase of excitability could spread over the two hemispheres through the hippocampal commissure, which is confirmed through the increased coherence between temporal regions associated with greater hippocampal atrophy.

MCI with hippocampal atrophy is characterized with decrease in frontoparietal coherence. Temporal areas wall is created as a result of increase in excitability in medial temporal areas subsequent to the hippocampal atrophy. This might cause impairment in long-range, frontoparietal functional connections in each hemisphere.

11.4.2 Multivariate Phase Synchronization

The degree of phase synchronization within a multivariate time series is measured through multivariate phase synchronization (MPS). It also allows synchronization mapping within spatially extended systems. A specific landscape of synchronization is revealed through whole-head mapping at early stage of Alzheimer's disease. Moreover, the synchronization is characterized through decrease in MPS over the frontotemporal region and an increase over the temporo-parieto-occipital region predominantly of the left hemisphere. There is significant correlation between abnormal MPS in both anterior and posterior clusters and the MMSE score, binding regional EEG synchronization that characterizes cognitive decline in Alzheimer's patients.

The early and possibly preclinical condition of Alzheimer's is characterized through increased EEG synchronization that is likely to result in loss of inhibitory interneurons. The ongoing degradation of anatomical connectivity in Alzheimer's results in hyposynchronization with progression of the disease condition. The hypoactive state of a region in early stage of Alzheimer's is manifested through increased electroencephalogram (EEG), indicating hypersynchronization in the posterior cortex (lateral and medial parietal and posterior cingulate cortices, extending into lateral occipital and medial temporal regions). At a preclinical stage, this is a very early event observed, where neither cognitive deficits nor cerebral atrophy are detected.

The lateral and medial temporal regions including parahippocampal and fusiform gyri of the left hemisphere and the uncus are the networks with decreased intra-regional synchronization that are influenced at early stage of Alzheimer's. A comparable spatial pattern of demyelination of juxtacortical white matter (U-fibers) signifies the asymmetry of hyposynchronization with greater effects in the left hemisphere. It has been shown that patients suffering from Alzheimer's exhibit longer characteristic path length over a wide range of thresholds.

11.4.3 Network Topology-Associated Alzheimer's

The decrease in clustering coefficient, reduced global and local efficiencies, and high path length characterizes the disruptions of network integrity and reductions of network efficiency among Alzheimer's patients (Wang et al. 2013). There is increase in nodal strength S (weighted degree) in the theta and delta bands exhibiting the bilateral medial orbitofrontal cortices (MOF). Theta activity is responsible of oscillating with medial frontal and cingulate cortex as it arises from the hippocampus that is associated with memory functioning. Concerning the low-frequency fluctuations, Alzheimer's patients possess stronger functional connectivity related to the bilateral medial orbitofrontal regions. This feature of Alzheimer's suggests that these patients are compensated for disruption toward the hippocampus-related connectivity. On the contrary, reduction on nodal strength within the delta band is exhibited by the right para-hippocampus. Moreover, the impairment of memory encoding depends on the decline of the nodal strength in parahippocampus gyrus. The regions experiencing significant increase in the nodal shortest path length include:

- Several central regions (preC and postC)
- Right middle frontal (CMF)
- Right inferior frontal regions (PT and PO)
- Bilateral parietal and occipital regions

The larger separations among the nodes in the network are indicated through its increased nodal shortest path length, indicating costs for information transmission and integration. The importance of a node for communication within the network is quantified through modal efficiency. Significant alterations for nodal efficiency have been observed in different cortical regions in the delta, theta, beta 2, and gamma bands. The regions with reduced nodal efficiency are likely to be distributed in the right hemisphere in the low-frequency delta band that covers the following parts:

- Right inferior frontal region
- Temporal region
- Parietal region
- Occipital region

The disrupted connections between the thalamus and multiple cortical regions are reflected through the distributed regions with significantly decreased nodal efficiency lateralized to the right hemisphere in the delta band.

11.5 Chapter Summary

Alzheimer's disease is considered as the most common form of dementia, and it is likely to worsen over time; however, it is not a normal part of aging. The onset of Alzheimer's results in the prevention of brain cell to perform well. It has been shown that plaques and tangles play an important role in blocking communication and disrupting processes between the nerve cells. The development of Alzheimer's depends on the expression of three amyloid precursor protein (APP) genes and two presenilin genes (PSEN-1 and PSEN-2) resulting in Alzheimer's. The appearance of pathological emblems results in the establishment of Alzheimer's, which involve the neurofibrillary tangles and senile plaques. The neuronal death and synaptic failure occur as a result extracellular deposits of amyloid beta and intracellular self-gathered clumps of tau proteins. The neurons are likely to suffer oxidative stress, when there is interaction between Aβ and receptor at advanced glycation stage, which enhances the inflammatory responses within the microglia.

References

Abraha I, Rimland JM, Trotta FM, Dell'Aquila G, Cruz-Jentoft A, Petrovic M, Cherubini A (2017) Systematic review of systematic reviews of non-pharmacological interventions to treat behavioural disturbances in older patients with dementia. The SENATOR-OnTop series. BMJ Open 7(3):e012759

Alzheimer's Association, 2005. Basics of Alzheimer's disease

Amjad H, Roth DL, Samus QM, Yasar S, Wolff JL (2016) Potentially unsafe activities and living conditions of older adults with dementia. J Am Geriatr Soc 64(6):1223–1232

Ashraf J, Ahmad J, Ali A, Ul-Haq Z (2018) Analyzing the behavior of neuronal pathways in Alzheimer's using petri net modeling approach. Front Neuroinform 12:26

Austrom MG, Boustani M, LaMantia MA (2018) Ongoing medical management to maximize health and well-being for persons living with dementia. The Gerontologist 58(suppl_1):S48–S57

Bekris LM, Yu CE, Bird TD, Tsuang DW (2010) Genetics of Alzheimer disease. J Geriatr Psychiatry Neurol 23(4):213–227

Bennett DA, Schneider JA, Bienias JL, Evans DA, Wilson RS (2005) Mild cognitive impairment is related to Alzheimer disease pathology and cerebral infarctions. Neurology 64(5):834–841

Bertram L, Lill CM, Tanzi RE (2010) The genetics of Alzheimer disease: back to the future. Neuron 68(2):270–281

Bunn F, Burn AM, Goodman C, Rait G, Norton S, Robinson L, Brayne C (2014) Comorbidity and dementia: a scoping review of the literature. BMC Med 12(1):192

Cacace R, Sleegers K, Van Broeckhoven C (2016) Molecular genetics of early-onset Alzheimer's disease revisited. Alzheimers Dement 12(6):733–748

Castello MA, Soriano S (2014) On the origin of Alzheimer's disease. Trials and tribulations of the amyloid hypothesis. Ageing Res Rev 13:10–12

De Strooper B, Karran E (2016) The cellular phase of Alzheimer's disease. Cell 164(4):603–615

Du AT, Schuff N, Kramer JH, Rosen HJ, Gorno-Tempini ML, Rankin K, Weiner MW (2007) Different regional patterns of cortical thinning in Alzheimer's disease and frontotemporal dementia. Brain 130(4):1159–1166

ElSharkawy AM (2014) Breast cancer. In: Al-Naggar RA (ed) Principles and practice of cancer prevention and control. OMICS Group eBooks, Foster City

Fazio S, Pace D, Kallmyer B, Pike J (2018) Alzheimer's Association towards guidelines for dementia care practice: recommendations with emphasis on high-quality, person-centered care in long-term and community-based care settings. Alzheimers Dement 14:520–521

Gentier RJ, Verheijen BM, Zamboni M, Stroeken M, Hermes DJ, Küsters B, Van Leeuwen FW (2015) Localization of mutant ubiquitin in the brain of a transgenic mouse line with proteasomal inhibition and its validation at specific sites in Alzheimer's disease. Front Neuroanat 9:26

Hardy J, Bogdanovic N, Winblad B, Portelius E, Andreasen N, Cedazo-Minguez A, Zetterberg H (2014) Pathways to Alzheimer's disease. J Intern Med 275(3):296–303

Heppner FL, Ransohoff RM, Becher B (2015) Immune attack: the role of inflammation in Alzheimer disease. Nat Rev Neurosci 16(6):358

Huang C, Wahlund LO, Dierks T, Julin P, Winblad B, Jelic V (2000) Discrimination of Alzheimer's disease and mild cognitive impairment by equivalent EEG sources: a cross-sectional and longitudinal study. Clin Neurophysiol 111(11):1961–1967

Koppel J, Campagne F, Vingtdeux V, Dreses-Werringloer U, Ewers M, Rujescu D, Greenwald BS (2011) CALHM1 P86L polymorphism modulates CSF Aβ levels in cognitively healthy individuals at risk for Alzheimer's disease. Mol Med 17(9–10):974

Loy CT, Schofield PR, Turner AM, Kwok JB (2014) Genetics of dementia. Lancet 383(9919):828–840

Mace NL, Rabins PV (2011) The 36-hour day: a family guide to caring for people who have Alzheimer disease, related dementias, and memory loss. JHU Press, Baltimore

Maslow K, Fortinsky RH (2018) Nonphysician care providers can help to increase detection of cognitive impairment and encourage diagnostic evaluation for dementia in community and residential care settings. The Gerontologist 58(suppl_1):S20–S31

Masters CL, Bateman R, Blennow K, Rowe CC, Sperling RA, Cummings JL (2015) Alzheimer's disease. Nat Rev Dis Primers 15(1):15056

National Institute on Aging. Alzheimer's disease Fact Sheet. 2016. Available from: https://www.nia.nih.gov/health/alzheimers-disease-fact-sheet

Poil SS, De Haan W, van der Flier WM, Mansvelder HD, Scheltens P, Linkenkaer-Hansen K (2013) Integrative EEG biomarkers predict progression to Alzheimer's disease at the MCI stage. Front Aging Neurosci 5:58

Putcha D, Brickhouse M, O'Keefe K, Sullivan C, Rentz D, Marshall G, Sperling R (2011) Hippocampal hyperactivation associated with cortical thinning in Alzheimer's disease signature regions in non-demented elderly adults. J Neurosci 31(48):17680–17688

Quinodoz SA, Ollikainen N, Tabak B, Palla A, Schmidt JM, Detmar E, Aznauryan E (2017) Higher-order inter-chromosomal hubs shape 3-dimensional genome organization in the nucleus. bioRxiv. https://doi.org/10.1101/219683

Schaefer EJ, Bongard V, Beiser AS, Lamon-Fava S, Robins SJ, Au R, Wolf PA (2006) Plasma phosphatidylcholine docosahexaenoic acid content and risk of dementia and Alzheimer disease: the Framingham Heart Study. Arch Neurol 63(11):1545–1550

Wang J, Zuo X, Dai Z, Xia M, Zhao Z, Zhao X, He Y (2013) Disrupted functional brain connectome in individuals at risk for Alzheimer's disease. Biol Psychiatry 73(5):472–481

World Health Organization. WHO|Dementia. WHO [Internet]. World Health Organization. 2017. Available from: http://www.who.int/mediacentre/factsheets/fs362/en/

Biotechnology and Bioinformatics Applications in Alzheimer's Disease

12

Mahmoud A. Ali, Athanasios Alexiou, and Ghulam Md Ashraf

Abstract

Alzheimer's disease is one of the most severe types of dementia that causes problems with memory, thinking, and behavior. Biotechnology and bioinformatics are nowadays involved in the establishment of advanced methods of diagnosis and treatment, including molecular medicine, personalized medicine, gene identification and manipulation, as well as neural engineering. Next-generation sequencing is one of the strongest tools for studying genetic diseases and gene mutations. Additionally, brain-computer interface could be used in the near future to assist people with paralysis or other related disorders and physical injuries to move toward into a better way of life, restoring memory or improving the way of everyday life. This chapter aims to provide an overview of the most common and an advanced application of biotechnology and bioinformatics in Alzheimer's including the genome-wide association studies and the role of microbiome detection in Alzheimer's disease.

M. A. Ali (✉)
Biotechnology Program, Faculty of Agriculture, Cairo University, Giza, Egypt
e-mail: mahmoud.bioinfo@gmail.com

A. Alexiou
Novel Global Community Educational Foundation, Hebersham, NSW, Australia

AFNP Med, Wien, Austria

G. M. Ashraf
King Fahd Medical Research Center, King Abdulaziz University, Jeddah, Saudi Arabia

Department of Medical Laboratory Technology, Faculty of Applied Medical Sciences, King Abdulaziz University, Jeddah, Saudi Arabia

G. M. Ashraf, A. Alexiou (eds.), *Biological, Diagnostic and Therapeutic Advances in Alzheimer's Disease*, https://doi.org/10.1007/978-981-13-9636-6_12

Keywords
Algorithms in biology · Alzheimer's disease · Bioinformatics · Brain-computer interface · Genome-wide association study · Next-generation sequencing · Neural engineering

12.1 Introduction

The human brain is the most important and complicated part of the human body, and the central nervous system (CNS) gathers information and controls all functions of the body simultaneously. Neuroengineering is a discipline within biomedical engineering that uses engineering techniques to understand, repair, replace, enhance, and exploit the properties of the neural system. Neural engineering is based on computational neuroscience, experimental neuroscience, clinical neurology, electrical engineering, and signal processing of living neural tissue and encompasses elements from robotics, cybernetics, computer engineering, neural tissue engineering, materials science, and nanotechnology. Additionally, discrete mathematics techniques, dynamic programming (DP) methods, and greedy (GA) and hybrid algorithms are mainly used to address optimization genetic problems and to demonstrate a multitude of acceptable solutions (Levitin 2011). The most widespread problems of dynamic programming have the following attributes:

1. Simple subproblems where the original problem is severed in a sum of simpler subproblems.
2. Principle of optimality: The optimum total solution is the result of the combination of optimum subproblem solutions.
3. Overlapping subproblems: Although they seem to be uncorrelated, it is possible that the subproblems share optimal solutions of common subproblems.

There are a few of studies concerning the application of DP methods in biology and bioinformatics in general. The DP idea runs for many problems such as the coin-row problem, the change-making problem, the coin-collecting problem, the knapsack problem and memory functions, the optimal binary search trees, the hidden Markov models, and the various alignment techniques.

12.2 Neuroinformatics and Brain-Computer Interface

Electroencephalography (EEG) is an electrophysiological monitoring method to record the electrical activity of the brain and use it to our advantage. EEG records the electrical signals of the brain's neurons and measures the voltage fluctuations of the ionic current in the neurons of the brain. Clinically, EEG refers to the recording of the brain's spontaneous electrical activity over a period.

Electroencephalography is recorded from multiple electrodes placed on the scalp. Usually, EEG is used to diagnose neurodegeneration, which causes abnormalities in EEG readings (Alexiou et al. 2017). Additionally, EEG can be used to diagnose sleep disorders, coma, encephalopathy, and brain death (Lau et al. 2012). A brain-computer interface (BCI) is a direct connection between a wired brain and an external device. BCIs usually are targeting at mapping, augmenting, or repairing human cognitive or sensory-motor functions. Nowadays, BCI research and development is focused on neuroprosthetics applications that aim at restoring damaged hearing, sight, and movement. Due to the outstanding cortical plasticity of the brain, signals from implanted prostheses can, after adaptation, be handled by the brain like a natural sensor (Herwig et al. 2003). BCI can be used in a variety of ways. It can enable for people with paralysis to do actions as typing via direct brain control at the highest speeds and accuracy levels. It can also make people, who are unable to move, feel like they can move using a hybrid treatment solution of BCI combined with virtual reality (VR). Studies have come to the conclusion that BCI can improve the quality of life for people with paralysis or accidents. Recently, researchers have made innovated steps in the art therapy research, by using noninvasive methods like the sonogenetics method where low-pressure ultrasound is applied for the activation of neurons even in the deeper brain regions (Ibsen et al. 2015) or by generating sounds as a direct biofeedback from the brain signals to understand real-time temporal patterns related to behavior and cognitive abnormal conditions (Chatzichronis et al. 2019).

12.3 Alzheimer's Disease Diagnosis

Current approaches for the diagnosis of various neurological diseases employ a variety of different methodologies. Apart from the standard paper and pencil neuropsychological tests used for the detection of AD-related cognitive impairments, researchers have identified reliable biomarkers through cerebrospinal fluid analyses, molecular neuroimaging, and structural and functional MRI associated with alterations in certain brain areas (Mantzavinos and Alexiou 2017). For instance, a "timeline" of the unobserved progress of AD before the appearance of its symptoms has recently been framed by scientists. This task included spinal fluid and blood test, assessments of mental ability, and brain scans. This finding strongly suggests that some important changes occur in the brain of patients suffering from inherited AD, these changes occur years before the appearance of actual symptoms, and this may have significant implications for the diagnostics and therapeutics of AD. Despite such remarkable progress in the early and accurate classification of the predementia stage, the currently employed methodologies are expensive, invasive, impractical, and sometimes painful. Also, these methods fail to incorporate the comprehensive set of behavioral patterns that are spontaneously generated by populations with or at the risk of developing MCI while engaged in everyday activities. While there is a consensus that dementia-related brain pathology causes a diverse set of behavioral alterations, there is still no systematic attempt to explore these complex alterations, relate them to existing biomarkers, compare and contrast them with normal behaviors, and extract patterns and algorithms that can effectively differentiate between the normal and pathological behavior.

12.4 Genome-Wide Association Studies

In genome-wide association studies (GWAS), if potential biomarkers do not provide clear information about the disease mechanism or have very small effect on disease risk and very small portion on the total estimated genetic variance of the disease, they may be proved inefficient. Researchers in order to detect genetic factors that are highly associated with AD have developed a multi-SNP GWAS to detect interactions between different loci and compute linkage disequilibrium (LD) between each SNP and AD (Figs. 12.1 and 12.2) (Bodily et al. 2016).

Several studies have focused on the role of CSF biomarkers in predicting the possibility of patients affected by MCI to convert to AD. The results seem to indicate that patients with MCI who have a higher incidence to develop AD have lower CSF levels of Aβ42 compared with healthy controls and stable MCI cases (Mattsson 2009, 2010). While CSF levels of Aβ42 and tau proteins are the most common and efficient potential AD biomarkers, in a recent GWAS study, researchers in order to reveal sex-specific genetic AD predictors identified loci that show many disparate associations between males and females (Deming et al. 2018). Specifically, in CSF Aβ42 and total tau tests on female and male patients, researchers identified sex-specific associations with amyloidosis in two loci in chromosomes 4 and 6, suggesting a crucial role for SERPINB1 in females and for OSTN and CLDN16 in tau pathology (Deming et al. 2018). Another latest study revealed the role of the M20-domain-containing protein 1 (PM20D1) in neuroprotection against AD. The overexpression of PM20D1 reduced cell death and Aβ levels, improving simultaneously cognitive performance (Sanchez-Mut et al. 2018).

Fig. 12.1 Single-SNP GWAS. (The algorithm is reprinted from Bodily et al. 2016, distributed under the terms of the Creative Commons Attribution 4.0 International License)

```
Input: Genotypes for a set of SNP loci, L and phenotypes
       for a set of individuals, S
for SNP location l in L do
    initialize Counts
    for individual s ∈ S do
        B_j ← phenotype of s
        for genotype A_i at l do
            increment Counts(A_i, B_j)
        end for
    end for
    compute p(A_i ∧ B_j) for all values of A_i, B_j
    compute p(A_i) and p(B_j) for all values of A_i, B_j
    D ← p(A_1 ∧ B_1) − (p(A_1) × p(B_1))
    compute and report Pearson's r
end for
```

Fig. 12.2 Multi-SNP parallelized GWAS. (The algorithm is reprinted from Bodily et al. 2016, distributed under the terms of the Creative Commons Attribution 4.0 International License)

Input: Genotypes for a set of SNP loci, L and phenotypes for a set of individuals, S, and SNP IDs $START$, END

for SNP location l in L within partition $\{START, END\}$ **do**
    for SNP location m in L **do**
        initialize $Counts$
        for individual $s \in S$ **do**
            $C_k \leftarrow$ phenotype of s
            for genotype A_i at l **do**
                for genotype B_i at m **do**
                    increment $Counts(A_i, B_j, C_k)$
                end for
            end for
        end for
        compute $p(A_i \wedge B_j \wedge C_k)$ for all values of A_i, B_j, C_k
        compute $p(A_i)$, $p(B_j)$, and $p(C_k)$ for all values of A_i, B_j, C_k
        $D \leftarrow p(A_1 \wedge B_1 \wedge C_k) - (p(A_1 \wedge B_j) \times p(C_k))$
        compute and report Pearson's r
    end for
end for

S.No.	Gene Name	Ac. No.	Length (Amino acids)	Tissue Type
1	A2M	AAH26246	353	LIVER
2	ABCB1	AA130425	1280	CERABELLUM
3	ACHE	AAH36813	546	BRAIN
4	AD7C-NTP	AAC08737	375	NEURONAL
5	APLP1	AAH12889	650	OVARY
6	APOB	AAH51278	825	LIVER
7	APOD	AAH07402	189	MELANOTIC MELANOMA
8	APOE	BBA96060	63	BLOOD
9	APP	AAH65529	751	RETIOBLASTOMA
10	BACE1	AAH36084	501	BRAIN
11	ABCA2	AAH08755	867	EYE
12	ABAD	AAH08708	252	BRAIN
13	BCHE	AAH08396	64	BRAIN
14	BDNF	AAH29795	247	BRAIN
15	CASP6	AAH00305	293	LUNG
16	CCK	AAH08283	115	PANCRESS
17	CCR5	AAB09551	215	NOT SPECIFIED
18	CD36	CCA83662	472	NOT SPECIFIED
19	CD40LG	CAI42902	240	NOT SPECIFIED
20	CDH1	AAY68225	882	NOT SPECIFIED
21	CDK5	CAG33322	292	NOT SPECIFIED
22	CETP	AAB59388	425	LIVER
23	CFTR	NP_000483	1480	NOT SPECIFIED
24	CHAT	AAI30618	630	CERABELLUM
25	CHRNA7	AAH37571	321	BRAIN
26	CLU	AAH19588	449	BRAIN
27	CSF1	AAH21117	554	KIDNEY
28	CSNK1D	AAH15775	409	SPLEEN
29	CTNNA3	AAH65819	516	PERIPHERAL NERVOUS SYSTEM
30	CTSD	CAG33228	412	NOT SPECIFIED
31	CYCS	AAH67222	105	CARCINONA
32	CYP19A1	AAH56258	359	PLACENTA
33	DBN1	AAH07567	649	RHABDOMYOSARCOMA
34	DCN	AAH05322	359	LIVER
35	DSCR1	AAH02864	197	LUNG
36	ESR1	CAI42285	595	NOT SPECIFIED
37	ESR2	AAH24181	323	TESTIS
38	FGF2	NP_001997	288	NOT SPECIFIED
39	FN1	AAH16875	268	CARCINOMA
40	GAL	AAH30241	123	CARCINOMA
41	GLUL	AAH51726	373	BRAIN
42	GSK3B	AAM88578	74	FETAL BRAIN SYSTEM
43	HTR6	AAH74996	440	FETAL BRAIN SYSTEM
44	IGF1R	AAH10607	55	BRAIN
45	FE65	AAH10854	708	LEIMYOSARCOMA
46	IL18	CAG46798	193	NOT SPECIFIED
47	LRP1	AAH21204	439	RHABDMOMYOOSAREOMA
48	MAPK1	AAH99905	360	BRAIN
49	MAPT	AAH0558	352	BRAIN
50	NCSTN	AAH47621	689	TESTIS
51	PIN1	AAH31971	45	TESTIS
52	PLAU	AAH13575	431	CARCINOMA
53	PNMT	NP_002677	282	NOT SPECIFIED
54	MCP1	AAO75526	25	NOT SPECIFIED
55	NP1	AAK61283	473	NOT SPECIFIED
56	NgR	ABC69293	600	BRAIN
57	PAT1	AAC83973	585	NOT SPECIFIED
58	IVIg	CAC29069	110	B-CELL
59	LPL	CAG3335	475	NOT SPECIFIED
60	PSEN1	AAH11729	463	MELANOTIC MELONOMA
61	PSEN2	CAH73110	448	NOT SPECIFIED
62	S100B	AAH01766	92	MELANOTIC MELONOMA
63	SNCA	AAI08276	140	MELANOTIC MELONOMA
65	STH	AI30322	128	CEREBELLUM
66	UBB	AAH3899	229	BRAIN
67	VEGF	AAA35789	191	NOT SPECIFIED
68	PRND	AAH43644	176	TESTIS
69	PARP1	AAH14206	250	MELANOTIC MELONOMA
70	MAPK10	AAH35057	461	BRAIN
71	MAPK14	AAH31574	360	BRAIN
72	IL1RAPL2	AAI0478	686	BRAIN
73	IL2RA	CAI41071	200	NOT SPECIFIED
74	IDE	CAI132670	1019	NOT SPECIFIED

Fig. 12.3 Proteins involved in the pathogenesis of Alzheimer's disease (The algorithm is reprinted from Rao et al. 2008, distributed under the terms of the Creative Commons Attribution 4.0 International License)

Multiple sequence alignment and phylogenetic analysis underlie also the significant role of certain proteins to the development and progression of AD (Fig. 12.3) (Rao et al. 2008; Mantzavinos and Alexiou 2017). Presenilin 1 (PS1) carries out the major function of γ-secretase called proteolysis. Additionally, Presenilin 2 (PS2) cooperates in chemical signal transmission from the cell membrane into the nucleus, a very important task for the cell growth and maturation. Mutations in PS1 (chromosome 12), PS2 (chromosome 1), and APP (chromosome 21) genes may increase Aβ42 generation, which is currently recognized as a potential AD biomarker and risk factor (Rao et al. 2008).

Additionally, long noncoding RNAs (lnRNAs) which are defined as transcripts longer than 200 nucleotides that are not translated into protein can regulate gene expression and seem to be involved in many neurodegeneration disorders. BACE1-AS regulates the expression of BACE1 by increasing BACE1 mRNA stability and generating additional BACE1 through a posttranscriptional feed-forward mechanism. The same mechanism increases Aβ concentrations, the main constituent of senile plaques (Luo et al. 2016). A recent GWAS study shows that DNA methylation in the TOMM40 gene on chromosome 19 is associated with AD and revealed three main SNPs highly associated with AD: rs593742 (ADAM10, locus chr15:58873555–59120077); rs889555 210 (BCKDK/KAT8, locus chr16:30916129–31155458); and rs6504163 (ACE, locus 211 chr17:61545779–61578207) (Marioni et al. 2018).

PAB10 gene is related to AD resilience, with a significant role in neurotransmitter release, phagosome maturation, and GLUT4 translocation and in endocytosis. While RAB10 is involved in neuronal axon genesis, researchers assume that 50% knockdown of PAB10 may result in a 45% reduction in Aβ42 levels and a 61% reduction in the Aβ42/Aβ40 ratio. It seems that targeting RAB10 could provide a novel AD therapy (Ridge et al. 2017).

In a recent study concerning gut microbiome alterations in AD, researchers identified significant differences in the composition of gut microbiome between the control group and the AD group, where AD participants shown decreased abundance of Firmicutes and Actinobacteria and increased abundance of Bacteroidetes compared to the healthy participants (Figs. 12.4, 12.5 and 12.6) (Minter et al. 2017; Vogt et al. 2017). Additionally, observed correlations between levels of differentially abundant genera and CSF AD biomarkers (Fig. 12.7) suggest that gut microbiome is related with AD (Vogt et al. 2017).

12.5 Discussion

Any effort focused on early detection and longitudinal study of the individuals that might have a greater risk to develop AD over the course of their life will improve our understanding of the natural history of AD (Gardner et al. 2009). Improved knowledge of the early onset of AD may also be important for future AD therapies which may have the greatest impact if treatment is initiated already in the prodromal stage, while most of the ongoing drug tests failed to stop AD

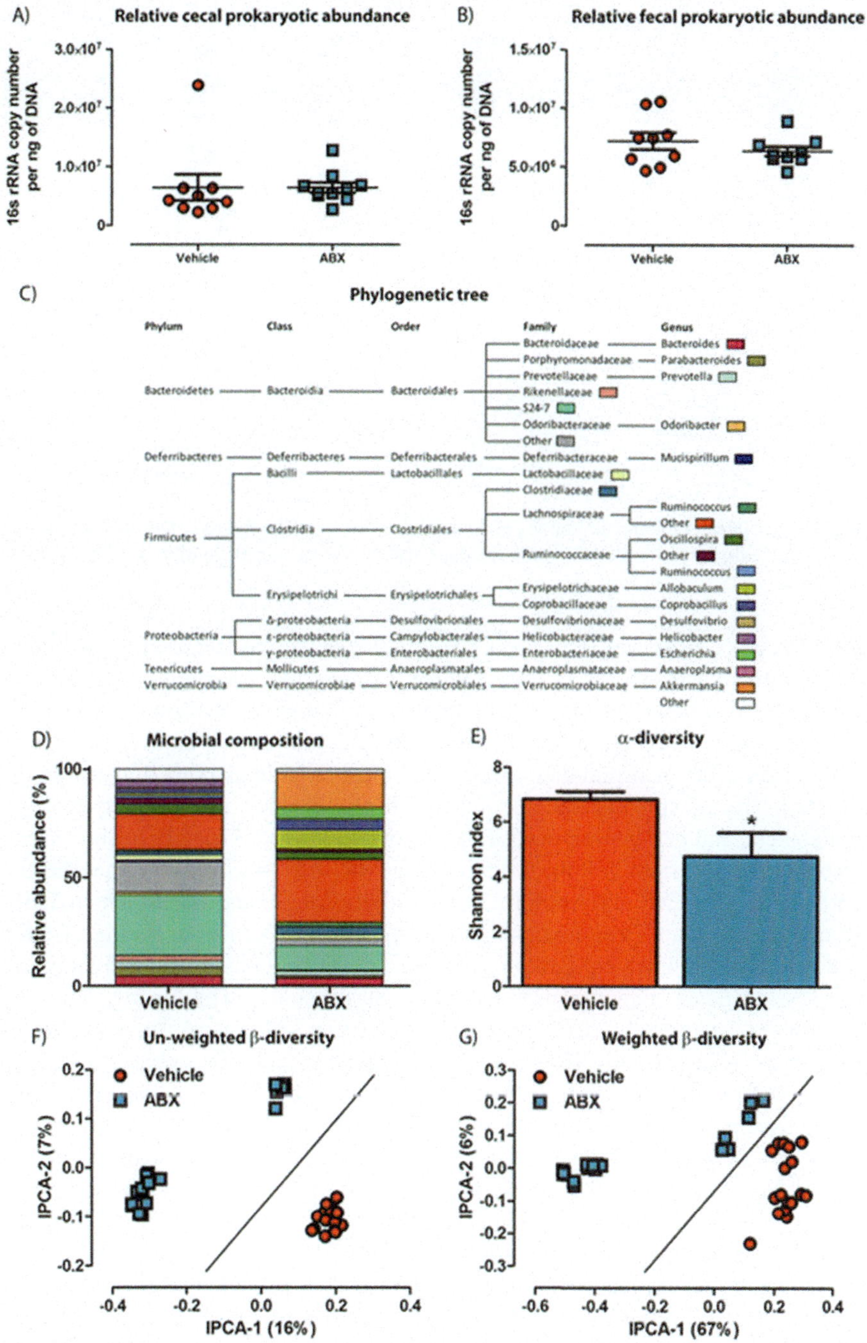

Fig. 12.4 Alterations in gut microbial diversity in an aged $APP_{SWE}/PS1_{\Delta E9}$ murine model of Alzheimer's disease (The figure is reprinted from Minter et al. 2017,)

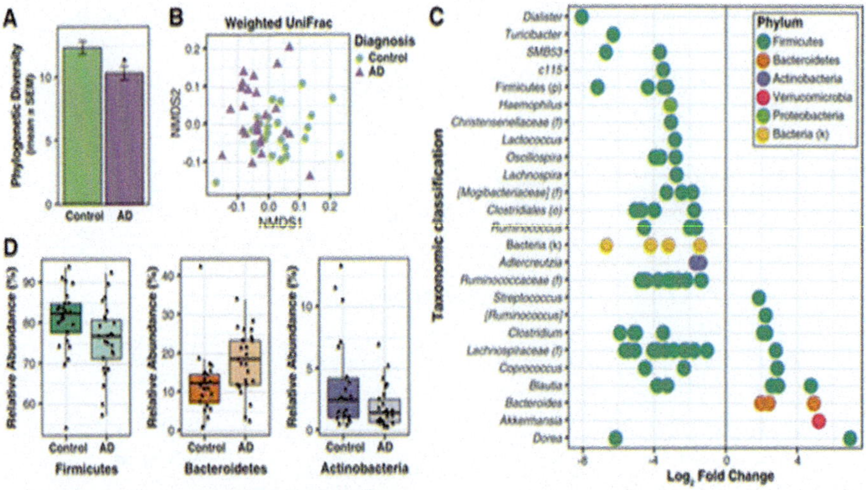

Fig. 12.5 Gut microbiome alterations in Alzheimer's disease (The figure is reprinted from Vogt et al. 2017, distributed under the terms of the Creative Commons Attribution 4.0 International License)

progression (Callaway 2012). Much focus has been directed on patients with mild cognitive impairment (MCI), to recognize markers that will predict the progression of the disease, while many patients with MCI display the same morphological changes with AD patients (Petersen et al. 2001; Blennow et al. 2006; Visser and Verhey 2008). Current data suggest that most reliable markers of disease progression are an episodic memory and semantic fluency for neuropsychology, delta, theta, and beta power, event-related potentials and P3 for EEG, whole brain atrophy and ventricle volume for structural MR, Aβ1–42, Aβ1–40 phospho-tau and total-tau for CSF, and ApoE for blood. Additional clinical assessment includes Mini-Mental State Examination (MMSE), Clinical Dementia Rating (CDR), Medical History, Physical Examination, Hachinski Ischemic Scale, the Geriatric Depression Scale, Functional Assessment Questionnaire, The Neuropsychiatric Inventory Questionnaire (NPI-Q), and the Alzheimer's Disease Assessment Scale-Cognitive (ADAS-COG). Nowadays, it is clear also that AD is characterized by Aβ overexpression, which in turn causes an alteration to the proteins, thus facilitating fission and fusion events in mitochondria (Wang et al. 2008). Moreover, this results in increased ROS production, mitochondrial dysfunction, reduced potential of the mitochondrial membrane, increased ATP generation, and enhanced mitochondrial fragmentation (Wang et al. 2008). In human AD brain, Aβ interacts with Aβ binding mitochondrial matrix protein (alcohol dehydrogenase) and has been suggested to facilitate oxidative stress and mitochondrial dysfunction (Lustbader et al. 2004). On the other hand, increased Aβ production has been proposed to be an outcome of neuronal apoptosis (Chen and Chan 2009; Martin 2010). Moreover, defects in fission and fusion-mediated mitochondrial dynamics have been reported to cause a decrease in mitochondrial movement (Chen et al.

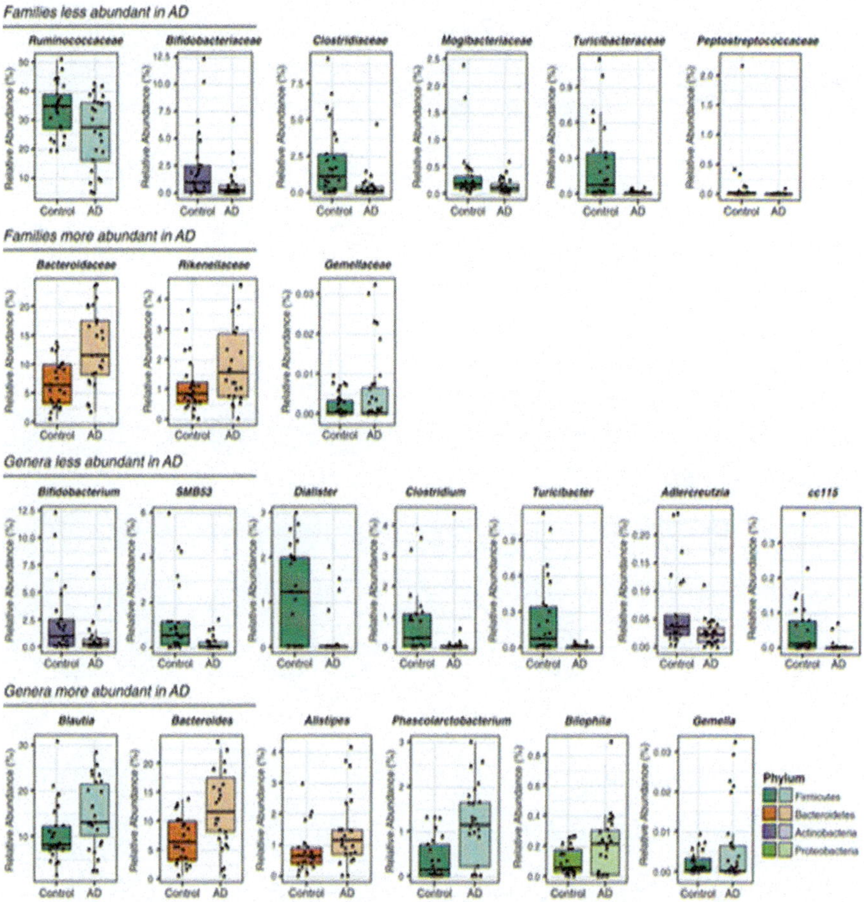

Fig. 12.6 Gut microbiome alterations in Alzheimer's disease – taxonomy by families and genera (The figure is reprinted from Vogt et al. 2017, distributed under the terms of the Creative Commons Attribution 4.0 International License)

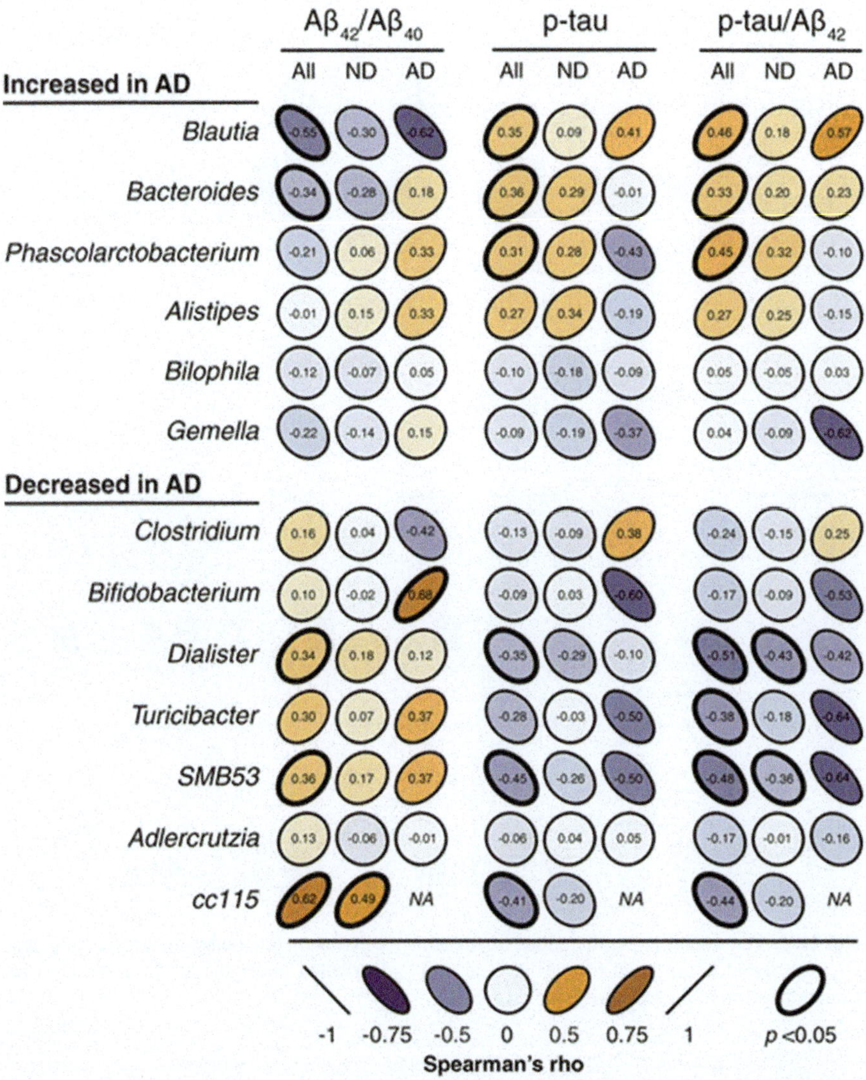

Fig. 12.7 Genera and cerebrospinal fluid biomarkers of AD and genera (The figure is reprinted from Vogt et al. 2017, distributed under the terms of the Creative Commons Attribution 4.0 International License)

2007). In fusion-deficient neurons, swelling and aggregation-induced increase in mitochondrial diameter have been reported to obstruct their smooth passage into neuritis, thus inducing apoptosis in axons and dendrites of mitochondria (Chen et al. 2007). These deficiencies further lead to abnormal neuronal development, thus causing neurodegeneration gradually (Chen et al. 2007).

References

Alexiou A, Mantzavinos VD, Greig NH, Kamal MA (2017) A Bayesian model for the prediction and early diagnosis of Alzheimer's disease. Front Aging Neurosci 9:77. https://doi.org/10.3389/fnagi.2017.00077

Blennow K, deLeon MJ, Zetterberg H (2006) Alzheimer's disease. Lancet 368(9533):387–403

Bodily PM, Fujimoto MS, Page JT, Clement MJ, Ebbert MT, Ridge PG, Alzheimer's Disease Neuroimaging Initiative (2016) A novel approach for multi-SNP GWAS and its application in Alzheimer's disease. BMC Bioinf 17(Suppl 7):268. https://doi.org/10.1186/s12859-016-1093-7

Callaway E (2012) Alzheimer's drugs take a new tack. Nature 489:13–14. https://doi.org/10.1038/48901

Chatzichronis S, Alexiou A, Simou P, Matzavinos V, Tsiamis V et al (2019) Neurocognitive assessment software for enrichment sensory environments. J Proteomics Bioinform 12:018–028. https://doi.org/10.4172/0974-276X.1000492

Chen H, Chan DC (2009) Mitochondrial dynamics – fusion, fission, movement, and mitophagy-in neurodegenerative diseases. Hum Mol Genet 18(R2):R169–R176

Chen H, McCaffery JM, Chan DC (2007) Mitochondrial fusion protects against neurodegeneration in the cerebellum. Cell 130(3):548–562

Deming Y, Filipello F, Cignarella F, Hsu S, Mikesell R, Li Z, Del-Aguila JL, Dube U, Farias FG, Bradley J, Cantoni C, Benitez B, Budde J, Ibanez L, Fernandez MV, Blennow K, Nellgard B, Zetterberg H, Heslegrave A, Johansson PM, Svensson J, Lleo A, Alcolea D, Clarimon J, Rami L, Molinuevo JL, Suarez-Calvet M, Morenas-Rodriguez E, Kleinberger G, Ewers M, Brett TJ, Haass C, Harari O, Karch CM, Piccio L, Cruchaga C (2018) The MS4A gene cluster is a key regulator of soluble TREM2 and Alzheimer disease risk. bioRxiv 352179. https://doi.org/10.1101/352179

Gardner A et al (2009) Differences at brain SPECT between depressed females with and without adult ADHD and healthy controls: etiological considerations. Behav Brain Funct 5(1):37

Herwig U, Satrapi P, Schönfeldt-Lecuona C (2003) Using the international 10-20 EEG system for positioning of transcranial magnetic stimulation. Brain Topogr 16:95–99. https://doi.org/10.1023/B:BRAT.0000006333.93597.9d

Ibsen S et al (2015) Sonogenetics is a non-invasive approach to activating neurons in Caenorhabditis elegans. Nat Commun 6:8264. https://doi.org/10.1038/ncomms9264

Lau T, Gwin J, Ferris D (2012) How many electrodes are really needed for EEG-based mobile brain imaging? J Behav Brain Sci 2(3):387–393. https://doi.org/10.4236/jbbs.2012.23044

Levitin A (2011) Introduction to the design and analysis of algorithms, 3rd edn. Pearson Education Limited, Harlow

Luo S, Lu JY, Liu L (2016) Divergent lncRNAs regulate gene expression and lineage differentiation in pluripotent cells. Cell Stem Cell 18(5):637–652

Lustbader JW et al (2004) ABAD directly links Abeta to mitochondrial toxicity in Alzheimer's disease. Science (New York, NY) 304(5669):448–452

Mantzavinos V, Alexiou A (2017) Biomarkers for Alzheimer's disease diagnosis. Curr Alzheimer Res 14(11):1149–1154

Marioni RE, Harris SE, McRae AF et al (2018) GWAS on family history of Alzheimer's disease. Transl Psychiatry 8:99

Martin LJ (2010) Mitochondrial and cell death mechanisms in neurodegenerative diseases. Pharmaceuticals 3(4):839–915

Mattsson N (2009) CSF biomarkers and incipient Alzheimer disease in patients with mild cognitive impairment. JAMA 302:385

Mattsson N (2010) CSF biomarkers and incipient Alzheimer disease in patients with mild cognitive impairment. JAMA 302(4):385–393. https://doi.org/10.1001/jama.2009.1064

Minter MR, Hinterleitner R, Meisel M, Zhang C, Leone V, Zhang X, Oyler-Castrillo P, Zhang X, Musch MW, Shen X, Jabri B, Chang EB, Tanzi RE et al (2017) Antibiotic-induced perturbations in microbial diversity during post-natal development alters amyloid pathology in an aged APPSWE/PS1ΔE9 murine model of Alzheimer's disease. Sci Rep 7(1):10411. https://doi.org/10.1038/s41598-017-11047-w

Petersen RC, Doody R, Kurz A et al (2001) Current concepts in mild cognitive impairment. Arch Neurol 58(12):1985–1992

Rao AA, Reddi KK, Thota H (2008) Bioinformatic analysis of Alzheimer's disease using functional protein sequences. J Proteomics Bioinform 1:036–042. https://doi.org/10.4172/jpb.1000007

Ridge PG, Karch CM, Hsu S, Arano I, Teerlink CC, Ebbert M, Gonzalez Murcia JD, Farnham JM, Damato AR, Allen M, Wang X, Harari O, Fernandez VM, Guerreiro R, Bras J, Hardy J, Munger R, Norton M, Sassi C, Singleton A, Younkin SG, Dickson DW, Golde TE, Price ND, Ertekin-Taner N, Cruchaga C, Goate AM, Corcoran C, Tschanz J, Cannon-Albright LA, Kauwe J, Alzheimer's Disease Neuroimaging Initiative (2017) Linkage, whole genome sequence, and biological data implicate variants in RAB10 in Alzheimer's disease resilience. Genome Med 9(1):100. https://doi.org/10.1186/s13073-017-0486-1

Sanchez-Mut JV et al (2018) PM20D1 is a quantitative trait locus associated with Alzheimer's disease. Nat Med 24:598–603

Visser PJ, Verhey FR (2008) Mild cognitive impairment as predictor for Alzheimer's disease in clinical practice: effect of age and diagnostic criteria. Psychol Med 38(1):113–122

Vogt NM, Kerby RL, Dill-McFarland KA, Harding SJ, Merluzzi AP, Johnson SC, Carlsson CM, Asthana S, Zetterberg H, Blennow K, Bendlin BB et al (2017) Gut microbiome alterations in Alzheimer's disease. Sci Rep 7(1):13537. https://doi.org/10.1038/s41598-017-13601-y

Wang X et al (2008) Amyloid-beta overproduction causes abnormal mitochondrial dynamics via differential modulation of mitochondrial fission/fusion proteins. Proc Natl Acad Sci U S A 105(49):19318–19323

Neurobiological Mechanisms Involved in the Pathogenesis of Alzheimer's Disease

13

Fayaz Ahmad Mir and Zaigham Abbas Rizvi

Abstract

Alzheimer's disease (AD) is one of the many neurodegenerative disorders which is characterized by progressive loss of neurons due to the extracellular accumulation of misprocessed and aggregated amyloid beta (Aβ)-plaques and appearance of intracellular neurofibrillary tangles containing hyperphosphorylated tau protein which ultimately leads to loss of synapses and cognitive decline. Aggregation of amyloid beta (Aβ)-plaques is the hallmark of AD. Aβ is the proteolytic cleavage product of amyloid precursor protein (APP) which is cleaved by β- and γ-secretase enzymes into $A\beta_{1-42}$ and $A\beta_{1-40}$ isoforms where the former readily aggregate more rapidly than the latter. Tau protein, the major component of neurofibrillary tangles, is a microtubule-associated protein which is usually soluble but becomes insoluble as it forms tangles of oligomers which is thought to be initiated by toxic concentrations of Aβ-plaques. Recent studies have shown that some genetic mutations, genomic instability and other factors like head injuries, depression, imbalanced diet and age progression all contribute to the development and progression of AD. The most important gene, for which a role in ageing-related late-onset AD has been established since a decade, is APOE where different variants of the gene differently predispose the individuals to the development of AD. In this chapter, we will be highlighting well-established molecular and cellular mechanisms behind the development and progression of AD, the regions in the brain that are affected and the known genetic basis behind the onset and pathophysiology of AD. In the later section, we will address some of

F. A. Mir
School of Life Sciences, Jawaharlal Nehru University, New Delhi, Delhi, India

Z. A. Rizvi (✉)
Centre for Microbial Ecology, Translational Health Science and Technology Institute, NCR Biotech Science Cluster, Faridabad, Haryana, India
e-mail: rizviza@thsti.res.in

G. M. Ashraf, A. Alexiou (eds.), *Biological, Diagnostic and Therapeutic Advances in Alzheimer's Disease*, https://doi.org/10.1007/978-981-13-9636-6_13

the current and prospective therapeutic interventions based on our current understanding of neurobiological mechanisms underlying AD.

Keywords
Alzheimer's disease · Aβ-plaques · Neurodegeneration · Cytotoxicity · Neuroinflammation tau protein · Axonal transport

13.1 Introduction

Alzheimer's disease is a type of progressive neurodegenerative disorder characterized by dementia, which constitutes a group of symptoms including problems in learning and memory, defects in language, speech and other cognitive functions that affect individual's ability to carry out his/her routine activities (Fig. 13.1) (Association, A.s 2018). AD was first described almost a century ago by Dr. Alois Alzheimer (German psychiatrist) at a conference of Bavarian Psychiatrists, where Dr. Alois Alzheimer presented an intriguing case of a patient named Auguste D, a lady in mid-50s, suffering from dementia, paranoia and many other psychological changes. Dr. Alois, based on autopsy reports, explained that there were lesions in and around some specific nerve cells of her brain. With the advent of an electron microscope, which allowed the scientists to magnify the neuronal cells up to a

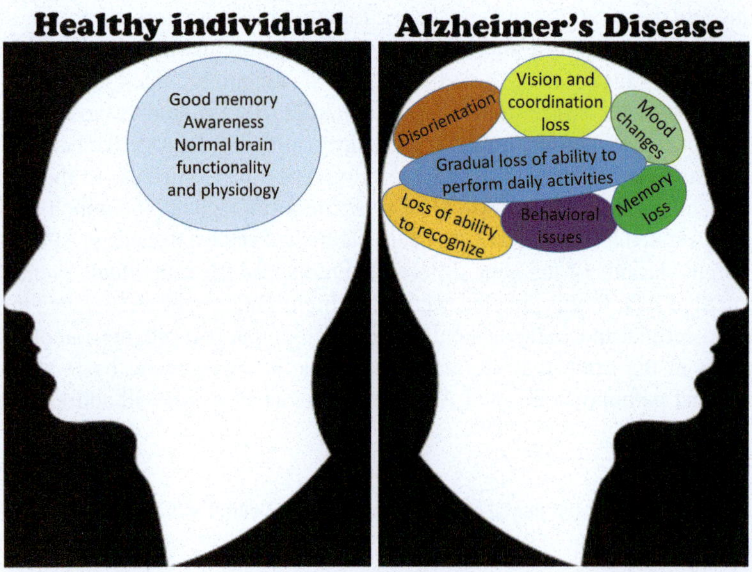

Fig. 13.1 Comparison of brain functionality and physiology of healthy and Alzheimer's disease individual. Alzheimer's disease is characterized by progressive loss of neurons affecting normal brain physiology. Alzheimer's patients suffer from disorientation, memory and behavioural loss and impairments in coordination and suffer an inability to perform day-to-day activity

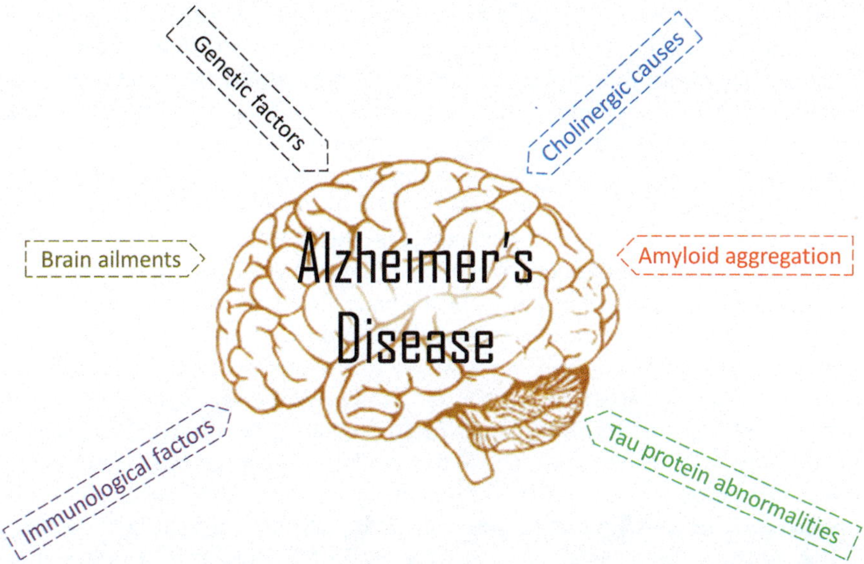

Fig. 13.2 Pathophysiological causes in Alzheimer's disease. Alzheimer's disease could be caused due to a number of underlying biological factors. While aggregation of amyloid beta protein and tau protein abnormalities are believed to be the major contributing factors, other factors involving gene mutations, inflammatory responses and decline in neurotransmitters levels such as acetylcholine may also contribute to the disease onset and progression

million times, further paved the way for understanding the deeper aspects of Alzheimer's disease. By the 1980s protein components of senile plaques and neurofibrillary tangles, which are the hallmarks of Alzheimer's disease, were identified along with the identification and characterization of β-amyloid and microtubule-associated tau proteins (Sisodia and Tanzi 2007).

Current understandings suggest that the causative agents behind Alzheimer's disease vary from brain ailments, cholinergic factors to cellular as well as molecular causes such as genetic susceptibility, protein aggregation/dysfunction and inflammatory factors (Fig. 13.2). In the USA alone, it is estimated that 5.7 million people suffer from Alzheimer's disease, and the number is projected to grow to around 13.8 million by the year 2050. Data recorded from official death certificates in the year 2015 showed that 110,561 Americans had died due to Alzheimer's disease making it the sixth leading cause of deaths in the USA (Association, A.s 2018). Several reports suggest that there is a correlation between Alzheimer's disease and age-specific occurrences which in the last three decades have significantly decreased (Prince et al. 2016; Satizabal et al. 2016; Matthews et al. 2013). Moreover, some studies have found that there is a higher rate of prevalence and incidence of Alzheimer's-related dementia in African-American minorities as compared to European-Americans (Steenland et al. 2016). Thus, an understanding of the genetic, cellular and molecular causes and risk factors of Alzheimer's disease would prolong the life of the patients suffering from it and would pave the way for its early detection so that it can be treated at early stages.

Alzheimer disease is a central nervous system pathology but its consequences and manifestations extend beyond the brain. Alzheimer's pathology affects many brain regions involved in higher brain functions like speech, cognition, decision-making, etc. Moreover, various types of neurons especially acetyl-cholinergic and glutamatergic are affected in the pathophysiology of AD. There is widespread neuronal degeneration in AD which is caused by senile plaques of aggregated amyloid β-protein and neurofibrillary tangles (NFTs) mainly composed of hyperphosphorylated microtubule-associated tau protein, the two histopathological hallmarks of the disease (Sisodia and Tanzi 2007) which finally causes the loss of synapses and neurons leading to gross atrophy of the patient's brain. Although these two hallmarks of AD differentially yet significantly contribute to the development of AD, neurobiologists are divided in their opinion on whether beta-amyloid cascade initiates AD (beta-amyloid hypothesis) or it is the tau protein which gets aggregated due to hyperphosphorylation and misfolding thereby initiating the disease (Mudher and Lovestone 2002). Nevertheless, amyloid beta (Aβ)-protein aggregation and senile plaque formation is still considered to be the distinct and early pathological hallmark in early-onset AD which further induces the activation of sequential events of cell death by hyperphosphorylated tau protein in the form of NFTs (Sun et al. 2015a). In the amyloidogenic pathway, amyloid plaques are formed due to excessive production and aggregation of Aβ-peptide due to proteolysis of the amyloid precursor protein (APP) by α-secretase and β-secretase enzymes respectively. Mutations in the APP gene have been found as the prominent causes responsible for the development of familial AD (Ohshima et al. 2018). Moreover, mutations in other genes encoding for enzymes of APP processing complex like presenilins (PS-I/II), α-secretase and β-secretases cause the impairments in APP processing leading to deleterious accumulation of toxic Aβ-peptides which ultimately cluster into soluble oligomers initiating cell toxicity (Chow et al. 2010; Kelleher and Shen 2017). Moreover, microtubule-associated protein tau when hyperphosphorylated aggregates into NFTs leading to cellular degeneration because of impairments in cell signalling and axonal transport processes causing brain atrophy. Various cell signalling pathways like Wnt/β-catenin and MAPK pathways act downstream of Aβ and tau-mediated effects finally modulating cellular physiology and initiating apoptosis, thereby augmenting neurodegeneration.

13.2 Brain Areas Affected in AD

The human brain contains around a hundred billion nerve cells which collect, process and transmit the information through electrical and chemical communications. The central nervous system (CNS) in higher vertebrates is divided into four distinct parts which include the cerebrum, cerebellum, diencephalon and brainstem. The average weight of an adult human brain is approximately 1300–1400 grams and is composed mainly of neurons and glial cells. Furthermore, different glial cells such as astrocytes, oligodendrocytes and microglia carry out specialized functions like support, nutrition and immune functions within the CNS. Different brain regions have diverse functions and are affected differently in various CNS pathophysiologies as suggested by studies utilizing functional magnetic resonance imaging (fMRI), nuclear magnetic resonance (NMR) imaging and post-mortem autopsy reports.

Some brain regions have been found to be more susceptible to the pathophysiological and metabolic features specific to the AD (Liang et al. 2008). The major brain areas where changes in gene expression, metabolic alterations as well as Aβ-plaque formation have been found include the hippocampus, entorhinal cortex, superior frontal gyrus, middle temporal gyrus, posterior cingulate cortex and primary visual cortex, among others (Liang et al. 2008; Valla et al. 2001; Vogt et al. 1992; Li et al. 2012; Cui et al. 2007). Although AD pathology does not affect multiple brain areas simultaneously, gene expression analysis and brain imaging studies have revealed that AD progressively affects the regions one after the other (Thal et al. 2002). AD has its various stages of progression like stage 1 with no impairment, stage 2 with very mild impairment, stage 3 with a mild decline of cognition, stage 4 and 5 with the moderate decline with difficulty in arithmetic calculations to short-term memory loss and significant confusions. In the final stages of 6 and 7, there is severe to the very severe decline of cognitive abilities with reduced ability to respond to the environment and nearing death.

Convincing reports came out in the late 1990s that neuronal atrophy in selective brain regions was highly correlated with the cognitive deficits characteristic of AD. One of the brain regions called the medial temporal lobe (MTL) show the highest density of histopathological markers akin to AD including Aβ-amyloid plaques and neurofibrillary tangles of tau protein (Smith 2002; Jobst et al. 1992). In longitudinal studies of Alzheimer's patients, it was found by various imaging techniques and autopsies that their fluid-filled brain ventricles were enlarged as the brain tissues are progressively lost along with enhanced neuropsychological deterioration in these patients (Luxenberg et al. 1987). Significantly higher loss of tissues from MTL was reported by various CT studies done in AD patients with a rate of 15% of MTL tissue loss per year, therefore, shrinking the MTL rapidly thereby confirming the higher susceptibility of MTL to such catastrophic histopathological events of AD pathology (Jobst et al. 1994). Moreover, in normal healthy subjects, MTL shrunk only minutely (1/10th of the rate as in AD) thereby pointing towards a conclusion that AD cannot be solely occurring due to the rigorous or accelerated ageing but it must be the consequences of a disease process and associated pathology (Smith 2002). Furthermore, reports of serial MRI scans of AD patients suggest that regional atrophy from MTL nucleus spreads progressively to other brain areas in a highly specific manner engulfing most of the brain (Scahill et al. 2002).

As the hippocampus, which is the most important brain area involved in learning and storing memory, is a part of MTL, it is imperative that neuronal atrophy and synapse loss from these important brain areas involved in higher cognitive functions result in cognitive dysfunctions in AD patients. Hippocampal subfields like CA1, CA3, CA2 and dentate gyrus have been differently implicated in AD pathogenesis by various studies so far. High-resolution MRI imaging of AD patients at 3 teslas (magnetic flux density) have shown that CA1 anterior-dorsal region of the hippocampus is severely affected by Alzheimer's-associated atrophy but not in the normal ageing process (Smith 2002). Some studies have suggested that a specific part of the hippocampus called cornu ammonis (CA), of a normal elderly human, occupies about 1.5 ml volume and harbours about nine million neurons, which shrinks by 66% of the normal volume and the number of neurons in this area of hippocampus significantly drops by 84% in the terminal stage of AD (Bobinski et al. 1996).

Another important brain area which is severely affected in the early stages of AD is the brainstem nucleus locus coeruleus (LC). In the past decades, various studies provided strong evidence by analysing post-mortem autopsy reports of brain samples obtained from AD patients, concluding that in Alzheimer's disease, there is a predominant neuronal loss within LC (German et al. 1992). Moreover, reports suggest that LC is the brain region where the early formation of neurofibrillary tangles occurs in young adults and aged people which serves as the platform for further propagation of the disease (Braak and Del Tredici 2011; Grudzien et al. 2007). Interestingly, reports suggest that in Alzheimer's pathological condition, LC neurons show ectopic expression of cell cycle-regulating proteins like proliferating cell nuclear antigen (PCNA), cyclin-dependent kinase-4 (CDK-4), cyclin D and cyclin B1 reminiscent of cell death by atrophy (Busser et al. 1998). Additionally, one study reports that LC neurons are lost more extensively than cholinergic cells from nucleus basalis of Meynert and corroborated better with the duration of the Alzheimer's disease (Zarow et al. 2003), pointing to the sensitivity of LC neurons to the pathological insults than other brain regions. In addition, LC-NE (norepinephrine) suppresses neuroinflammation via modulation of microglia. Microglia are key players in neuroinflammation-mediated neurodegeneration in AD by suppressing Aβ-42-induced cytokine and chemokine release and NE-activated microglia phagocytose Aβ, thereby preventing their accumulation (Heneka et al. 2010). Therefore, neuronal loss from LC declines the NE content in the brain which might interfere in Aβ-42 clearance by activated microglia leading to its accumulation up to the toxic levels (Chalermpalanupap et al. 2013). LC degeneration-induced NE reductions in the brain promote pro-inflammatory responses while subjugating the anti-inflammatory responses and reducing the clearance of Aβ-42 poses a triple threat in the development of AD (Chalermpalanupap et al. 2013). However, it is not known whether neuronal loss and a concomitant decline in NE in the brain precedes AD pathology or is followed by the onset of AD pathology.

Amygdala, which is a part of the limbic system located in the temporal lobe of the brain, is mainly responsible for emotional and fearful information processing and regulates the fear-based responses, memory and survival instincts. Regression analysis studies have shown that in mild cognitive impairment early stage of AD, amygdala atrophy was severe and comparable to that of the hippocampus (Poulin et al. 2011). Moreover, in various post-mortem reports of AD patients, it has been well documented that β-amyloid senile plaques, as well as neurofibrillary tangles, were significantly present in amygdala along with the prominent loss of amygdalar neurons (Herzog and Kemper 1980; Scott et al. 1991, 1992). However, the comparative studies of neuronal loss and atrophy between amygdala and hippocampus are to some extent inconsistent, but nevertheless, amygdalar atrophy in AD contributes to the progression of neuropsychiatric and emotional disturbances in AD patients.

Recent advances in medical science especially live imaging techniques like fMRI and positron emission tomography (PET) have helped the doctors to locate the brain regions having atrophy in living patients unlike in the past where after death the autopsy reports were main confirmatory evidence of AD. Moreover, in some typical cases of rare AD forms, the atrophy may not begin from hippocampus and memory is not affected in the early stages. One form of atypical AD displays posterior cortical atrophy where early signs of atrophy appear in occipital and parietal lobes which affects the visual and spatial awareness abilities of the patient. Patients usually find it difficult to read or identify objects even if their eyes are healthy. It is still not understood as to why and how the origin of atrophy in the brain differs among individual patients and why the course of progression pattern varies. Future research about coordinated brain areas and their functions could provide us with clues about these unresolved mysteries.

13.3 Cellular Mechanism of Amyloid Beta (Aβ)-Plaque Formation

The cellular, as well as genetic factors contributing to the pathophysiology of AD, have been rigorously investigated, but the molecular mechanisms underlying the progression and epistemology of the symptoms of the disease remained a haunting mystery for scientists so far. As per the reports and research conducted so far in in vitro as well as on various animal models of AD and human autopsies, β-amyloid plaques are said to be the central players involved in the pathogenesis of AD (Fig. 13.3). The amyloid cascade hypothesis was first time proposed by Hardy and Allsop in 1991, and it is still the best explanatory hypothesis supported by a large number of experiments (Carrillo-Mora et al. 2014). According to this hypothesis, the starting events which drive the progression of neural damage in AD include the production and excessive accumulation/aggregation of Aβ-peptides in intracellular as well as in extracellular spaces (Hardy and Allsop 1991; Eckman and Eckman 2007). Since amyloid precursor protein (APP) is the integral membrane receptor glycoprotein of unknown function present on neurons and from which β-amyloid peptides are formed after proteolysis by β-site APP cleaving enzyme-1 (BACE-1) and γ-secretase enzymes, it has remained a prominent target for AD-related research. Amyloid-β (Aβ) peptides formed after proteolysis of APP are mostly short peptides containing 38–43 amino acids. APP, the transmembrane protein, belongs to an evolutionarily conserved protein family which also include APP-like protein-1 (APLP-1) and APLP-2 proteins (Wasco et al. 1993).

Physiologically, cellular APP processing takes place through two major pathways known as amyloidogenic and non-amyloidogenic. Non-amyloidogenic pathway results in the formation of soluble extracellular fragment known as sAPPα after the extracellular domain of APP is cleaved. The extracellular domain of APP is cleaved by α-secretase enzymes in association with disintegrin and metalloproteinases (ADAM, which also includes ADAM 9, ADAM 10 and ADAM 17) all of which exhibit neuroprotective and neurotrophic activities within the CNS (Pietri et al. 2013;

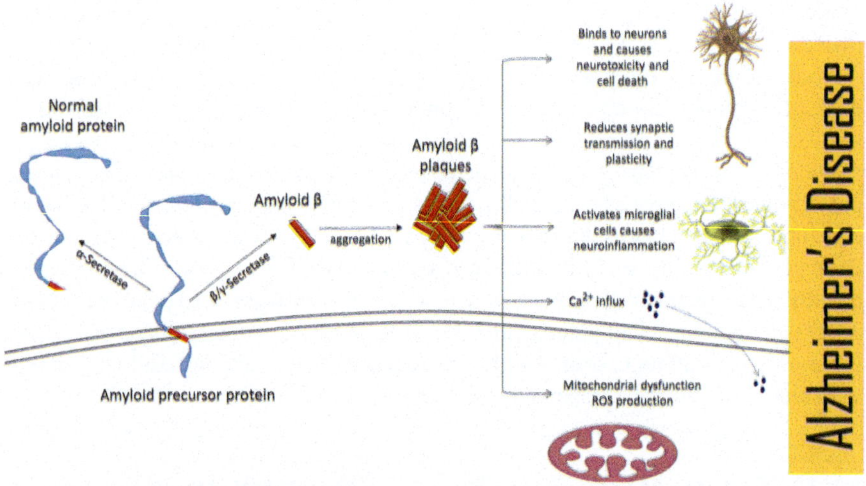

Fig. 13.3 Amyloid beta protein aggregation leads to Alzheimer's disease. According to the most popular hypothesis, amyloid beta aggregation is the major underlying factor behind Alzheimer's disease. Amyloid beta arises by the proteolysis of amyloid precursor membrane protein at β site. Amyloid beta peptides then self-aggregate by direct interaction forming plaques causing cytotoxicity. Generation of these amyloid beta plaques then triggers a number of pathophysiological responses that lead to the development of Alzheimer's disease

Edwards et al. 2008). Next, γ-secretase enzymes residing at the cytosolic side of plasma membrane act on APP resulting in the formation of an intracellular fragment known as APP intracellular C-terminal domain (AICD) (Chang and Suh 2010). However, in the amyloidogenic pathway which is central to the pathogenesis of AD, APP is cleaved by β-secretase enzyme usually called BACE-1, which upon cleavage generates two fragments sAPP-β (N-terminal) and CT99 (Zhang et al. 2012; de Paula et al. 2009). Aβ-peptide is ultimately found to be the main constituent of the senile plaques which are formed by the proteolytic cleavage of APP (Wilquet and De Strooper 2004). Among Aβ-peptides produced after the cleavage of APP, the peptides whose carboxyl terminals end with 40th and 42nd amino acid of APP are designated as Aβ 1–40 and Aβ 1–42, respectively, and these are main constituents of senile plaques responsible for neuronal cell death and are the highest markers in the brain of AD patients (Roher et al. 1993; Miller et al. 1993). Further, CT99 fragment is again cleaved by γ-secretase enzyme complex which also includes nicastrin, presenilin 1 and 2, anterior pharynx defective-1 and presenilin enhancer-2, all of which are located in the plasma membrane as a complex of protein-cleaving enzymes called intra-membrane-cleaving proteases (I-CLiPs). The enzyme, β-secretase or BACE-1, acts on specific β-sites of APP containing aspartic acid-1 and glutamic acid-11 creating C-terminal membrane-embedded fragment of C89 or C99 (Vassar et al. 1999; Gouras et al. 1998). Elevated activity of β/γ-secretase produces enhanced levels of sAPP-β and C99 which are further processed by γ-secretase, ultimately resulting into the formation of Aβ 1–40/Aβ 1–40 and AICD which consequently leads to plaque formation and apoptosis via activation of caspase-6 enzymes (de Paula et al. 2009; Gupta and Goyal 2016). Because of two additional amino acids alanine and isoleucine present in Aβ 1–42, these peptides become more hydrophobic and vigorously aggregate than Aβ 1–40 peptides (Grimm and Hartmann 2012). All the four components of an intra-membrane complex of γ-secretase mentioned above are vital for processing of APP, but the major contribution to AD pathology comes from presenilins as more than 150 point mutations of autosomal type within these presenilin genes have been linked to familial early-onset AD (Duering et al. 2005; St George-Hyslop and Petit 2005). Furthermore, these point mutations in presenilin genes, especially PS-1, have been found to accelerate the generation of Aβ 1–42 type of peptides by the gain of deleterious functions (Kowalska 2004; Duff et al. 1996).

All these proteolytic activities by intra-membranous γ-secretase complex acting on APP are shown to occur at special membrane microdomains called lipid rafts which are cholesterol and sphingolipid-enriched membrane patches where γ-secretase subunits reside as well. Apart from this, these lipid rafts also serve as a platform for various signal transduction pathways as well as cell adhesion and protein trafficking (Grimm and Hartmann 2012; Vetrivel et al. 2004, 2005). Production of Aβ-peptides is a normal physiological process within the CNS, and in AD, the balance between the rate of Aβ-peptide production and clearance is disturbed which results in aggregation of amyloid plaques due to excessive accumulation of Aβ 1–40 and Aβ 1–42 peptides which ultimately leads to neurotoxicity, oxidative stress, microglial activation and alteration of protein kinase/phosphatase activity which eventually leads to cellular degeneration (Fig. 13.3) (Wildsmith et al. 2013; Selkoe and Hardy 2016; Sun et al. 2015b). The exact mechanisms behind the excessive production or lower rate of clearance of Aβ-peptides and how do these senile plaques induce neuronal cytotoxicity still remains to be elucidated. However, recent reports have provided important clues about the nature of these Aβ-peptides and their interaction with other biomolecules within the cells.

13.3.1 Aβ-Induced Cellular Cytotoxicity

As mentioned earlier, the functions of Aβ-peptides under normal physiological conditions are not well understood and these peptides seem to be present even in healthy individuals and are produced throughout life as normal metabolic or cellular by-products. However, some of the animal model studies fail to assign any particular loss of physiological function to the absence of Aβ-peptides (Sadigh-Eteghad et al. 2014; Luo et al. 2003). Interestingly, exogenously applied Aβ 1–40 peptides to cell cultures at picomolar ranges have displayed neurotrophic and neuroprotective roles (Yankner et al. 1990; Plant et al. 2003). Nevertheless, it is beyond doubt that aggregated Aβ-peptides mainly constituting peptides below 50 amino acid residues forming the amyloid senile plaques are the major hallmarks of the AD pathophysiology impairing synaptic plasticity and memory (Shankar et al. 2008; Kayed et al. 2003). As the disease progresses, the ratio of Aβ 1–42/Aβ 1–40 increases in the brain mainly due to either higher rate of Aβ 1–42 formation or reduced levels of Aβ 1–40 and other soluble oligomers of Aβ-peptides which results in the formation of senile plaques as Aβ 1–42 are more toxic because of the higher rate of aggregation (Glabe 2005; Walsh and Selkoe 2007). Additionally, Aβ-plaques also interact with some of the cell membrane receptors and other proteins inducing programmed cell death and hence immensely contribute to the neurodegeneration processes (Fig. 13.3) (Zhu et al. 2015; Small et al. 2001). Looking at the molecular levels, oligomerization of Aβ-peptides caused by misfolding of APP-derived peptides could generate cellular ionic imbalances leading to various aspects of pathophysiological outcomes of AD (Lal et al. 2007). This effect is likely mediated by either reactive oxygen species (ROS) generated by Aβ-peptide aggregation or the interaction of these plaques with cellular membranes and associated receptors/ion channels (Fig. 13.3) (Lal et al. 2007).

Interestingly, Aβ-peptides behave as a double-edged sword because on one side, they display neurotrophic and neuroprotective effects and antagonistically they cause diverse toxic effects (Atwood et al. 2003). These differential effects of Aβ-peptides within CNS might be determined by the relative ratio of Aβ 1–40/Aβ 1–42, and it is not well understood how these peptides switch their functionality from being neurotrophic like increasing neurogenesis in the hippocampus and enhancing memory consolidation to being toxic, depressing Long-term potentiation (LTP) and initiating apoptosis (Lopez-Toledano and Shelanski 2004). Some studies have suggested that it is the imbalance between its excessive production and a lower rate of clearance from CNS which makes Aβ-peptides the most toxic by-products impairing other vital cellular signalling processes, altering membrane permeability, inducing excitotoxicity via interaction with many neurotransmitter receptors and creating a flux of ROS with protein oxidation (Carrillo-Mora et al. 2014; Lin et al. 2001; Canevari et al. 1999; Rosales-Corral et al. 2004a; Butterfield et al. 2007). Moreover, human beings are consistently exposed to trace amounts of various metals like aluminium, copper, iron, zinc, etc., and for decades, these metals are suspected to play rather a blurred role as a risk factor in the progression of AD. Interestingly enough, Aβ-peptides are efficient ion chelators, and it has been found that zinc-chelated Aβ-oligomers are more toxic to brain tissues. Further evidence for their role in AD came from various ex vivo studies where it has been shown that zinc-Aβ-oligomers potentially inhibit LTP in the hippocampus via activation of microglia (Lee et al. 2018). Moreover, the size and shape of Aβ-oligomers are correlated with their intensity of toxicity with reports suggesting that an increase in the size of Aβ-oligomer assembly decreases the intensity of deleterious effects (Sengupta et al. 2016). Aβ dimmers have been found to be more toxic than other assemblies and they provide building blocks for the formation of other intermediate Aβ-oligomers (O'Nuallain et al. 2010; Mc Donald et al. 2015).

13.3.2 Aβ-Plaques and Oxidative Stress

Pathophysiology induced by Aβ-senile plaques is mediated by several different mechanisms like defective axonal transport, apoptosis, oxidative stress, calcium dyshomeostasis, neuroinflammation and mitochondrial malfunctioning (Rosales-Corral et al. 2004b; Moreira 2018). The main source of free radicals (molecular species with one or more unpaired electrons in their outer shell) is the reduction of molecular oxygen in water molecules yielding superoxide radicals which finally produces hydrogen peroxide (H_2O_2) by accepting an electron. Further, H_2O_2 is further reduced to highly reactive hydroxyl radicals called reactive oxygen species (ROS) reacting with biomolecules including proteins, DNA/RNA as well as lipids and damaging them by oxidation-reduction reactions. Enhanced oxidative stress results from an imbalance between the levels of anti-oxidants and oxidants in favour of the later. Recently, there has been a rising interest in the role played by oxidative stress in various neurodegenerative disorders like AD, Parkinson's disease, cerebral ischaemia-reperfusion and Down's syndrome. Elevated oxidative stress is another hallmark of AD, and the recently held notion is that methionine amino acid residue present at the 35th position of Aβ-peptide is crucial for the neurotoxicity induced by Aβ-oligomers by generating oxidative stress (Butterfield et al. 2013). Various lines of evidence indicate that oxidative stress load develops from the cellular lipid bilayers where Aβ 1–42 peptide inserts itself as oligomers initiating lipid peroxidation thereby serving as the source of ROS generation (Markesbery 1997; Mark et al. 1997; Butterfield et al. 2001).

Recent studies have suggested that lofty levels of oxidative stress occur in brain regions where Aβ 1–42 levels are high compared to the brain regions having lower levels of Aβ 1–42 oligomers (Butterfield and Boyd-Kimball 2018). Furthermore, studies on cortical synaptosomes, cultured hippocampal neurons and primary neuronal/astrocytes cultures indicate that Aβ-peptides of various lengths induces protein oxidation as measured by elevated levels of protein carbonyls (Perluigi et al. 2006a; Varadarajan et al. 1999, 2001). In the presence of many anti-oxidants including vitamin E, Aβ 1–42 oligomers failed to generate ROS in neuronal cultures and protein oxidation was prevented (Yatin et al. 1999). Interestingly, in vivo administration of anti-oxidants like D609 (a tricyclodecanol xanthic acid derivative) and ferulic acid ethyl ester (FAEE) independently abrogated protein carbonyl production (protein oxidation) induced by Aβ 1–42 oligomers (Perluigi et al. 2006b; Ansari et al. 2006). These studies indicate that Aβ-oligomers induce oxidative stress via peroxidation of proteins and lipids rendering them unstable and dysfunctional leading to the progression of AD.

Neuronal membranes are most dynamic in function as they are electrically excitable and any functional or structural alteration of neuronal membranes could lead to disastrous consequences for their survival. Aβ 1–42 oligomers interact with neuronal membranes and jeopardize their functional integrity by various mechanisms including insertion into the membrane and subsequent pore formation leading to changes in membrane permeability (Butterfield and Lashuel 2010). Apart from pore formation, neuronal membranes could be potential targets for ROS resulting in disruption of their function via several mechanisms like lipid peroxidation reactions which have been shown to induce loss of asymmetry in phospholipids in membranes of synaptosome which serves as an early signal for apoptosis. Among the products of lipid peroxidation which are elevated in AD pathology, 4-hydroxynonenal (4HNE) and acrolein (an unsaturated aldehyde) are highly reactive and both induce apoptosis and disrupt ionic homeostasis of neuronal cells (Castegna et al. 2004). The enzyme amino-phospholipid translocase or flippase, which is ATP dependent, maintains the asymmetry of lipid bilayers, and 4HNE generated by lipid peroxidation oxidizes a cysteine residue in its catalytic domain rendering the flippase enzyme dysfunctional and thereby disrupting the bilayer asymmetry. Furthermore, 4HNE lipid peroxidation product can alter the conformational structure and function of various membrane proteins by conjugating with them and subsequently causing neurotoxicity and neural degeneration in AD (Subramaniam et al. 1997). Another important mechanism by which ROS accelerates the pathophysiology induced by Aβ-peptides is acting via membrane receptor known as low-density lipoprotein receptor-related protein 1 (LRP1) which is also known by other names such as apolipoprotein E receptor (APOER). LRP1, a protein receptor known to play an important role in cellular endocytosis, apolipoprotein metabolism and cell signalling, was recently shown to undergo oxidation by ROS resulting in impaired clearance of Aβ-peptides from the brain (Cheignon et al. 2018). It is well established that LRP-1 is a multifunctional receptor protein that plays an important role in the efflux of toxic Aβ-peptides from CNS to blood via blood-brain barrier, and this activity is significantly reduced under AD pathology (Jeynes and Provias 2008; Sagare et al. 2007; Ito et al. 2007).

As mentioned earlier, Aβ-peptides has metal chelating properties and interact with metals like zinc, copper, and iron which enhances the cytotoxicity of Aβ-peptides via ROS and biomolecular oxidation. It has been suggested that elevated ROS load in AD brain is highly correlated with increased concentrations of iron (Fe) and copper (Cu) in the brain, both of which are responsible for ROS generation via Fenton reaction (Jomova et al. 2010). Recent results obtained from Rutherford backscattering spectrometry and scanning transmission ion microscopy substantiate the facts that toxicity induced by Aβ-peptides is elevated in presence of metal ions as they deposit in and around Aβ-plaques as compared with surrounding tissues resulting in fibril assembly as well as β-sheet formation in Aβ-peptides (Rajendran et al. 2009). As mentioned earlier that Aβ-induced cytotoxicity depends on the conformational structure and length of the Aβ-peptides where non-β sheets are non-fibrillar and β sheets are fibrillar and toxic. Recent studies suggest that the aggregation of Aβ-peptides is also controlled by pH, the concentration of Aβ-peptides as well as levels of metal ions including copper, zinc, and iron (Jomova et al. 2010). In natural healthy state, generation of ROS is followed by elevated levels of anti-oxidant enzymes like haem oxygenase-1, catalase and superoxide dismutase (SOD), and interestingly the mRNA levels of these enzymes are reportedly elevated in a consistent manner under AD pathology (Premkumar et al. 1995), but the activity of these anti-oxidant enzymes is significantly reduced, and the causes for this reduced activity are not clear (Omar et al. 1999). Some studies showed that ribonucleic acid (RNA) is one of the major targets of ROS-induced oxidation in the pathology of AD, whereas mitochondria serve as the main source of ROS generation (Nunomura et al. 2009). However, the exact mechanism behind amyloid plaque deposition and elevated ROS load is not fully understood yet some evidence conclude that aggregated Aβ-plaques initiate a chronic inflammatory response in AD pathology and enormous free radicals are released from activated microglial cells.

Strong evidence supports the view that the cytotoxicity of Aβ-peptides is not inherent and requires the presence of ROS to induce such toxic cellular effects leading to AD pathology. Indeed, it was shown that in the absence of redox metal ions like copper, zinc or iron, Aβ-peptides did not induce toxic effects as the strong binding capacity of Aβ-peptides to these metals to reduce them subsequently producing hydrogen peroxide as well ROS load (Huang et al. 1999). As mentioned earlier, methionine 35 of Aβ-peptides is strongly implicated in AD pathogenesis as this residue of Aβ-peptide is the most susceptible to oxidation reactions in vivo (Jomova et al. 2010). Reports suggest that methionine-35 gets oxidized to methionine-sulphoxide which significantly reduces the pro-apoptotic and toxic effects of Aβ-peptides on isolated mitochondria (Jomova et al. 2010). Furthermore, lipid peroxidation due to ROS results in the formation of various carbonyl products which can damage DNA and proteins. Various types of advanced glycation end products (AGE) and advanced lipid peroxidation end products (ALE) are produced during ROS reactions. These AGEs and ALEs further interact with receptors like a receptor for AGEs (RAGE) activating downstream signalling pathways which subsequently leads to the production of pro-inflammatory cytokines such as interleukin-6 (Valko et al. 2005). Recent research reports suggest that glyceraldehyde-derived AGEs mediated by ROS reactions elevate the APP and Aβ levels in mice models, and in the presence of AGEs, the toxicity of Aβ-peptides is significantly heightened (Ko et al. 2015). The mechanism behind the role played by AGEs in AD pathology is still debatable, but new reports are emerging which suggest that AGEs up-regulate the expression of the proteins associated with the processing of APP including BACE1 and PS1 and reduce the expression of neuroprotective as well as antioxidative sirtuin-1 via ROS action. AGEs are also reported to up-regulate the expression of GRP78 (an endoplasmic reticulum chaperone protein) and elevate the cell-death signalling pathways including proteins like p53 and pro-apoptotic caspases (Ko et al. 2015). Therefore, ROS generated by Aβ-peptide aggregation alter the chemical nature of vital biomolecules generating secondary toxic products which ultimately leads to neuronal death via different cellular signalling pathways.

Normal ageing has its effects on the functioning of microglia like heightened activation but reduced function and proliferation as a result of cellular senescence (Miller and Streit 2007), and these processes contribute to the age-associated neurodegenerative diseases. In pursuit of phagocytosis of Aβ-plaques, activated microglia release pro-inflammatory chemokines and are clustered around amyloid plaques mounting a sustained inflammatory response which significantly damage neurons (Meda et al. 1995). Furthermore, as mentioned above, ROS load is an important hallmark of AD; microglia activation-induced neuronal damage also involves ROS production in AD, and it is still not clear how Aβ induces ROS generation in microglia, but it likely involves the binding of Aβ-peptides to the cell surface receptors and ion channels on microglia. Recent reports suggest that Aβ-oligomers stimulate microglial cells for ROS production by acting on certain ion channels and one such ion channel has been identified as voltage-gated transient receptor potential ion channel (TRPV) based on inhibition by drugs which aborted ROS production in stimulated microglia (Schilling and Eder 2011). These cation channels modulating Aβ-induced microglial ROS production could serve as potential therapeutic drug targets for controlling AD pathogenesis. Generation of ROS inside microglia is derived from various sources acted upon by peroxidase enzymes like NADPH oxidase which plays an indispensable role in the generation of oxygen free radicals from molecular oxygen which further takes part in the formation of ROS consequently modulating morphology and pro-inflammatory gene expression in activated microglial cells (Gao et al. 2002; Qin et al. 2004). It is interesting to note that the expression of microglial NADPH oxidase is up-regulated in neurodegenerative diseases like AD and Parkinson's disease (PD) pointing towards the importance of this enzyme in neurodegeneration diseases where it could be targeted for treatment of AD and PD because it is the main ROS-producing enzyme (Wu et al. 2003; Block 2008).

Therefore, an important link between AD pathology and oxidative stress load is through protein oxidation and lipid peroxidation induced by ROS which accelerates AD pathology via secondary effects of these ROS on cells. Finding early markers in AD has been the toughest task for neurobiologists and oxidative stress as an early biomarker in the aetiology of mild cognitive impairment (MCI) stage of AD would provide an important diagnostic methodology for early detection of AD (Butterfield et al. 2006). Interestingly Aβ-peptide itself undergoes oxidative damage in the presence of ROS, and oxidized Aβ-peptide further coordinates metal ions consequently creating more ROS load (Cassagnes et al. 2013) as well as accelerates the aggregation of Aβ-peptide. Hence, the possibility of therapeutic strategies targeting ROS pathways derived from better and deeper insights of mechanisms involved in ROS generation could be a step forward towards treating AD. Moreover, it is known that Aβ-induced-oxidative stress is not global in the brain of AD patients; rather Aβ-poor regions like the cerebellum do not show signs of ROS production in AD pathology pointing to the fact that Aβ-formation is upstream of ROS generation and strategies targeting excess production of Aβ-peptide shall be able to restrict ROS production also (Cheignon et al. 2018). However, it is still unclear whether oxidative stress associated with Aβ inducing neurodegeneration is the primary or secondary to the alternative sources of ROS generation.

13.4 Role of Tau Protein in the Development of AD

Another important pathological hallmark of AD is the excessive accrual of hyperphosphorylated microtubule-associated protein (MAP) called tau protein which forms the intracellular neurofibrillary tangles (NFTs) inside the neurons leading to a plethora of physiological dysfunctions (Buee et al. 2000). We will now discuss the probable cellular and molecular mechanisms behind the generation and pathology induced by NFTs.

Tau, an unfolded microtubule-associated protein which binds to microtubules of the cytoskeleton, is involved in their assembly as well as stabilization (Grundke-Iqbal et al. 1986). Tau proteins are mostly located in axons and in small quantities are also found in dendrites of neurons (Ittner et al. 2010). Apart from these two places, tau proteins within the nucleus play a vital role in the regulation of transcriptional activity as well as in genomic maintenance under non-pathological physiological conditions (Violet et al. 2015). Under normal physiological conditions, tau provides stability to microtubules and also regulates the intracellular trafficking of vesicles and other transport processes within neurons which makes it an important player in the axonal transport process and defects therein during AD pathologies (Vershinin et al. 2007; Talmat-Amar et al. 2018). However, during AD, the normal physiological function of tau is disrupted which consequently leads to the development of neurofibrillary tangles (NFT) (Fig. 13.4). The gene which encodes for tau protein is located on chromosome 17 consisting of 16 exons. Alternative splicing of exons including 2, 3 and 10 leads to six possible combinations, for six tau protein isoforms, mainly found in neurons and mostly located in axonal regions. These six isoforms of tau usually contain 352–441 amino acid residues. The three isoforms which do not have exon 10 contain three binding domains with the microtubules (3R) and the other three have four (4R). The latter isoforms having 4R have stronger interaction and better stability (Buee et al. 2000). Alternative splicing of a single gene called microtubule-associative protein tau (MAPT) generates numerous isoforms of tau protein. Primarily, tau plays an important role in maintaining overall

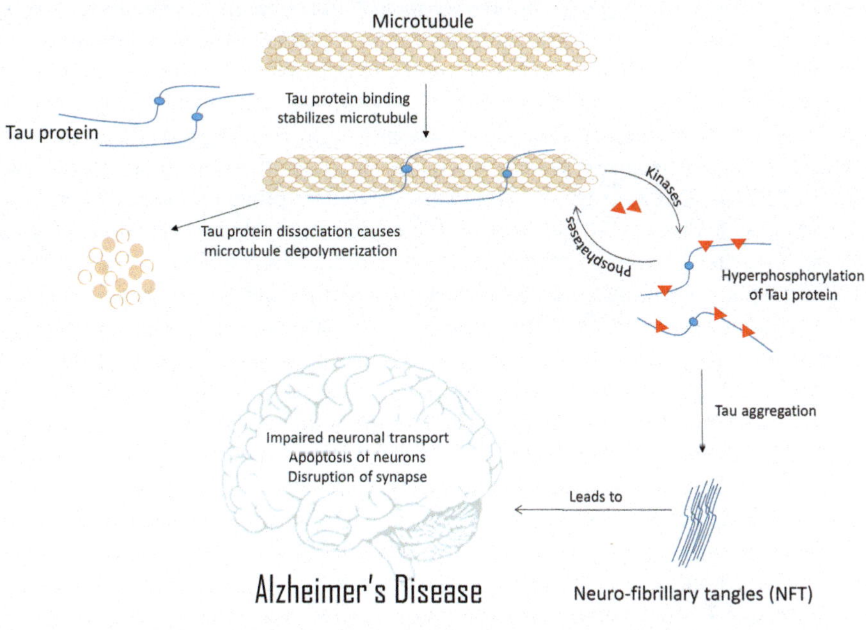

Fig. 13.4 Tau protein-mediated development of Alzheimer's disease. Tau proteins are important for the stability and functionality of microtubules. Hyperphosphorylation of tau protein causes its dissociation from microtubules, which then promotes pairing of hyperphosphorylated tau threads. Many phosphorylated tau threads associate together to generate neurofibrillary tangles (NFT) which could block synaptic junctions thereby causing blockage of axonal transport and neuronal cell deaths

neuronal morphology as it regulates certain vital cellular processes including neurite outgrowth, microtubule assembly as well as stability which are indispensable for cytoskeleton maintenance and axonal transport processes (Wang and Mandelkow 2016).

Additionally, tau is tightly regulated during homeostasis as well as under stress-induced cellular responses by a cascade of post-translational modifications such as glycation, glycosylation, cleavage, nitration, ubiquitination and phosphorylation. Post-translational interferences severely affect tau-microtubule binding and consequently promote its misfolding causing it to aggregate into NFTs. Normal site-specific phosphorylation of the tau protein by various kinases regulates its ability to bind tubulin and promote microtubule assembly (Lindwall and Cole 1984). Optimal levels of tau phosphorylation are required for the proper functioning of tau protein, whereas excessive phosphorylation disrupts optimal functions involving tau protein which also enhance tau-tau protein aggregation resulting in the formation of NFTs (Fig. 13.4). Tau contains 85 putative serine or threonine phosphorylation sites mainly located in the microtubule-binding region (MBR) and the proline-rich domain of the protein (Hanger et al. 2009). Phosphorylation of tau protein is mediated by several types of protein kinases such as the serine/threonine/tyrosine kinases and most of the sites are located on either side of MBR.

Hyperphosphorylation of tau proteins causes the helical and straight filaments to form neurofibrillary tangles. The loss of microtubules-binding capacity provokes cytoskeleton destabilization, affecting neuronal transport processes like synaptic transmission and depriving neurons of trophic factors which eventually causes neurodegeneration and neuronal death (Fig. 13.4) (Montoliu-Gaya and Villegas 2015). The major tau kinases that play an important role in its phosphorylation include GSK3β, cyclin-dependent protein kinase 5 (CDK5), cAMP-dependent protein kinase (PKA), MAPK, calcium-calmodulin-dependent protein kinase II (CaMk II) (Baudier and Cole 1988) and microtubule affinity-regulating kinase (MARK) (Noble et al. 2003). Various in vivo and in vitro studies suggested that GSK-3β plays a vital role in controlling tau phosphorylation at various specific amino acid residue sites including $\mathrm{Ser}^{199,\,404,396,400}$, Thr^{231} and Ser^{413} residues that are mostly phosphorylated in tau which subsequently aggregates into toxic paired helical filaments (PHF) (Liu et al. 2002). Among various kinases which take part in phosphorylation cascade of tau, GSK-3β plays a major role in phosphorylating tau under in vivo conditions. Furthermore, phosphorylation of GSK-3β kinase at Thr^{231} residue induces a local conformational rearrangement resulting in increased accessibility of GSK-3β or other kinases to excessively phosphorylate tau protein at various sites leading to its hyperphosphorylated state aggregating into PHFs. On the other hand, in vitro studies have found a set of protein phosphatases which play a complementary and opposite role in de-phosphorylation of tau including protein phosphatase 1(PP1), PP2A, PP2B and PP2C that might act as therapeutic targets in near future (Avila et al. 2004). Among these protein phosphatases, PP2A has a stronger role in dephosphorylation of abnormal tau as compared to other phosphatases (Gong et al. 1994). Reports suggest that overall tau protein phosphorylation status is three to four times higher in AD-affected brain samples as compared to samples from a healthy brain.

Strong evidence suggests that dysregulated and high phosphorylation status of tau is central to the AD progression and pathogenesis, however, under normal conditions, all tau proteins are associated and bound to the microtubules. Apart from tau hyperphosphorylation, numerous other neuronal proteins, such as MAP 1B, neurofilaments, β-tubulin, as well as β-catenin, are also hyperphosphorylated under AD pathology in the brain. Therefore, it can be concluded that decreased PP2A phosphatase activity, which removes the phosphate groups from these proteins including tau, might be the underlying cause for abnormal hyperphosphorylation of tau in AD leading to cytotoxicity and progression of the disease.

The histopathological indices of AD patient's brain samples are characterized not only by the presence of intracellular NFTs but also by the presence of Aβ-plaques. These neuritic plaques would further stimulate the production of more NFTs disrupting synaptic topography, consequently reducing the rate of neurotransmission as well as inducing atrophy of tangle-containing neurons ultimately resulting into the behavioural pathophysiological signs including dementia (Hardy and Selkoe 2002). The role played by Aβ-oligomers in inducing hyper-phosphorylation of tau protein through activation of GSK-3β kinase has been identified as connecting link between Aβ-plaques and tau pathologies in AD progression (Stancu et al. 2014). Accumulation of Aβ-oligomers affects downstream Akt-cell survival signalling pathways through inhibition of phosphatidylinositol-3-kinase (PI-3K) as well as by activating GSK-3β, subsequently causing tau hyperphosphorylation (Magrane et al. 2005). Additionally, studies have found that Aβ-peptides up-regulates calcineurin-1 (RCAN1) expression which induces long-term depression at synapses and facilitates increased tau phosphorylation through two different mechanisms. Firstly, RCAN1 hinders the activity of calcineurin, which is a phosphatase and takes part in tau dephosphorylation, and secondly, RCAN1 up-regulates the activity of GSK-3β, a kinase which hyperphopshorylates tau causing PHF-induced toxicity. Therefore, these reports suggest that elevated expression of RCAN1 has a major role in AD progression and it also elevates the pathophysiology of AD by causing mitochondrial dysfunctions which ultimately activate many apoptotic pathways leading to neurodegeneration (Lloret et al. 2011).

Studies cited above reveal that hyperphosphorylation of tau is central to the development and progression of AD via NFT-induced cellular toxicity leading to neuronal degeneration. Abnormally hyperphosphorylated tau protein, apart from losing its biological activity by disassociating from microtubules, also promotes its polymerization into PHFs. The soluble abnormal tau and/or its oligomers are in many ways toxic to neurons probably due to the defence mechanisms employed by neurons. Abnormally, hyper-phosphorylated tau via self-aggregation further polymerizes into highly aggregated PHFs/NFTs which finally choke the affected neurons by impairing their axonal transport processes thereby depriving them of neurotrophic factors inducing cell death (Fig. 13.4). A large number of reports have revealed a role for abnormally phosphorylated tau protein in the neuronal cytoskeleton collapse in ageing and neurodegenerative tauopathies in AD. Control of tau phosphorylation by targeting tau kinases such as GSK-3β might be a feasible therapeutic strategy to prevent tau aggregation and its associated pathological effects in controlling AD progression. However, excessive phosphorylation of tau appears to be a prerequisite but not sufficient alone to induce tau aggregation. Therefore, other tau post-translational modifications might also be involved in tau pathologies. However, tau proteins forming NFTs regardless of its post-translational modifications can also lead to cellular toxicity, and furthermore, it has been shown that suppression of tau pathology also blocks Aβ-induced toxicity and reduces memory deficit in AD animal models. These data provide a future perspective suggesting that reduction of the overall tau phosphorylation status may constitute a neuroprotective strategy to prevent tauopathy-induced AD. However, some studies suggest depleting axonal tau protein would hinder active axonal transport processes, and, as exemplified for cholinergic neurons, reducing axonal transport processes would ultimately deplete neurotrophic growth factors (NFG) and their receptors inducing neuronal degeneration. Moreover, the collapse of the neuronal cytoskeleton may have deleterious outcomes for a number of cellular processes including axonal and dendritic transport systems, affecting the distribution of proteins within a neuron, irregular distribution of signalling molecules and organelles throughout the cell. Maintenance of neuronal morphology, as well as contacts with neighbouring neurons through synaptic connectivity, will also be affected. Deterioration of these indispensable cellular physiological processes ultimately leads to neurodegeneration and cognitive impairments in AD patients. Therefore, amelioration of tau dysfunctions and maintenance of the neuronal cytoskeleton by restoring axonal transport processes could be important therapeutic strategies for the treatment of AD in the near future.

13.5 Axonal Transport Dysfunctions in AD

In AD animal models especially in rodents, it has been found that death of the neurons occurs via multiple pathways involved in atrophy or apoptosis. Long before in 1939, Huxley and Hodgkin using the squid giant axon demonstrated how an action potential is carried along the axons, and consequently, it was found later that these action potentials mediate the release of neurotransmitter vesicles into the synapse for neuronal communication (Hodgkin and Huxley 1939; Vicario-Orri et al. 2015). A large number of studies have provided convincing evidence that structural, as well as functional, defects in axons play a crucial role in the pathophysiology of AD. The intriguing hypothesis is that the main distinctive feature that makes neurons more prone to AD-associated atrophy is their widely extended cellular morphology consisting of about 99% of cellular volume in axonal compartments alone (Morfini et al. 2009). Neurons within the nervous system communicate with each other by sending the information via electric or chemical signal down their axons to their axon terminals which are received by postsynaptic neurons via their dendrites. Proper functioning of axons and their respective terminals depend heavily on the axonal transport processes carrying various proteins, neurotransmitter vesicles, organelles and other macromolecules from the cell body to the axon terminals. Axonal transport is one of the main cellular processes by which neuronal function is maintained and it requires cytoskeleton proteins like motor and adaptor proteins for trafficking and transportation of cargoes within the neuronal cells. Reports suggest that axonopathy was found to precede other well-established AD-related pathologies by more than a year in mouse models of AD and also in early stages of human AD patients (Stokin et al. 2005). Axonal transport defects involve axonal swelling due to excessive accumulation of vesicles, microtubule-associated and motor proteins and cellular organelles which block the neuronal communication by hindering synaptic transmission due to impaired vesicular transport (Yagishita 1978). Cytoskeleton motor proteins like dyneins, myosins and kinesins carry axonal cargoes from soma to the axonal terminals via tracks made up of microtubules and are therefore vital to understand the role of imbalanced and dysregulated transport processes occurring in axons and its involvement in the pathogenesis of AD and related neurodegenerative diseases (Cash et al. 2003). Microtubules in the axons of neurons have a discrete polarity with the faster growing plus end laid towards the nerve terminals and slower growing minus end pointed towards the soma. Furthermore, various genetic and biochemical studies have shown that the kinesin superfamily of proteins move towards the plus end and carry out fast anterograde (from cell body to the nerve terminals) transport in neurons, whereas most of the retrograde transport (from nerve terminals to the cell body) is carried by dynein proteins (Goldstein and Yang 2000; Hirokawa 1998). A study carried in *Drosophila* genetically disrupted cytoplasmic dynein heavy chain (cDHC) and dynactin subunit p150 (Glued) resulted in a phenotype of the fly similar to that observed when genetic disruption of kinesin-1 subunits was carried which consequently resulted in posterior paralysis of larva and tremendous axonal swellings filled by membranous cargoes in nerves (Martin et al. 1999). Additionally, *roadblock*-a new *Drosophila* mutant of axonal transport having same posterior paralysis and axonal swellings with accumulations of cargoes was employed to identify a new dynein light chain, again supplementing the existing evidence that dynein plays an essential role in retrograde axonal transport system (Bowman et al. 1999).

Axonal transport is classified into two types: slow transport involving slower rates of cargo movements which mainly consist of anterograde movement of cytoplasmic proteins and cytoskeleton elements involved in neuronal morphogenesis and fast anterograde transport (FAT) which is bidirectional in movement (Hirokawa 1998; Brown 2003). Additionally, FAT is involved in the transport of synaptic vesicles and neurotrophic factors which ensure synaptic activity as well as neuronal health simultaneously; therefore, FAT pathways are severely affected in AD pathology. Various primary mutations found in genes encoding proteins for axonal transport machinery play a critical role in an imbalance of FAT and subsequent progress of various neurodegenerative diseases of CNS. For example, in vitro studies have found that mutations in kinesin protein KIF5A resulted in dysfunctional and pathogenic anterograde axonal transport in a neurodegenerative disease called hereditary spastic paraplegia (HSP) which primarily affects motoneurons (Ebbing et al. 2008). Similarly, there are other CNS diseases and pathologies associated or caused by mutations in kinesin superfamily of proteins. Anterograde vesicular transport is rendered dysfunctional in Charcot-Marie-Tooth hereditary neuropathy type-2 (CMT-2) disease by the mutation in another kinesin called KIF1β causing distal muscle weakness and atrophy (Zhao et al. 2001; Xu et al. 2018). Interestingly, in another neurodegenerative disorder called spinal and bulbar muscular atrophy (SBMA) which is also known as Kennedy's disease, one of the proteins taking part in axonal retrograde transport, namely, dynactin, was found mutated at p150 subunit (Puls et al. 2003). Dynactin-1 mutations are also responsible for other neurodegenerative diseases like Perry syndrome characterized by Parkinsonian-neuronal loss and dementia (Farrer et al. 2009; Newsway et al. 2010). Apart from these primary mutations in axonal transport machinery-related genes, there are various other mutations that cause secondary malfunctioning in the FAT process like mutations of SGP4 encoding spastin protein which is an ATPase-regulating microtubule dynamics and is involved in around 40% of autosomal dominant forms of hereditary spastic paraplegia (HSP) disease (Baas and Qiang 2005). Furthermore, knocking-out of SGP4 gene in mice resulted in the tremendous accumulation of cellular organelles along with cytoskeleton proteins and subsequent swelling of the axons characterizing the impairment of axonal transport machinery in these animal models (Tarrade et al. 2006). Axonal transport also helps the neurons in meeting energy demands by transporting mitochondria in both the directions and mutations in mitofusin-2 gene which encodes an outer mitochondrial membrane GTPase involved in mitochondrial transport along the axon disrupt both anterograde as well as retrograde transport of mitochondria in Charcot-Marie-Tooth disease (CMT) type 2A (Misko et al. 2010; Baloh et al. 2007). Although it is an established fact that axonal transport dysfunctions precede the progression of various neurodegenerative diseases, it is however not well understood how exactly the defects in FAT cause the development of these diseases as their mechanism are yet to be established clearly. However, identification and characterization of primary mutations in proteins involved in axonal transport machinery provides strong evidence that FAT defects play a crucial role in the development of these neurodegenerative diseases.

Unlike above neurodegenerative diseases, there is no particular known mutation in axonal transport machinery that could lead to the progression of AD. However, most of the studies suggest that FAT defects precede some of the pathogenic hallmarks in the progression of AD in human patients as well as in transgenic mice models for AD (Stokin et al. 2005). Most of the proteins taking part in the processing of APP which also leads into the formation of toxic Aβ-peptides undergo axonal transport including APP, BACE1 (β-secretase) and PS1, and all of them have played an important role in the aetiology and progression of AD. In AD, due to dysfunctional axonal transport, neurotrophins have been found to be irregularly distributed and dysregulated leading to decreased survival of neurons (Vicario-Orri et al. 2015). Although the exact functions of APP remains a mystery, some studies have implicated APP in fast axonal transport by acting as a receptor for anterograde motor kinesin (Kamal et al. 2001). Furthermore, tau which is a microtubule-associated protein that binds and stabilizes microtubules is hyperphosphorylated and entangled in AD; therefore, in its absence, microtubules become destabilized, and axonal transport is disrupted in AD pathology indicating exciting possibility to treat AD by the use of microtubule-stabilizing drugs including taxanes, epothilones as well as peloruside as shown by recent reports (Higuchi et al. 2002; Lee et al. 1994). It is a well-established fact that APP is transported in axons via FAT and this process is dependent on kinesin-1 protein (Kamal et al. 2001; Koo et al. 1990; Ferreira et al. 1993). Furthermore, it has been suggested that APP may be transported by a vesicular complex also containing BACE-1 and presenilins which are critical for processing of APP into amyloidogenic/toxic Aβ-peptides which later aggregate into oligomers (Kamal et al. 2001; Gunawardena and Goldstein 2001). These studies altogether suggest that misregulation of APP directly by mutations like in familial AD or indirectly via proteins interacting with APP during its processing and transport could lead to defects in FAT and deprive the neurons of critical components essential for survival thereby initiating neurodegeneration (Gunawardena and Goldstein 2004).

As presenilin proteins help in the processing of APP, mutations in presenilin genes have been observed in most cases of early familial AD. Several studies suggest that presenilin interacts with glycogen synthase kinase-3β (GSK3β) and its substrate tau (Takashima et al. 1998). GSK3β play various roles and it phosphorylates kinesin light chains which lead to detachment of motor from the cargo thereby preventing further transport of cargoes (Morfini et al. 2002). Early studies conducted on transgenic mice overexpressing wild-type APP showed that before the appearance of Aβ-plaques, FAT defects, characterized by axonal swellings due to the accumulation of cellular organelles, cytoskeleton proteins and vesicles, were found in the brain which suggested that FAT events may play an important role in AD progression (Stokin et al. 2005). Furthermore, these disruptions in FAT happened in absence of phosphorylated tau protein suggesting that Aβ-peptide is the main trigger of such alterations in axonal transport (Wirths et al. 2006). Additionally, it has been shown in transgenic mice overexpressing mutated form of APP (Tg2576) that the FAT rates of olfactory nerve considerably decrease even before the plaque formation appears (Smith et al. 2007). Further, APP is carried down to axon by FAT through an indirect interaction with JNK interacting protein-1 known as JIP1 (Matsuda et al. 2001; Muresan and Muresan 2005). Scaffolding protein JIP1, in turn, binds to kinesin light chain (KLC) at its C-terminal at a conserved 11 amino acid motif and helps in the bidirectional transport of APP, and knocking out JIP1 produces anterograde and retrograde transport deficits of APP in neurons which further leads to axonopathy (Verhey et al. 2001). Apart from Aβ-induced axonal transport impairments, NFTs equally play an important role in this dysfunction as tau protein binds and stabilizes the microtubules which are indispensable for transport processes like rail tracks of cellular trafficking. Hyperphosphorylated tau destabilizes microtubules shattering off the axonal transport process and depriving neurons of trophic factors needed for its survival. The most deleterious effects of impairments in axonal transport are caused by roadblocks in organelle transport like mitochondria to the axonal terminals where the intense energy demanding processes like vesicular trafficking and exocytosis needs immense ATP generated by mitochondria. Apart from this axonal transport impairments also affect neurotransmission as neurons are unable to move neurotransmitter vesicles inside them causing loss of synapses and apoptosis.

13.5.1 Aβ-Induced Secondary Impairments in Axonal Transport

As Aβ-peptides occur in monomers, dimers, oligomers and fibrillar structures, their toxicity levels are correlated with their size and structures. Both monomeric and fibrillar Aβ-peptides have been shown to affect the axonal transport machinery in general and mitochondrial transport in particular in cultured hippocampal neurons (Rui et al. 2006; Hiruma et al. 2003). The role of soluble oligomers of Aβ in FAT defects has been assessed most because of the strong evidence that these soluble oligomers are most toxic causing profound neurodegeneration. These soluble oligomers of Aβ when applied to squid and murine hippocampal neurons disrupted their FAT machinery probably mediated by NMDA receptor-dependent mechanisms involving GSK3β (Pigino et al. 2009; Decker et al. 2010). Although, Aβ-oligomers have been found to disrupt both the types of transport processes, anterograde and retrograde, recent studies suggest that they affect anterograde transport more significantly than retrograde transport (Rui et al. 2006; Wang et al. 2010) by which synaptic vesicle and mitochondrial anterograde transport is prohibited which subsequently disrupt synaptic activity and ultimately lead to neural degeneration. Furthermore, it is not known how exactly Aβ-oligomers disrupt FAT, but new insights point towards various plausible mechanisms including inhibitory effects of Aβ-oligomers on actin polymerization and aggregation (Hiruma et al. 2003). Studies suggest that both high concentration and low concentration of Aβ-oligomers cause defects in axonal transport and such defects are reversible upon washout of Aβ-oligomers (Hiruma et al. 2003; Tang et al. 2012).

As axonal transport is a fundamental need for the survival and maintenance of neuronal health, its disruption is bound to lead a cascade of pathologies subsequently leading to various types of neuronal degenerative diseases. Many studies have proposed that FAT disruption is one of the main causes behind the progression of AD and it involves both Aβ-soluble oligomers as well as hyperphosphorylated tau proteins. However, the precise mechanism of how FAT leads to the development of diseases like AD remains unknown and additionally it is not well understood if FAT defects are the primary causes of the disease or they are simply the consequences of AD pathology.

13.6 Mitochondrial Dysfunctions in the Pathophysiology of AD

The energy demands of the brain are significantly high and account for around 20% of the energy consumption of the whole body (Magistretti and Allaman 2015). Mitochondria are the key organelle responsible for cellular energy production, and its loss of function may result in moderate to severe fatigue and diminished production of ATP at the cellular level. As neurons exclusively use glucose as the primary energy source unless starved, this depicts their higher dependence on mitochondria for aerobic oxidative phosphorylation for their energetic needs. The energy requirements of neurons are largely driven to maintain the ionic gradients across the plasma membranes which are critical for the generation of action potentials (nerve impulses) and hence neuronal communication. Therefore, it is not surprising that malfunctioning of mitochondria could lead to plethora of deleterious consequences including generation of free radicals, modulation of mitochondrial permeability transition and secondary excitotoxicity, impaired calcium buffering and oxidative damage which all are well reported in a number of neurodegenerative diseases such as Alzheimer's, Parkinson's, Huntington's and amyotrophic lateral sclerosis (McInnes 2013).

Since mitochondria are vulnerable to oxidative damage, the interaction between oxidative stress and mitochondrial dysfunction may initiate or accelerate the generation of reactive oxygen species that are critical for the pathogenesis of AD (Wang et al. 2014). Oxidative stress-induced damage involves functional alterations in mitochondria which are early observed events in AD prior to the appearances of Aß-plaques. In AD pathogenesis, the impaired activity of three important enzymes involved in the tricarboxylic acid (TCA) including pyruvate dehydrogenase, isocitrate dehydrogenase and α-ketoglutarate dehydrogenase is observed due to mitochondrial dysfunctioning (Gibson et al. 2012). Reports suggest that impaired ATP generation within a neuron also leads to activation of kinases such as ERK1 and ERK2 which further phosphorylates tau proteins in paired helical filaments, like a state similar in AD pathogenesis. A higher degree of β-amyloid deposition, overexpression of oxidative stress markers, mitochondrial DNA (mtDNA) deletions and mitochondrial structural abnormalities are found in animal models of AD as compared to normal control subjects.

13.6.1 Alterations in Metabolic Functions and mtDNA Defects in AD

As mitochondria are indispensable for providing the energy source to the neurons for their maintenance of ionic gradients across the plasma membrane and constant release of synaptic vesicles whenever action potential is generated, defects in mitochondrial enzymatic machinery have been reported in AD pathology. Studies of transcriptome and genomic analysis revealed that glucose metabolism in cerebral tissues declined in AD which was found to be associated with down-regulation of nuclear genes encoding subunits of the mitochondrial electron transport chain (Godoy et al. 2014). A number of mitochondrial key enzymes like cytochrome oxidase are reported to have reduced expression in the brain under AD pathology. Additionally, elevated levels of oxidative stress likely cause a further decline in enzymatic activities and expression profiles of some of the important metabolic enzymes such as α-ketoglutarate dehydrogenase complex, cytochrome oxidase (COX) and pyruvate dehydrogenase complex (García et al. 2013). Furthermore, utilizing electron micrography, cytochrome oxidase (COX) histochemical staining and mtDNA in situ hybridization techniques, various reports suggest that neurons which displayed ROS-induced damage also had increased mtDNA content and elevated expression of COX in AD biopsy samples (Bonda et al. 2010). Additionally, it has been suggested that in AD, there is enhanced oxidative damage in nuclear and mtDNA in neurons under oxidative stress indicated by the presence of higher levels of multiple oxidized bases in nuclear and mtDNA from AD patient samples. Since key enzymatic complexes of the citric acid cycle as well as of electron transport chain (ETC) act as oxidizing or reducing mediators/agents, any dysfunctions in these complexes result in the aberrant transfer of electrons consequently generating oxidative free radicals which further lead to mitochondrial stress and initiation of apoptosis. Interestingly, mitochondrial ETC complex-I and complex-IV deregulations have been found tau-toxicity and Aβ-dependent processes respectively (Eckert et al. 2010). Apart from displaying lower expression/activity of α-ketoglutarate in AD pathology, AD brain has significant loss of α-ketoglutarate enriched neurons in cortical layer 2/4 of which is mostly affected in AD.

As neuroscientists are constantly searching for early detectable biomarkers for AD diagnosis at initial stages, indicators like decline in brain glucose metabolism are considered useful measures to consider cognition status and functionality in AD and are useful for early diagnosis and to predict future cognitive decline as reduced glucose metabolism in the diseased brain is one of the significant abnormalities under AD pathogenesis. Deficits in mitochondrial functions and glucose metabolism are well-established pathological hallmarks of AD and brain ageing contributing to neurodegeneration. In fact, in early AD, the brain shows region-specific hypometabolism of glucose and mitochondrial dysfunctions having harmful consequences for neurons by producing large ROS, depleting ATP levels and caspase-3-mediated programmed cell death (Fig. 13.5). Progressive accumulations of Aβ-peptides in mitochondria have been reported to alter mitochondrial dynamics including fission-fusion equilibrium and Aβ-induced activation of NMDAR's along with the excessive release of calcium from the lumen of endoplasmic reticulum

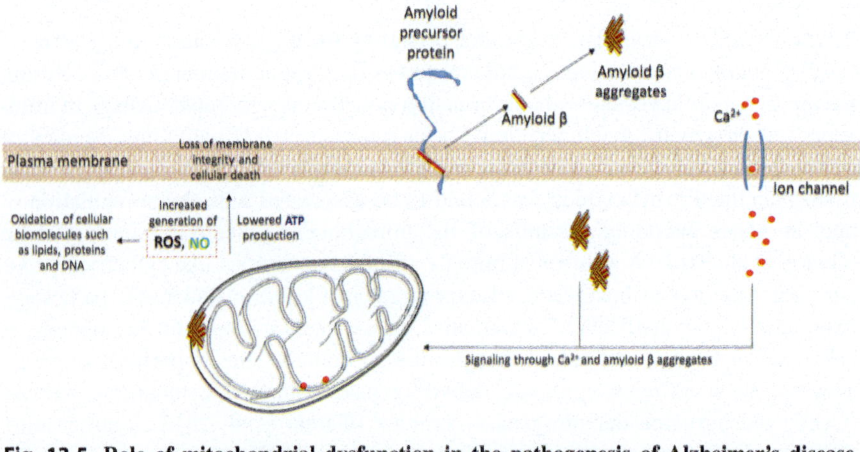

Fig. 13.5 Role of mitochondrial dysfunction in the pathogenesis of Alzheimer's disease. Aggregation of amyloid beta proteins and an increased influx of calcium ions caused a rapid generation of reactive oxygen species and nitric oxide molecules. These increased levels of ROS and NO species then trigger a number of secondary responses such as oxidation of DNA, proteins and lipids, loss of membrane integrity and cellular death, features which are the hallmark of neuronal degradation and development of Alzheimer's disease

causes calcium dyshomeostasis in mitochondria resulting in neuronal injury (Xu et al. 2017). Substantiated by positron emission tomography (PET) imaging, most dementia patients display decreased glucose metabolism in many brain areas, however, severe changes have been reported in parieto-temporal association cortex and in the frontal lobes while brainstem, cerebellum and basal ganglia are not affected. Various reports suggest that Aβ-peptides are inserted and clog many mitochondrial protein importing channels thereby depriving this organelle of various nuclear genome-encoded proteins/enzymes like COX which finally degenerate these organelles and ultimately the whole neurons are killed. Therefore, restoring metabolic defects in mitochondria and rescuing the reduced expression of glycolytic and oxidative phosphorylation enzymes could potentially act as one of the most important therapeutic targets for AD treatment.

Furthermore, increased oxidative damage to mtDNA causing mitochondrial dysfunctions to exacerbate the pathology in AD further links the ROS cascade with AD pathology (Fig. 13.5). Elevated levels of sporadic mutations are found in the mitochondrial DNA control regions which are unique to AD. As a result of such mutations, there are reduced levels of protective proteins like histones and other DNA repair machinery proteins; thus the hotspots of ROS attack increase and make mtDNA vulnerable (Xu et al. 2017). Analysis of the oxidized nucleotide levels in mitochondria shows almost three times higher oxidative damage in AD brain which might be the cause of increased mutation. Many of these mutations occur at the important transcription as well as at replication regulatory elements that causes reduced expression of vital mitochondrial proteins subsequently causing deleterious effects on mitochondrial function. Reports suggest that mtDNA deletions are responsible for deficiency of various mitochondrial enzymes like COX which is a well-characterized marker of mitochondrial dysfunction atypical in AD pathology (Moreira 2018). Moreover, recent data on mtDNA rearrangement from post-mortem AD brain tissues of AD patients suggest that there are three kinds of DNA rearrangements including deletions, F-type rearrangements (duplication) and R-type rearrangements, and these rearrangements are 2.7-fold higher in AD pathology which may be the driving force for AD pathology rather than consequences of it (Chen et al. 2016). As the mitochondria are versatile organelle and neurons are highly energy demanding, the continuous DNA replication in mitochondria for distribution to its daughter organelle makes their genome prone to replication errors resulting in the accumulation of a large pool of mutations. Apart from these mutations, mtDNA rearrangements including deletions, tandem duplications as well as insertions subsequently alter the protein-coding function of this organelle DNA which is deleterious for neuronal physiology and ultimately result in cell death.

13.6.2 Mitochondrial Dysfunction and Neuronal Apoptosis in AD

Apoptosis is a programmed cell death that is involved in the selective demise of neurons during early development, toxic insults to cells and various neurodegenerative diseases including AD. Mitochondria are versatile cell organelles which coordinate many cellular processes like redox signalling, calcium homeostasis, energy production, synaptic plasticity and arbitration of cell survival or death. Mitochondria lie in the centre of the intrinsic apoptotic pathway which predominates in the nervous system and is initiated by receiving cell death signals like DNA damage, depletion of trophic factors and oxidative stress (Mattson et al. 2008). Intrinsic pathway involves activation of various kinases like JNK and other transcription factors like p53 which induce the gene expression and mitochondrial translocation of various pro-apoptotic proteins such as Bax and Bak which belong to Bcl-2 family members, forming a pore in the outer mitochondrial membrane. Cytochrome-c escapes from mitochondria through these membrane pores and once in the cytosol, it binds to apoptosis protease activating factor-1 as well as ATP to form apoptosome which then activates caspase-9 and that in turn cleaves and activates caspase-3 executing programmed cell death (Mattson et al. 2008).

In neurodegenerative pathology such as AD, oxidative stress and malfunctioning of mitochondria inappropriately initiate apoptosis accelerating neurodegeneration. Aβ-peptides affect mitochondrial physiology in many ways, including inhibiting α-subunit of ATP synthase, declining activities of many enzymes of Kreb's cycle, reducing respiratory state and inhibiting the transportation of nuclear proteins to mitochondria thereby rendering the mitochondria dysfunctional. Furthermore, pore formation by Aβ in mt-membrane releases cytochrome-c thereby initiating apoptosis contributing to the neurodegeneration process. Additionally, under AD pathology, cytochrome-c release via mt-membrane pores activates caspase 9, and further it cleaves and activates caspase 3 which has been shown to cleave tau protein at D421 site which assists in nucleation-dependent NFT formation (Rissman et al. 2004). Therefore, one of the main targets in combating mitochondrial dysfunction-induced cell death is to overexpress the anti-apoptotic genes, for example, overexpression of Bcl-2 genes may protect neurons against Aß-induced cell death.

13.7 Conclusion

Advances made in molecular biological tools during the last few decades have essentially contributed to the understanding of cellular as well as molecular mechanisms including cell signalling pathways underlying AD pathogenesis (Fig. 13.6). Present understanding of the underlying complex molecular mechanisms involved in the development of AD involves various hypothesis where a plethora of factors interrelates with each other but none of these hypotheses fully explains the complex dogma of cellular pathology involved in AD pathogenesis thereby demanding further investigations. Less is known about the mechanisms responsible for initial triggering as well as early factors causing AD including abnormal accumulation of Aβ oligomers, oxidative stress and tau pathology remain poorly understood and it remains a challenge as to how these factors contribute to neuronal killing creating memory deficits and other behavioural consequences of AD. Genetic factors drastically contribute to the development and progression of AD but still, the cause of these pathological factors and their interaction with environmental factors is still unknown.

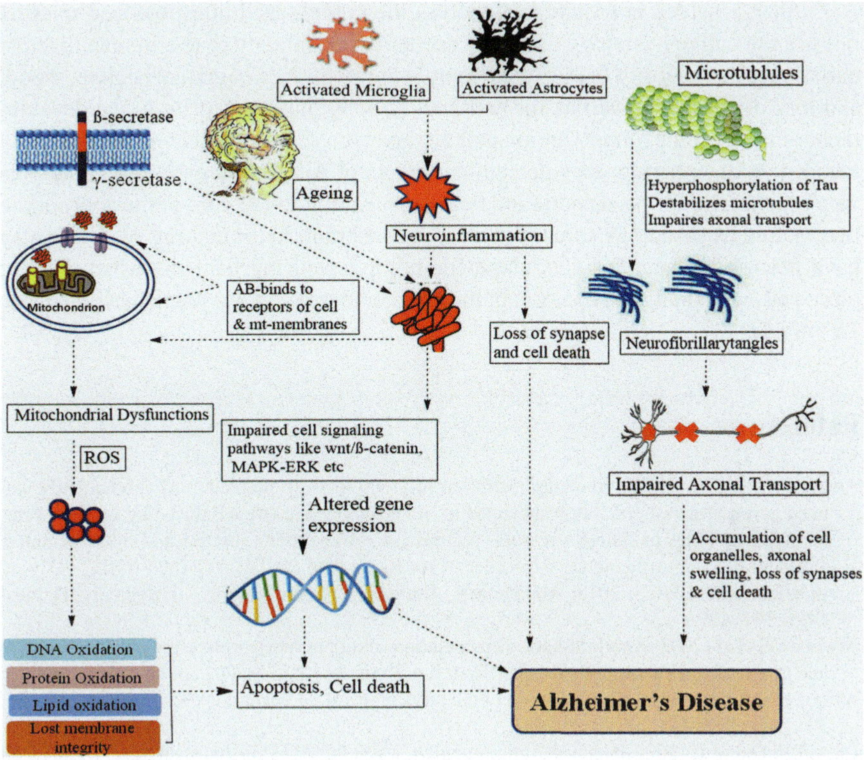

Fig. 13.6 Neurobiological mechanisms involved in the pathogenesis of Alzheimer's disease. A schematic model illustrating diverse neurobiological mechanisms that leads to the onset and development of Alzheimer's disease

A deeper understanding of the imbalances between the rate of Aβ production and its clearance as well as enzymatic activities in tau hyperphosphorylation will provide some clues about the possible treatment strategies for AD. Targeting rate-limiting enzymes for Aβ production such as BACE1 for therapeutic potential has not provided much help; therefore, new molecular targets such as Wnt/β-catenin and MAPK pathways as well as increasing scavenging capacities of biological scavengers to decrease oxidative stress may prove helpful. Oxidative stress has been found to induce tau hyperphosphorylation via enhanced activation of GSKβ-3 enzyme, and furthermore, this shatters off the axonal transport in neurons leading to their apoptosis; therefore, new anti-oxidant compounds shall be screened for scavenging ROS. Mitochondrial dysfunctions significantly contribute to the development and progression of AD and is an early feature in AD pathology. These abnormalities in mitochondria function like oxidative stress, aberrant calcium buffering and inappropriate initiating of the intrinsic pathway of apoptosis together lead to neuronal atrophy in AD brain.

Further, research is required to address the complicated questions and missing links in the pathophysiology of AD. A deeper understanding of the complex factors underlying the majority of AD behavioural symptoms like cognitive decline, mood, and loss of speech as well as for neuronal atrophy may contribute to the development of novel clinical intervention and therapeutics. Current researches are majorly focused on deciphering early detection markers of AD pathology which could provide ample time for therapeutic interventions leading to the cure at the appropriate time before its pathology spreads globally in the brain. In the coming years, we may have new and deeper insights about the complicated mechanisms which lead to these pathophysiological changes in the brain, and treating AD would become more procurable.

References

Ansari MA et al (2006) In vivo administration of D609 leads to protection of subsequently isolated gerbil brain mitochondria subjected to in vitro oxidative stress induced by amyloid beta-peptide and other oxidative stressors: relevance to Alzheimer's disease and other oxidative stress-related neurodegenerative disorders. Free Radic Biol Med 41(11):1694–1703

Association, A.s (2018) 2018 Alzheimer's disease facts and figures. Alzheimers Dement 14(3):367–429

Atwood CS et al (2003) Amyloid-beta: a chameleon walking in two worlds: a review of the trophic and toxic properties of amyloid-beta. Brain Res Brain Res Rev 43(1):1–16

Avila J et al (2004) Role of tau protein in both physiological and pathological conditions. Physiol Rev 84(2):361–384

Baas PW, Qiang L (2005) Neuronal microtubules: when the MAP is the roadblock. Trends Cell Biol 15(4):183–187

Baloh RH et al (2007) Altered axonal mitochondrial transport in the pathogenesis of Charcot-Marie-Tooth disease from mitofusin 2 mutations. J Neurosci 27(2):422–430

Baudier J, Cole RD (1988) Interactions between the microtubule-associated tau proteins and S100b regulate tau phosphorylation by the Ca2+/calmodulin-dependent protein kinase II. J Biol Chem 263(12):5876–5883

Block ML (2008) NADPH oxidase as a therapeutic target in Alzheimer's disease. BMC Neurosci 9(Suppl 2):S8

Bobinski M et al (1996) Neurofibrillary pathology–correlation with hippocampal formation atrophy in Alzheimer disease. Neurobiol Aging 17(6):909–919

Bonda DJ et al (2010) Mitochondrial dynamics in Alzheimer's disease: opportunities for future treatment strategies. Drugs Aging 27(3):181–192

Bowman AB et al (1999) Drosophila roadblock and Chlamydomonas LC7: a conserved family of dynein-associated proteins involved in axonal transport, flagellar motility, and mitosis. J Cell Biol 146(1):165–180

Braak H, Del Tredici K (2011) The pathological process underlying Alzheimer's disease in individuals under thirty. Acta Neuropathol 121(2):171–181

Brown A (2003) Axonal transport of membranous and nonmembranous cargoes: a unified perspective. J Cell Biol 160(6):817–821

Buee L et al (2000) Tau protein isoforms, phosphorylation and role in neurodegenerative disorders. Brain Res Brain Res Rev 33(1):95–130

Busser J, Geldmacher DS, Herrup K (1998) Ectopic cell cycle proteins predict the sites of neuronal cell death in Alzheimer's disease brain. J Neurosci 18(8):2801–2807

Butterfield DA, Boyd-Kimball D (2018) Oxidative stress, amyloid-β peptide, and altered key molecular pathways in the pathogenesis and progression of Alzheimer's disease. J Alzheimers Dis 62(3):1345–1367

Butterfield SM, Lashuel HA (2010) Amyloidogenic protein–membrane interactions: mechanistic insight from model systems. Angew Chem Int Ed 49(33):5628–5654

Butterfield DA et al (2001) Evidence of oxidative damage in Alzheimer's disease brain: central role for amyloid beta-peptide. Trends Mol Med 7(12):548–554

Butterfield DA et al (2006) Redox proteomics identification of oxidatively modified hippocampal proteins in mild cognitive impairment: insights into the development of Alzheimer's disease. Neurobiol Dis 22(2):223–232

Butterfield DA et al (2007) Roles of amyloid β-peptide-associated oxidative stress and brain protein modifications in the pathogenesis of Alzheimer's disease and mild cognitive impairment. Free Radic Biol Med 43(5):658–677

Butterfield DA, Swomley AM, Sultana R (2013) Amyloid β-peptide (1–42)-induced oxidative stress in Alzheimer disease: importance in disease pathogenesis and progression. Antioxid Redox Signal 19(8):823–835

Canevari L, Clark JB, Bates TE (1999) β-Amyloid fragment 25–35 selectively decreases complex IV activity in isolated mitochondria. FEBS Lett 457(1):131–134

Carrillo-Mora P, Luna R, Colin-Barenque L (2014) Amyloid beta: multiple mechanisms of toxicity and only some protective effects? Oxidative Med Cell Longev 2014:795375

Cash AD et al (2003) Microtubule reduction in Alzheimer's disease and aging is independent of tau filament formation. Am J Pathol 162(5):1623–1627

Cassagnes L-E et al (2013) The catalytically active copper-amyloid-Beta state: coordination site responsible for reactive oxygen species production. Angew Chem Int Ed 52(42):11110–11113

Castegna A et al (2004) Modulation of phospholipid asymmetry in synaptosomal membranes by the lipid peroxidation products, 4-hydroxynonenal and acrolein: implications for Alzheimer's disease. Brain Res 1004(1–2):193–197

Chalermpalanupap T et al (2013) Targeting norepinephrine in mild cognitive impairment and Alzheimer's disease. Alzheimers Res Ther 5(2):21

Chang KA, Suh YH (2010) Possible roles of amyloid intracellular domain of amyloid precursor protein. BMB Rep 43(10):656–663

Cheignon C et al (2018) Oxidative stress and the amyloid beta peptide in Alzheimer's disease. Redox Biol 14:450–464

Chen Y et al (2016) Mitochondrial DNA rearrangement Spectrum in brain tissue of Alzheimer's disease: analysis of 13 cases. PLoS One 11(6):e0154582

Chow VW et al (2010) An overview of APP processing enzymes and products. NeuroMolecular Med 12(1):1–12

Cui JG et al (2007) Expression of inflammatory genes in the primary visual cortex of late-stage Alzheimer's disease. Neuroreport 18(2):115–119

de Paula VDJR et al (2009) *Neurobiological pathways to Alzheimer's disease: Amyloid-beta, TAU protein or both?* Dementia & Neuropsychologia **3**(3):188–194

Decker H et al (2010) Amyloid-beta peptide oligomers disrupt axonal transport through an NMDA receptor-dependent mechanism that is mediated by glycogen synthase kinase 3beta in primary cultured hippocampal neurons. J Neurosci 30(27):9166–9171

Duering M et al (2005) Mean age of onset in familial Alzheimer's disease is determined by amyloid beta 42. Neurobiol Aging 26(6):785–788

Duff K et al (1996) Increased amyloid-beta42(43) in brains of mice expressing mutant presenilin 1. Nature 383(6602):710–713

Ebbing B et al (2008) Effect of spastic paraplegia mutations in KIF5A kinesin on transport activity. Hum Mol Genet 17(9):1245–1252

Eckert A et al (2010) Convergence of amyloid-beta and tau pathologies on mitochondria in vivo. Mol Neurobiol 41(2–3):107–114

Eckman CB, Eckman EA (2007) An update on the amyloid hypothesis. Neurol Clin 25(3):669–682

Edwards DR, Handsley MM, Pennington CJ (2008) The ADAM metalloproteinases. Mol Asp Med 29(5):258–289

Farrer MJ et al (2009) DCTN1 mutations in Perry syndrome. Nat Genet 41(2):163–165

Ferreira A, Caceres A, Kosik KS (1993) Intraneuronal compartments of the amyloid precursor protein. J Neurosci 13(7):3112–3123

Gao HM et al (2002) Microglial activation-mediated delayed and progressive degeneration of rat nigral dopaminergic neurons: relevance to Parkinson's disease. J Neurochem 81(6):1285–1297

García-Escudero V et al (2013) Deconstructing mitochondrial dysfunction in Alzheimer disease. Oxidative Med Cell Longev 2013:13, Article Id 162152

German DC et al (1992) Disease-specific patterns of locus coeruleus cell loss. Ann Neurol 32(5):667–676

Gibson GE et al (2012) *Deficits in the mitochondrial enzyme α-ketoglutarate dehydrogenase lead to Alzheimer's disease-like calcium dysregulation*. Neurobiology of aging **33**(6):1121.e13–1121.e24

Glabe CC (2005) Amyloid accumulation and pathogenesis of Alzheimer's disease: significance of monomeric, oligomeric and fibrillar Abeta. Subcell Biochem 38:167–177

Godoy JA et al (2014) Signaling pathway cross talk in Alzheimer's disease. Cell Commun Signal 12(1):23

Goldstein LS, Yang Z (2000) Microtubule-based transport systems in neurons: the roles of kinesins and dyneins. Annu Rev Neurosci 23:39–71

Gong CX, Grundke-Iqbal I, Iqbal K (1994) Dephosphorylation of Alzheimer's disease abnormally phosphorylated tau by protein phosphatase-2A. Neuroscience 61(4):765–772

Gouras GK et al (1998) Generation and regulation of beta-amyloid peptide variants by neurons. J Neurochem 71(5):1920–1925

Grimm M, Hartmann T (2012) Recent understanding of the molecular mechanisms of Alzheimer's disease. J Addict Res Ther 5:1–27

Grudzien A et al (2007) Locus coeruleus neurofibrillary degeneration in aging, mild cognitive impairment and early Alzheimer's disease. Neurobiol Aging 28(3):327–335

Grundke-Iqbal I et al (1986) Abnormal phosphorylation of the microtubule-associated protein tau (tau) in Alzheimer cytoskeletal pathology. Proc Natl Acad Sci U S A 83(13):4913–4917

Gunawardena S, Goldstein LS (2001) Disruption of axonal transport and neuronal viability by amyloid precursor protein mutations in Drosophila. Neuron 32(3):389–401

Gunawardena S, Goldstein LS (2004) Cargo-carrying motor vehicles on the neuronal highway: transport pathways and neurodegenerative disease. J Neurobiol 58(2):258–271

Gupta A, Goyal R (2016) Amyloid beta plaque: a culprit for neurodegeneration. Acta Neurol Belg 116(4):445–450

Hanger DP, Seereeram A, Noble W (2009) Mediators of tau phosphorylation in the pathogenesis of Alzheimer's disease. Expert Rev Neurother 9(11):1647–1666

Hardy J, Allsop D (1991) Amyloid deposition as the central event in the aetiology of Alzheimer's disease. Trends Pharmacol Sci 12:383–388

Hardy J, Selkoe DJ (2002) The amyloid hypothesis of Alzheimer's disease: progress and problems on the road to therapeutics. Science 297(5580):353–356

Heneka MT et al (2010) Locus ceruleus controls Alzheimer's disease pathology by modulating microglial functions through norepinephrine. Proc Natl Acad Sci U S A 107(13):6058–6063

Herzog AG, Kemper TL (1980) Amygdaloid changes in aging and dementia. Arch Neurol 37(10):625–629

Higuchi M, Lee VM, Trojanowski JQ (2002) Tau and axonopathy in neurodegenerative disorders. NeuroMolecular Med 2(2):131–150

Hirokawa N (1998) Kinesin and dynein superfamily proteins and the mechanism of organelle transport. Science 279(5350):519–526

Hiruma H et al (2003) Glutamate and amyloid beta-protein rapidly inhibit fast axonal transport in cultured rat hippocampal neurons by different mechanisms. J Neurosci 23(26):8967–8977

Hodgkin AL, Huxley AF (1939) Action potentials recorded from inside a nerve fibre. Nature 144:710

Huang X et al (1999) Cu(II) potentiation of alzheimer abeta neurotoxicity. Correlation with cell-free hydrogen peroxide production and metal reduction. J Biol Chem 274(52):37111–37116

Ito S et al (2007) Cerebral clearance of human amyloid-β peptide (1–40) across the blood–brain barrier is reduced by self-aggregation and formation of low-density lipoprotein receptor-related protein-1 ligand complexes. J Neurochem 103(6):2482–2490

Ittner LM et al (2010) Dendritic function of tau mediates amyloid-beta toxicity in Alzheimer's disease mouse models. Cell 142(3):387–397

Jeynes B, Provias J (2008) Evidence for altered LRP/RAGE expression in Alzheimer lesion pathogenesis. Curr Alzheimer Res 5(5):432–437

Jobst KA et al (1992) Detection in life of confirmed Alzheimer's disease using a simple measurement of medial temporal lobe atrophy by computed tomography. Lancet 340(8829):1179–1183

Jobst KA et al (1994) Rapidly progressing atrophy of medial temporal lobe in Alzheimer's disease. Lancet 343(8901):829–830

Jomova K et al (2010) Metals, oxidative stress and neurodegenerative disorders. Mol Cell Biochem 345(1–2):91–104

Kamal A et al (2001) Kinesin-mediated axonal transport of a membrane compartment containing beta-secretase and presenilin-1 requires APP. Nature 414(6864):643–648

Kayed R et al (2003) Common structure of soluble amyloid oligomers implies common mechanism of pathogenesis. Science 300(5618):486–489

Kelleher RJ 3rd, Shen J (2017) *Presenilin-1 mutations and Alzheimer's disease*. Proc Natl Acad Sci U S A 114(4):629–631

Ko SY et al (2015) The possible mechanism of advanced Glycation end products (AGEs) for Alzheimer's disease. PLoS One 10(11):e0143345

Koo EH et al (1990) Precursor of amyloid protein in Alzheimer disease undergoes fast anterograde axonal transport. Proc Natl Acad Sci U S A 87(4):1561–1565

Kowalska A (2004) Genetic aspects of amyloid beta-protein fibrillogenesis in Alzheimer's disease. Folia Neuropathol 42(4):235–237

Lal R, Lin H, Quist AP (2007) Amyloid beta ion channel: 3D structure and relevance to amyloid channel paradigm. Biochim Biophys Acta 1768(8):1966–1975

Lee VM, Daughenbaugh R, Trojanowski JQ (1994) Microtubule stabilizing drugs for the treatment of Alzheimer's disease. Neurobiol Aging 15(Suppl 2):S87–S89

Lee M-C et al (2018) Zinc ion rapidly induces toxic, off-pathway amyloid-β oligomers distinct from amyloid-β derived diffusible ligands in Alzheimer's disease. Sci Rep 8(1):4772

Li Y et al (2012) Analysis of hippocampal gene expression profile of Alzheimer's disease model rats using genome chip bioinformatics. Neural Regen Res 7(5):332–340

Liang WS et al (2008) Altered neuronal gene expression in brain regions differentially affected by Alzheimer's disease: a reference data set. Physiol Genomics 33(2):240–256

Lin H, Bhatia R, Lal R (2001) Amyloid β protein forms ion channels: implications for Alzheimer's disease pathophysiology. FASEB J 15(13):2433–2444

Lindwall G, Cole RD (1984) Phosphorylation affects the ability of tau protein to promote microtubule assembly. J Biol Chem 259(8):5301–5305

Liu F et al (2002) Role of glycosylation in hyperphosphorylation of tau in Alzheimer's disease. FEBS Lett 512(1–3):101–106

Lloret A et al (2011) Amyloid-beta toxicity and tau hyperphosphorylation are linked via RCAN1 in Alzheimer's disease. J Alzheimers Dis 27(4):701–709

Lopez-Toledano MA, Shelanski ML (2004) Neurogenic effect of beta-amyloid peptide in the development of neural stem cells. J Neurosci 24(23):5439–5444

Luo Y et al (2003) BACE1 (β-secretase) knockout mice do not acquire compensatory gene expression changes or develop neural lesions over time. Neurobiol Dis 14(1):81–88

Luxenberg JS et al (1987) Rate of ventricular enlargement in dementia of the Alzheimer type correlates with rate of neuropsychological deterioration. Neurology 37(7):1135–1140

Magistretti PJ, Allaman I (2015) *A cellular perspective on brain energy metabolism and functional imaging*. Neuron 86(4):883–901

Magrane J et al (2005) Intraneuronal beta-amyloid expression downregulates the Akt survival pathway and blunts the stress response. J Neurosci 25(47):10960–10969

Mark RJ et al (1997) A role for 4-hydroxynonenal, an aldehydic product of lipid peroxidation, in disruption of ion homeostasis and neuronal death induced by amyloid beta-peptide. J Neurochem 68(1):255–264

Markesbery WR (1997) Oxidative stress hypothesis in Alzheimer's disease. Free Radic Biol Med 23(1):134–147

Martin M et al (1999) Cytoplasmic dynein, the dynactin complex, and kinesin are interdependent and essential for fast axonal transport. Mol Biol Cell 10(11):3717–3728

Matsuda S et al (2001) c-Jun N-terminal kinase (JNK)-interacting protein-1b/islet-brain-1 scaffolds Alzheimer's amyloid precursor protein with JNK. J Neurosci 21(17):6597–6607

Matthews FE et al (2013) *A two-decade comparison of prevalence of dementia in individuals aged 65 years and older from three geographical areas of England: results of the Cognitive Function and Ageing Study I and II*. Lancet (London, England) **382**(9902):1405–1412

Mattson MP, Gleichmann M, Cheng A (2008) Mitochondria in neuroplasticity and neurological disorders. Neuron 60(5):748–766

Mc Donald JM et al (2015) The aqueous phase of Alzheimer's disease brain contains assemblies built from approximately 4 and approximately 7 kDa Abeta species. Alzheimers Dement 11(11):1286–1305

McInnes J (2013) Insights on altered mitochondrial function and dynamics in the pathogenesis of neurodegeneration. Transl Neurodegener 2(1):12

Meda L et al (1995) Activation of microglial cells by beta-amyloid protein and interferon-gamma. Nature 374(6523):647–650

Miller KR, Streit WJ (2007) The effects of aging, injury and disease on microglial function: a case for cellular senescence. Neuron Glia Biol 3(3):245–253

Miller DL et al (1993) Peptide compositions of the cerebrovascular and senile plaque core amyloid deposits of Alzheimer's disease. Arch Biochem Biophys 301(1):41–52

Misko A et al (2010) Mitofusin 2 is necessary for transport of axonal mitochondria and interacts with the Miro/Milton complex. J Neurosci 30(12):4232–4240

Montoliu-Gaya L, Villegas S (2015) Protein structures in Alzheimer's disease: the basis for rationale therapeutic design. Arch Biochem Biophys 588:1–14

Moreira PI (2018) Sweet mitochondria: a shortcut to Alzheimer's disease. J Alzheimers Dis 62(3):1391–1401

Morfini G et al (2002) Glycogen synthase kinase 3 phosphorylates kinesin light chains and negatively regulates kinesin-based motility. EMBO J 21(3):281–293

Morfini GA et al (2009) Axonal transport defects in neurodegenerative diseases. J Neurosci 29(41):12776–12786

Mudher A, Lovestone S (2002) Alzheimer's disease-do tauists and baptists finally shake hands? Trends Neurosci 25(1):22–26

Muresan Z, Muresan V (2005) Coordinated transport of phosphorylated amyloid-beta precursor protein and c-Jun NH2-terminal kinase-interacting protein-1. J Cell Biol 171(4):615–625

Newsway V et al (2010) Perry syndrome due to the DCTN1 G71R mutation – a distinctive L-DOPA responsive disorder with behavioural syndrome, vertical gaze palsy and respiratory failure. Mov Disord 25(6):767–770

Noble W et al (2003) Cdk5 is a key factor in tau aggregation and tangle formation in vivo. Neuron 38(4):555–565

Nunomura A et al (2009) RNA oxidation in Alzheimer disease and related neurodegenerative disorders. Acta Neuropathol 118(1):151–166

O'Nuallain B et al (2010) Amyloid β-protein dimers rapidly form stable Synaptotoxic Protofibrils. J Neurosci 30(43):14411–14419

Ohshima Y et al (2018) Mutations in the β-amyloid precursor protein in familial Alzheimer's disease increase Aβ oligomer production in cellular models. Heliyon 4(1):e00511–e00511

Omar RA et al (1999) Increased expression but reduced activity of antioxidant enzymes in Alzheimer's disease. J Alzheimers Dis 1(3):139–145

Perluigi M et al (2006a) In vivo protection by the xanthate tricyclodecan-9-yl-xanthogenate against amyloid beta-peptide (1-42)-induced oxidative stress. Neuroscience 138(4):1161–1170

Perluigi M et al (2006b) In vivo protective effects of ferulic acid ethyl ester against amyloid-beta peptide 1-42-induced oxidative stress. J Neurosci Res 84(2):418–426

Pietri M et al (2013) PDK1 decreases TACE-mediated alpha-secretase activity and promotes disease progression in prion and Alzheimer's diseases. Nat Med 19(9):1124–1131

Pigino G et al (2009) Disruption of fast axonal transport is a pathogenic mechanism for intraneuronal amyloid beta. Proc Natl Acad Sci U S A 106(14):5907–5912

Plant LD et al (2003) The production of amyloid beta peptide is a critical requirement for the viability of central neurons. J Neurosci 23(13):5531–5535

Poulin SP et al (2011) Amygdala atrophy is prominent in early Alzheimer's disease and relates to symptom severity. Psychiatry Res 194(1):7–13

Premkumar DR et al (1995) Induction of heme oxygenase-1 mRNA and protein in neocortex and cerebral vessels in Alzheimer's disease. J Neurochem 65(3):1399–1402

Prince M et al (2016) Recent global trends in the prevalence and incidence of dementia, and survival with dementia. Alzheimers Res Ther 8(1):23

Puls I et al (2003) Mutant dynactin in motor neuron disease. Nat Genet 33(4):455–456

Qin L et al (2004) NADPH oxidase mediates lipopolysaccharide-induced neurotoxicity and proinflammatory gene expression in activated microglia. J Biol Chem 279(2):1415–1421

Rajendran R et al (2009) A novel approach to the identification and quantitative elemental analysis of amyloid deposits—insights into the pathology of Alzheimer's disease. Biochem Biophys Res Commun 382(1):91–95

Rissman RA et al (2004) Caspase-cleavage of tau is an early event in Alzheimer disease tangle pathology. J Clin Invest 114(1):121–130

Roher AE et al (1993) Beta-amyloid-(1-42) is a major component of cerebrovascular amyloid deposits: implications for the pathology of Alzheimer disease. Proc Natl Acad Sci U S A 90(22):10836–10840

Rosales-Corral S et al (2004a) Kinetics of the neuroinflammation-oxidative stress correlation in rat brain following the injection of fibrillar amyloid-β onto the hippocampus in vivo. J Neuroimmunol 150(1–2):20–28

Rosales-Corral S et al (2004b) Kinetics of the neuroinflammation-oxidative stress correlation in rat brain following the injection of fibrillar amyloid-beta onto the hippocampus in vivo. J Neuroimmunol 150(1–2):20–28

Rui Y et al (2006) Acute impairment of mitochondrial trafficking by beta-amyloid peptides in hippocampal neurons. J Neurosci 26(41):10480–10487

Sadigh-Eteghad S et al (2014) Beta-amyloid exhibits antagonistic effects on alpha 7 nicotinic acetylcholine receptors in orchestrated manner. J Med Hypotheses Ideas 8(2):49–52

Sagare A et al (2007) Clearance of amyloid-β by circulating lipoprotein receptors. Nat Med 13:1029

Satizabal CL et al (2016) Incidence of dementia over three decades in the Framingham heart study. N Engl J Med 374(6):523–532

Scahill RI et al (2002) Mapping the evolution of regional atrophy in Alzheimer's disease: unbiased analysis of fluid-registered serial MRI. Proc Natl Acad Sci U S A 99(7):4703–4707

Schilling T, Eder C (2011) Amyloid-beta-induced reactive oxygen species production and priming are differentially regulated by ion channels in microglia. J Cell Physiol 226(12):3295–3302

Scott SA, DeKosky ST, Scheff SW (1991) Volumetric atrophy of the amygdala in Alzheimer's disease: quantitative serial reconstruction. Neurology 41(3):351–356

Scott SA et al (1992) Amygdala cell loss and atrophy in Alzheimer's disease. Ann Neurol 32(4):555–563

Selkoe DJ, Hardy J (2016) The amyloid hypothesis of Alzheimer's disease at 25 years. EMBO Mol Med 8(6):595–608

Sengupta U, Nilson AN, Kayed R (2016) The role of amyloid-β oligomers in toxicity, propagation, and immunotherapy. EBioMedicine 6:42–49

Shankar GM et al (2008) Amyloid-beta protein dimers isolated directly from Alzheimer's brains impair synaptic plasticity and memory. Nat Med 14(8):837–842
Sisodia SS, Tanzi RE (2007) Alzheimer's disease: advances in genetics, molecular and cellular biology. Springer Science & Business Media, New York
Small DH, Mok SS, Bornstein JC (2001) Alzheimer's disease and Abeta toxicity: from top to bottom. Nat Rev Neurosci 2(8):595–598
Smith AD (2002) Imaging the progression of Alzheimer pathology through the brain. Proc Natl Acad Sci 99(7):4135
Smith KDB et al (2007) In vivo axonal transport rates decrease in a mouse model of Alzheimer's disease. NeuroImage 35(4):1401–1408
St George-Hyslop PH, Petit A (2005) *Molecular biology and genetics of Alzheimer's disease*. C R Biol 328(2):119–130
Stancu IC et al (2014) Models of beta-amyloid induced Tau-pathology: the long and "folded" road to understand the mechanism. Mol Neurodegener 9:51
Steenland K et al (2016) A meta-analysis of Alzheimer's disease incidence and prevalence comparing African-Americans and Caucasians. J Alzheimers Dis 50(1):71–76
Stokin GB et al (2005) Axonopathy and transport deficits early in the pathogenesis of Alzheimer's disease. Science 307(5713):1282–1288
Subramaniam R et al (1997) The lipid peroxidation product, 4-hydroxy-2-trans-nonenal, alters the conformation of cortical synaptosomal membrane proteins. J Neurochem 69(3):1161–1169
Sun X, Chen WD, Wang YD (2015a) *beta-Amyloid: the key peptide in the pathogenesis of Alzheimer's disease*. Front Pharmacol 6:221
Sun X, Chen W-D, Wang Y-D (2015b) *β-Amyloid: the key peptide in the pathogenesis of Alzheimer's disease*. Front Pharmacol 6:221
Takashima A et al (1998) Presenilin 1 associates with glycogen synthase kinase-3beta and its substrate tau. Proc Natl Acad Sci U S A 95(16):9637–9641
Talmat-Amar Y, Arribat Y, Parmentier M-L (2018) Vesicular axonal transport is modified in vivo by Tau deletion or overexpression in Drosophila. Int J Mol Sci 19(3):744
Tang Y et al (2012) Early and selective impairments in axonal transport kinetics of synaptic cargoes induced by soluble amyloid beta-protein oligomers. Traffic 13(5):681–693
Tarrade A et al (2006) A mutation of spastin is responsible for swellings and impairment of transport in a region of axon characterized by changes in microtubule composition. Hum Mol Genet 15(24):3544–3558
Thal DR et al (2002) Phases of A beta-deposition in the human brain and its relevance for the development of AD. Neurology 58(12):1791–1800
Valko M, Morris H, Cronin MT (2005) Metals, toxicity and oxidative stress. Curr Med Chem 12(10):1161–1208
Valla J, Berndt JD, Gonzalez-Lima F (2001) Energy hypometabolism in posterior cingulate cortex of Alzheimer's patients: superficial laminar cytochrome oxidase associated with disease duration. J Neurosci 21(13):4923–4930
Varadarajan S et al (1999) Methionine residue 35 is important in amyloid beta-peptide-associated free radical oxidative stress. Brain Res Bull 50(2):133–141
Varadarajan S et al (2001) Different mechanisms of oxidative stress and neurotoxicity for Alzheimer's A beta(1–42) and A beta(25–35). J Am Chem Soc 123(24):5625–5631
Vassar R et al (1999) Beta-secretase cleavage of Alzheimer's amyloid precursor protein by the transmembrane aspartic protease BACE. Science 286(5440):735–741
Verhey KJ et al (2001) Cargo of kinesin identified as JIP scaffolding proteins and associated signaling molecules. J Cell Biol 152(5):959–970
Vershinin M et al (2007) Multiple-motor based transport and its regulation by Tau. Proc Natl Acad Sci U S A 104(1):87–92
Vetrivel KS et al (2004) Association of gamma-secretase with lipid rafts in post-Golgi and endosome membranes. J Biol Chem 279(43):44945–44954
Vetrivel KS et al (2005) Spatial segregation of gamma-secretase and substrates in distinct membrane domains. J Biol Chem 280(27):25892–25900

Vicario-Orri E, Opazo CM, Munoz FJ (2015) The pathophysiology of axonal transport in Alzheimer's disease. J Alzheimers Dis 43(4):1097–1113
Violet M et al (2015) Prefibrillar Tau oligomers alter the nucleic acid protective function of Tau in hippocampal neurons in vivo. Neurobiol Dis 82:540–551
Vogt BA, Crino PB, Vogt LJ (1992) Reorganization of cingulate cortex in Alzheimer's disease: neuron loss, neuritic plaques, and muscarinic receptor binding. Cereb Cortex 2(6):526–535
Walsh DM, Selkoe DJ (2007) A beta oligomers – a decade of discovery. J Neurochem 101(5):1172–1184
Wang Y, Mandelkow E (2016) Tau in physiology and pathology. Nat Rev Neurosci 17(1):5–21
Wang X et al (2010) Amyloid-beta-derived diffusible ligands cause impaired axonal transport of mitochondria in neurons. Neurodegener Dis 7(1–3):56–59
Wang X et al (2014) Oxidative stress and mitochondrial dysfunction in Alzheimer's disease. Biochim Biophys Acta 1842(8):1240–1247
Wasco W et al (1993) Isolation and characterization of APLP2 encoding a homologue of the Alzheimer's associated amyloid beta protein precursor. Nat Genet 5(1):95–100
Wildsmith KR et al (2013) Evidence for impaired amyloid beta clearance in Alzheimer's disease. Alzheimers Res Ther 5(4):33
Wilquet V, De Strooper B (2004) Amyloid-beta precursor protein processing in neurodegeneration. Curr Opin Neurobiol 14(5):582–588
Wirths O et al (2006) Axonopathy in an APP/PS1 transgenic mouse model of Alzheimer's disease. Acta Neuropathol 111(4):312–319
Wu DC et al (2003) NADPH oxidase mediates oxidative stress in the 1-methyl-4-phenyl-1,2,3,6-tetrahydropyridine model of Parkinson's disease. Proc Natl Acad Sci U S A 100(10):6145–6150
Xu L-L et al (2017) Mitochondrial dynamics changes with age in an APPsw/PS1dE9 mouse model of Alzheimer's disease. Neuroreport 28(4):222–228
Xu F et al (2018) KIF1Bβ mutations detected in hereditary neuropathy impair IGF1R transport and axon growth. The Journal of Cell Biology 217:3480–3496
Yagishita S (1978) Morphological investigations on axonal swellings and spheroids in various human diseases. Virchows Arch A Pathol Anat Histol 378(3):181–197
Yankner B, Duffy L, Kirschner D (1990) Neurotrophic and neurotoxic effects of amyloid beta protein: reversal by tachykinin neuropeptides. Science 250(4978):279–282
Yatin SM, Aksenov M, Butterfield DA (1999) The antioxidant vitamin E modulates amyloid beta-peptide-induced creatine kinase activity inhibition and increased protein oxidation: implications for the free radical hypothesis of Alzheimer's disease. Neurochem Res 24(3):427–435
Zarow C et al (2003) Neuronal loss is greater in the locus coeruleus than nucleus basalis and substantia nigra in Alzheimer and Parkinson diseases. Arch Neurol 60(3):337–341
Zhang H et al (2012) Proteolytic processing of Alzheimer's beta-amyloid precursor protein. J Neurochem 120(Suppl 1):9–21
Zhao C et al (2001) Charcot-Marie-tooth disease type 2A caused by mutation in a microtubule motor KIF1Bbeta. Cell 105(5):587–597
Zhu N et al (2015) Huperzine A protects neural stem cells against Abeta-induced apoptosis in a neural stem cells and microglia co-culture system. Int J Clin Exp Pathol 8(6):6425–6433

Immunotherapy in Alzheimer's Disease 14

Jyoti Gupta and Mehboob Hoque

Abstract

Alzheimer's disease (AD) is the most prevalent proteopathy characterized by dementia that appeals for major concern worldwide. The causative factor is the imbalance between production and clearance of the toxic Aβ peptide (40–43 amino acid long) from the brain. Among all the therapeutic approaches, the most prominent one is immunotherapy. Both active and passive immunotherapies have been worked upon. The majority of active immunization-based methods suffered risks of autoimmune toxic inflammation due to cross-reactivity with the nontoxic form. However, conventional monoclonal antibody (mAb)-based strategies have been designed to reduce Aβ level in brain and neutralize toxic effects. A multitude of clinical trials are being conducted using the passive therapeutic approach. Recently, alternative approaches including the recombinant fragments have emerged as a tool for safer and more effective therapy. Promising results have been observed in studies employing antibody fragments which include ScFv, BsAb, Fab, gammabodies, and intrabodies. Although there have been failures in some of these clinical trials, experiences gained from them can be used for designing better therapeutics. Currently, there is an urgent need of therapeutics which can target and clear off the senile plaques with limited side effects and toxicity.

J. Gupta
Interdisciplinary Biotechnology Unit, Aligarh Muslim University, Aligarh, India

M. Hoque (✉)
Department of Biological Sciences, Aliah University, Kolkata, India
e-mail: m_hoque@aliah.ac.in

G. M. Ashraf, A. Alexiou (eds.), *Biological, Diagnostic and Therapeutic Advances in Alzheimer's Disease*, https://doi.org/10.1007/978-981-13-9636-6_14

14.1 Introduction

Alzheimer's disease (AD) is a complex neurodegenerative proteopathy clinically manifested by progressive dementia. The pathological characteristics of AD are deposition of extracellular amyloid-beta (Aβ) peptides as senile neuritic plaques and accumulation of hyperphosphorylated fibrillar Tau in the form of neurofibrillary tangles (NFTs) in brain, as well as cerebral amyloid angiopathy (CAA) in arterial walls (Hashimoto et al. 2003; Tiraboschi et al. 2004). The Aβ peptide (40–43 amino acids long) is derived from the proteolytic cleavage of Aβ precursor protein (APP), a ubiquitously expressed family of transmembrane proteins that function in neuron growth, survival, and post-injury repair (Priller et al. 2006; Turner et al. 2003). The APP is cleaved sequentially by two proteolytic enzymes, β-secretase and γ-secretase, into smaller Aβ peptide fragments. Normally, Aβ-40 (40 amino acids in length) is the most abundant Aβ peptide species in the brain, however the toxic fibrillar Aβ is composed of Aβ-42 peptide (42 amino acids in length) (Hooper 2005). These are usually soluble peptides and undergo conformational changes and multistep polymerization to form β-rich aggregates and finally accumulate as toxic microscopic plaques. The Aβ peptides first aggregate to form oligomers, which further cluster together and form fibrils (also called protofibrils) with a regular β-sheet structure. These fibrils stick to each other to form mats that clump together with other substances and finally form plaques. Initially, these senile plaques accumulate intraneuronally (Ohnishi and Takano 2004; Tiraboschi et al. 2004), and eventually lead to cell death and finally deposit outside the neurons in dense formations (D'andrea et al. 2001; Greeneld et al. 2000). Pathogenesis of AD is also associated with the formation of intraneuronal NFTs from hyperphosphorylated microtubule-binding protein Tau, and therefore AD is also considered a tauopathy. The Tau proteins undergo phosphorylation, pair with other threads to create NFTs, and then disintegrate the neuron's transport system (Hernandez and Avila 2007).

The physiological Aβ concentration of interstitial fluid of healthy brain is regulated by various processes like (1) rate of Aβ production from APP, (2) influx of Aβ into the brain across blood-brain barrier (BBB) via receptor for advanced glycation end products (RAGE), (3) rapid clearance from the brain across BBB via low-density lipoprotein receptor-related protein-1 (LRP1), and (4) enzymatic degradation within the brain. In AD, the RAGE expression in the endothelial cells is increased and the LRP expression is reduced at the BBB (Deane et al. 2009). This restricts the clearance of Aβ from the brain and leads to accumulation followed by gradual oligomerization and formation of toxic amyloid species in the brain. Therefore, it is very crucial to remove excess Aβ from the brain by transporting across BBB and metabolism to prevent accumulation of toxic amyloid species in brain.

Currently available treatments for AD mainly focus on the amelioration of AD symptoms, while the underlying causes of the disease remain unaffected. Recent advances in AD research lead to the inception of new therapeutic avenues that aim to target and interfere with the pathogenic steps blocking the disease progression at early phases including preclinical stage. These therapies are termed disease-modifying therapies. The first pathogenic event in AD progression is the amyloid deposition that activates a cascade of additional mechanisms. Therefore, majority of the new therapeutic approaches aim to intervene in Aβ peptide deposition or to remove already deposited amyloids (Galimberti et al. 2013). Immunotherapy is the most promising approach for clearance of Aβ burden from the brain. Aβ immunotherapy includes both active and passive immunities to reduce Aβ deposition in the brain and enhance Aβ clearance by administering anti-Aβ antibodies or Aβ as immunogen.

14.2 Proposed Mechanisms for Immunotherapy

Being an immune-privileged organ, brain has minimum immune surveillance. This posits a challenge for AD immunotherapy. However, approximately 0.1% of plasma IgG is normally present in the cerebrospinal fluid (CSF) which ranges from 5 to 50 mg/mL in concentration (Morgan 2011). Therefore, immunotherapy is a pertinent approach to target Aβ in the brain and consequently its clearance. There have been several mechanisms of immunotherapy proposed for the reduction of Aβ deposition in brain.

The first proposed model of immunotherapy relies upon the antigen opsonizing activity of antibody leading to macrophage-mediated phagocytosis and complement activation (Bard et al. 2000). This suggests the entry of sufficient peripheral antibodies into the CNS and binding the Aβ fibrils causing opsonization. This anti-Aβ antibody/Aβ complex is then recognized by the Fc receptors present on the surface of macrophages and trigger phagocytosis (Fig. 14.1a). The phagocytic action is performed either by resident microglia or by infiltrating monocytes/macrophages. Several studies have elegantly demonstrated that amyloid deposits are decorated with anti-amyloid antibodies after systemic administration (Morgan 2011). Stoichiometry of antibody to Aβ even less than 1:1 is sufficient for clearance of the amyloid deposition. Both active and passive immunization strategies act by phagocytic mechanism for efficient clearance of amyloid plaques (Gelinas et al. 2004).

Second mechanism is explained as "the peripheral sink" hypothesis. This mechanism advocates the efflux of Aβ into bloodstream down its concentration gradient and clearance by the circulating anti-Aβ antibodies serving as a peripheral sink (Fig. 14.1b). These circulating anti-Aβ antibodies alter the equilibrium across the BBB for Aβ to efflux owing to the reduced free Aβ concentration in blood (DeMattos et al. 2001). This hypothesis is based on the argument that antibody penetration across BBB is not a critical step. Several studies have supported this hypothesis showing increased Aβ concentration in plasma of nondemented patients peripherally infused with anti-Aβ antibodies (Dodel et al. 2003) as well as in the serum of mice after active immunization (Lemere 2013).

The third mechanism which was proposed by Beak Solomon (Solomon et al. 1996, 1997) is based on the catalytic ability of antibodies. The catalytic antibodies modify the Aβ secondary structure into less aggregate forming conformations in the monomers associated with oligomers or fibrils (Fig. 14.1c). Initially, they reported

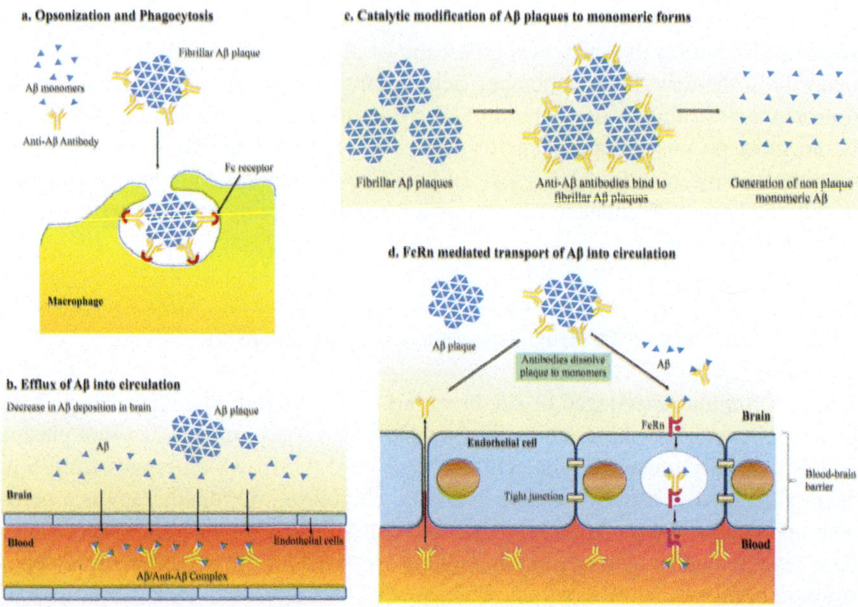

Fig. 14.1 *Mechanisms of immune clearance of amyloid from the brain.* (**a**) Antibody mediated opsonization and phagocytosis of Aβ aggregates. Anti-Aβ antibodies bind the Aβ fibrils and make available for phagocytosis by Fc receptor bearing macrophages. The phagocytozed amyloids are then digested and exported from the CNS. (**b**) Aβ efflux from the brain to circulation down its concentration gradient. Free Aβ in blood are bound by circulating anti-Aβ antibodies and block the influx of Aβ back to the brain from circulation. (**c**) Anti-Aβ antibodies bind to Aβ plaques and modify the secondary structure and dissolve the aggregates to monomeric units. (**d**) Anti-Aβ antibodies bind and dissolve the Aβ plaques. The antibody-antigen complexes are then bound to FcRn present on vascular epidermal cells and transported across BBB into circulation, thereby reducing brain plaque burden

that at stoichiometric ratio of 1:10, antibody specifically blocked fibril formation in vitro. Accordingly, this mechanism was later worked upon by Morgan and coworkers (Morgan 2011), and they demonstrated that antibodies could exhibit this activity at stoichiometry as low as 1:1000 (antibody: Aβ). Consequently, this hypothesis benefits from other adverse bystander effects involving macrophage activation, owing to such a low requirement of antibody.

Fourth mechanism is based on the FcRn (neonatal Fc receptor) mediated efflux of anti-Aβ antibody/Aβ complex across the BBB (Deane et al. 2005). When anti-Aβ monoclonal antibodies (mAbs) are administered intracranially, the anti-Aβ antibody/Aβ complex is formed and the Fc region of antibody binds to FcRn receptor on the epidermal cells of BBB and released into circulation by transcytosis as shown in Fig. 14.1d. Limitation of this mechanism lies on the use of limited antibodies, as excess antibodies may lead to competition for binding to FcRn receptor even in an uncomplexed form. Studies have demonstrated that clearance of amyloid plaques is inhibited by excess anti-Aβ mAbs administration (Gitter et al. 2002; Karlnoski et al. 2008).

However, none of the aforementioned mechanisms are mutually exclusive and restricted to a single antibody/fragment type. Any of the mechanisms can be adopted under a given set of conditions as the primary means of clearance or sequestration that would be influenced by several factors like the epitope, isotype, and the amyloid burden. Different effectors (antibody type/fragment) may operate via different mechanisms to different degrees, for instance, effectors unable to bind fibrillar amyloid and cross BBB may not induce phagocytosis, but function exceptionally well as a peripheral sink. Similarly, some effector molecules bind to Aβ fibrils but do not modify the secondary structure or modify to an extent that may not resolve the fibrils, still they can induce phagocytosis and FcRn-mediated transport.

14.3 Immunogenic Epitopes in Aβ

Interaction of antibody with antigen induces conformational changes in both the antigen and antibody. The degree of change may be insignificant to substantial (Amit et al. 1986; Bhat et al. 1990; Davies and Cohen 1996). When high affinity mAbs bind to the regions having high flexibility and antigenicity, it may induce alterations in the molecular dynamics of the whole protein chain or assembly (Frauenfelder et al. 1979; Karplus and Petsko 1990). Antibodies have the ability to recognize partially folded epitopes and induce proper folding to attain native conformation of the whole protein (Blond and Goldberg 1987; Carlson and Yarmush 1992; Solomon and Schwartz 1995). Antibodies also induce reversal of the toxic Aβ aggregates to its nontoxic forms. The mode of action is determined by the epitope specificity of anti-Aβ antibodies (Lichtlen and Mohajeri 2008). Based on the location of regions recognized by antibodies, four epitopic regions are found in Aβ peptide (Fig. 14.2).

14.3.1 The N-Terminal Region (Residues 1–16)

Amino acid residues 1–16 comprise the N-terminal epitope. This is an immunodominant region and antibodies specific to this region can recognize monomers, oligomers, protofibrils, fibrils, as well as APP. This indicates that the epitope is always exposed during Aβ fibrillogenesis (Ida et al. 1996; Robert et al. 2008). These antibodies can enter the CNS and display anti-aggregating properties both in vitro (Solomon et al. 1997) and in vivo (Bard et al. 2000). Site-directed mAbs bind to preformed amyloid fibrils and cause their disaggregation or solubilization at an equimolar ratio of mAb/Aβ peptide (Katzav-Gozansky et al. 1996; Solomon et al. 1996). The disaggregation effect depends on the epitope position on Aβ and the binding efficiency of mAbs. High affinity mAbs against N-terminal epitope of Aβ resulted in disaggregation and deterioration of amyloid fibril assembly even at lower concentration.

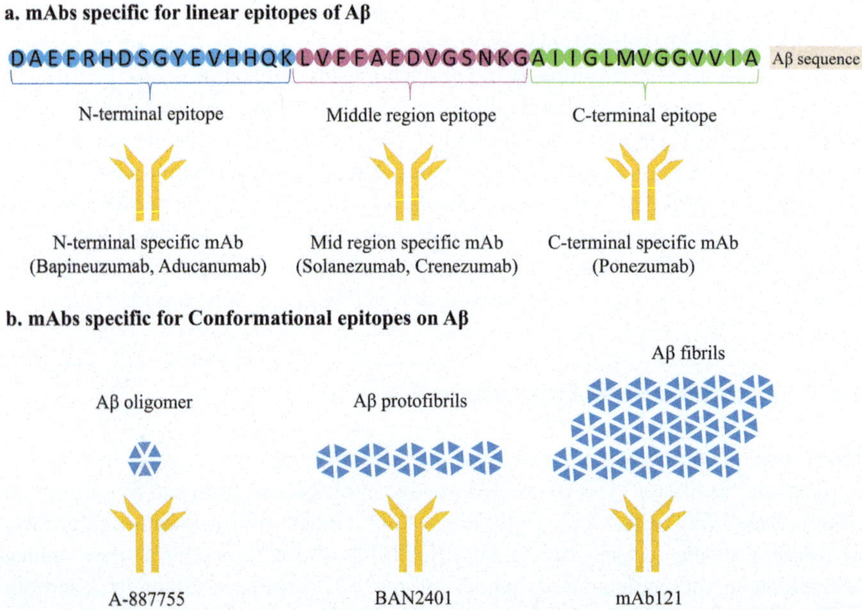

Fig. 14.2 *Various epitopes on the Aβ peptide and some of the mAbs directed toward them.* (**a**) Popular mAbs specific to linear epitopes of Aβ in the N-terminal region (blue shaded, residues 1–16), middle region (purple shaded, residues 17–32), and C-terminal region (green shaded, residues 33–42). (**b**) mAbs specific to conformational epitopes present in Aβ oligomers, Aβ protofibrils, or Aβ fibrils

14.3.2 The Middle Region (Residues 17–32)

The middle region epitope is present in the central portion of Aβ peptide spanning from 17 to 32 amino acids. Antibodies that recognize this epitope bind solely to Aβ monomers and do not enter into CNS (DeMattos et al. 2001). Therefore, antibodies directed toward this region are likely to perform the amyloid clearance by "peripheral sink effect."

14.3.3 The Dead End C-Terminal (Residues 33–42)

Amino acid residues 33–42 make up the epitope at C-terminal region of Aβ peptide. C-terminus-specific antibodies have the ability to enter CNS, and hence they are capable of Aβ clearance both by influx into the CNS and by the "peripheral sink effect." They are also associated with side effects that include CAA similar to antibodies against the N-terminal region (Gray et al. 2007; Lichtlen and Mohajeri 2008).

14.3.4 Conformational Epitope

The culprit for AD pathogenesis is not the monomeric rather the oligomeric form of Aβ (Lambert et al. 1998; Walsh et al. 2002). To disaggregate such oligomeric forms and protofibrils, the conformational antibodies would outstand the configurational antibodies raised against linear epitopes. Since the N-terminus is highly immunodominant, antibodies are raised against the entire Aβ peptide. To overcome this, an elegant strategy has been adopted in which synthetic oligomers lacking the N-terminal region are used as immunogen. These antibodies have been shown to have specificity for Aβ oligomers and help improve cognition in AD mice model (Hillen et al. 2010).

14.4 Active Immunotherapy

According to the amyloid cascade theory Aβ is the primary factor of AD pathogenesis. The neurodegeneration caused by Aβ species can be decreased by programming the immune system to generate anti-Aβ antibodies that can neutralize the complex and remove it from circulation. Synthetic full length or fragment or a modified fragment of the peptide is administered to evoke the production of anti-Aβ antibodies by B cells. This is called active Aβ immunotherapy.

Anti-Aβ antibodies were capable of preventing fibrillation, disrupting fibril formation, and thwarting the toxicity induced by fibrils as observed in cell culture-based assays (Solomon et al. 1997). Based on these encouraging results, Aβ was further investigated in vitro for its possible role as an active immunogen and whether it could prevent AD-related pathology. The first immunization trial came up with reduced plaque load without causing significant neurotoxicity after administration of full-length Aβ-42 in its aggregated form emulsified in Freund's adjuvant (Schenk et al. 1999). Similarly, when Freund's or alum adjuvant emulsified Aβ-42 or its homologous peptides were injected, it resulted in protection against development of cognitive deficits along with prevention of Aβ plaque pathology (Asuni et al. 2006; Janus et al. 2000; Lemere 2013; Morgan 2011; Sigurdsson et al., Sigurdsson et al. 2001, Sigurdsson et al. 2004). The principal epitope consisted of the N-terminal epitope of Aβ (residues 1 15). However, the mechanism of action and the mode of immune response elicited by these peptide-dependent immunogens have been a controversial debate till date. Peripheral injection of anti-Aβ mAbs could potentially reduce Aβ plaque load and improve behavior, suggesting the vaccine elicits humoral immune response (Bard et al. 2000; DeMattos et al. 2001; Lemere 2013; Wisniewski and Goñi 2012; Wisniewski and Goñi 2014). But the question then arises whether Fruend's adjuvant also induces Th1-mediated immune response. In another study, induction of Th2-mediated immune response was observed when immunized with nontoxic, non-fibrillogenic Aβ homologous peptides in adjuvants. Absence of Th1-mediated response reduced potential cytotoxicity (Lemere et al. 2001; Sigurdsson et al. 2001). These peptides were designed using the same first 15 amino acids with some appropriate amino acid substitutions. On the other hand,

T-cell epitopes are located at the mid- and C-terminal regions (Asuni et al. 2006; Sigurdsson et al. 2001, 2004).

These preclinical studies opened up an arena for clinical trials. The first clinical trial was launched by Elan/Wyeth's group in April 2000, in which a new adjuvant QS21 was used along with preaggregated Aβ-42 in the vaccine. This vaccine, AN1792, was designed to elicit a strong cell-mediated immune response as QS21 behaves similarly to Freund's adjuvants (Wisniewski and Frangione 2005). The study was conducted in the UK on 80 patients with mild to moderate AD (Bayer et al. 2005). More than half (53%) of the patients exhibited humoral immune response. In the same study, at a later stage of Phase I, polysorbate 80 was added to the preparation (Table 14.1). Use of this emulsifier improved Aβ-42 solubility. When the concentration of the emulsifier was increased, it led to a shift from the Th2 biased response toward Th1 proinflammatory response (Pride et al. 2008). The Phase II trial was conducted in October 2001, involving total 372 patients, out of which 300 were given higher formulation (AN1792 to placebo 4:1) of QS21 in polysorbate 80. This trial was withdrawn in January 2002 because a 6% of immunized patients developed ARIA (Amyloid Related Imaging Abnormality) and meningoencephalitis (Boche and Nicoll 2008; Wisniewski and Frangione 2005). The plausible explanation for such inflammatory process is the exposition of Aβ C-terminus region which activates Th2 response (Gilman et al. 2005). This is why the new second generation vaccines do not contain the culprit C-terminal fragment. A multitude of vaccination trials are currently in either Phase I or Phase II (www.clinicaltrials.gov).

The vaccine CAD106 has been designed for targeting only a B cell epitope, the small N-terminal fragment (residues 1–6) in conjunction with an adjuvant derived from bacteriophage Qβ coat proteins (Wisniewski and Goñi 2012). No cases of meningitis, meningoencephalitis, or vasogenic edema were diagnosed clinically or by imaging during the initial trial or two- year follow-up period.

Table 14.1 Anti-Aβ vaccines in clinical trial

Commercial name of vaccine	Company	Immunogen/ peptide region	Adjuvant	Clinical trial/ phase	Outcome
AN1792	Elan/Wyeth	Aggregated Aβ42	QS-21, polysorbate 80	IIa finished	No improvement
CAD106	Novartis	Aβ1–6	Bacteriophage Qβ protein	III	NR
ACC-001	Janssen and Pfizer	Aβ1–6	QS-21	II	NR
ACI-24	AC immune	Tetra-palmitoylated Aβ1–15 (β conformation)	Liposomes	II	NR
UB-311	United neuroscience, ltd.	Aβ1–14	Alum	II	NR

Another Janssen and Pfizer (AC-001) led Phase II clinical trial uses the same N-terminal Aβ (1–6) fragment combined with a carrier protein and surface-active saponin adjuvant QS21 (Ryan and Grundman 2009). Similarly, Affiris AG and GlaxoSmithKline used antigenic peptides, mimotopes, to target the N-terminal epitope of Aβ in their AD02 trials (Schneeberger et al. 2009). These peptides were generated using AFFITOME® technology (Wisniewski and Goñi 2015). Moreover, Affiris AG started one more Phase I trial targeting N-Terminus of pyroglutamate-modified Aβ. Usually these modified Aβ are found in the plaques and vascular amyloid deposits but not in CSF or plasma. Still these can be detected in the fluids only when the Aβ deposits are mobilized due to therapeutic interventions (DeMattos et al. 2002).

AC Immune also started Phase I/IIa clinical trials with ACI-24 that elicits humoral immune response against Aβ (trial identifier NCT02738450). The design of immunogen is based on tetra-palmitoylated amyloid 1–15 peptide which exists primarily in a β-sheet structure (Hickman et al. 2011; Muhs et al. 2007). Similarly, the other peptide-based vaccine UB311 consists of the N-terminal Aβ (1–14) peptide coupled with UBITh helper T-cell epitope, which specifically evokes Th1-mediated response. The vaccine is yet to be tested (identifier NCT02551809). The studies and clinical trials till date have demonstrated limited success as an effective approach toward AD pathology. The safety issues and the associated adverse effects still raise a major concern for dependence on active immunotherapeutic approaches.

14.5 Passive Immunotherapy

The active immunization was associated with low responsiveness and incidences of T-cell-dependent adversities. To overcome this, safer immunotherapeutic strategies have been adopted for clearance of brain amyloid deposits by using mAbs as well as recombinant antibody fragments. The therapeutic potential of antibody relies on its potential to prevent amyloid fibril formation from Aβ monomers and to dissolve the preformed fibrils. Standard in vitro assays which include electron microscopy, ThT binding, or MTT reduction assays are performed to assess the dissolution of aggregates and cytotoxicity (Levine 1993). Several forms of antibodies including antibody fragments have been investigated for passive immunotherapy of AD.

14.5.1 mAbs

Till date, a number of mAbs have been developed, and some are being considered for clinical application (Table 14.2).

14.5.1.1 Bapineuzumab

N-terminal epitopes are well displayed in aggregated forms of the Aβ; therefore antibodies directed against N-terminal epitopes also efficiently recognize the deposited fibrils. Bapineuzumab is one such humanized monoclonal antibody

Table 14.2 mAbs in clinical trial of AD

Monoclonal antibody	Company	Antigen(A)/epitope (E) and subclass of IgG	Target	Phase	Sample size	Age (years)	Dose and duration	Effects/ outcome	ARIA-E
Bapineuzumab	Janssen/ Pfizer	N-terminal Aβ1–5 (E)/ hIgG1	Monomers, oligomers fibrils, and amyloid plaques	III	2204	50–88*	0.5, 1, 2 mg/kg intravenous every 12 weeks for 78 weeks	Decreased cortical ^{11}C PiB and decreased pTau in CSF	High
Solanezumab	Eli Lilly	Aβ16–24(E)/hIgG1	Monomers	III	2129	55–90***	400 mg intravenous every 4 weeks for 78 weeks	No effect on brain Aβ or tau as diagnosed by PET	Low
Crenezumab	AC immune/ Genentech	Pyroglutamate-Aβ1–15 (A)/hIgG4	Monomers, oligomers, fibrils, and plaques	II	431	50–80**	300 mg subcutaneous every 2 weeks, 15 mg/kg intravenous every 4 weeks for 68 weeks	Elevated Aβ-42 in CSF	Low
Gantenerumab	Roche	N-terminal Aβ1–10 and central region Aβ18–27(E)/hIgG1	Oligomers and fibrils	III	799	50–85**	105 or 225 mg subcutaneous every 4 weeks for 104 weeks	Benefit in rapid progressors	High
Ponezumab	Janssen/ Pfizer	C-terminal Aβ40(E)/ hIgG2a	Monomers	II	1	>50*	10 mg/kg intravenous for 52 weeks	Elevated Aβ-42 in CSF	None
BAN2401	Biogen/ Eisai/ BioArctic	Aβ42 AM protofibrils(A)/ hIgG1		III	2	50–90**	78 weeks	*Ongoing*	
Aducanumab	Biogen	NT Aβ3–6 (E)/hIgG	Protofibrils	Ib		50–85**	75 weeks	*Ongoing*	High

Mod moderate, *pro* prodromal, *ARIA-E* amyloid related imaging abnormality-edema, *PiB* Pittsburgh compound B
Stage of disease: *Mild mod AD; **Mild-pro AD Aβ+; ***Mild-AD Aβ+

(Salloway et al. 2009) that binds five N-terminal residues. Studies conducted on animals indicate brain amyloid reduction (Sperling et al. 2012). However, it was associated with a risk of vasogenic edema and microhemorrhage (ARIA). The 3D6 is the murine precursor of humanized bapineuzumab, which has been reported to enter the brain of transgenic mice model, decorate the plaques, and induce Fc receptor-mediated phagocytosis (Bard et al. 2000). Several clinical studies have been conducted which indicate reduction in fibrillar amyloids in the brain of AD patients (Table 14.2). Meta-analysis based on clinical studies suggests lack of clinical efficacy of bapineuzumab and association with adverse effects (Abushouk et al. 2017). Therefore, use of bapineuzumab to treat AD patients is not recommended and can only be reconsidered after re-evaluating its efficacy in combinatorial formula.

14.5.1.2 Solanezumab

Solanezumab, another humanized IgG1 mAb, recognizes and targets a middle region (residues 16–26) amyloid peptide epitope (Farlow et al. 2012; Siemers et al. 2016). The murine precursor (m266) of antibody binds tightly to monomeric amyloid peptides, but not to aggregates or fibrils (DeMattos et al. 2001, 2002). Solanezumab is effective in reducing amyloid in transgenic mice and does not carry the risk of ARIA like the bapineuzumab. The proposed mechanism of action here is the peripheral sequestration of monomeric amyloid peptide, i.e., peripheral sink effect.

14.5.1.3 Crenezumab

Similar to solanezumab, crenezumab is another humanized mAb that recognizes and targets a midsequence (residues 13–24) amyloid peptide epitope. However, it differs from solanezumab in that it possesses an IgG4 backbone. The IgG4 isotype helps microglial phagocytic activity without raising cytokine storm (Adolfsson et al. 2012) which contributes to neurotoxicity as well as ARIA (Ultsch et al. 2016). In vitro studies report that crenezumab efficiently binds to amyloid fibrils as well as oligomers but to a lower extent to monomers (Adolfsson et al. 2012). The epitopes recognized by solanezumab and crenezumab overlap and therefore exhibit cross-reactivity. Aβ residues 21–26 adopt α-helical structure when bound to solanezumab, whereas residues 21–24 exhibit random coil structure when bound to crenezumab. The occurrence of α-helical epitope only in monomeric forms of Aβ but not in aggregates explains the preference of solanezumab to bind monomer.

14.5.1.4 Gantenerumab

Gantenerumab, the first fully human IgG1 anti-Aβ mAb, has the ability to bind to conformational epitopes which encompasses both the N-terminus (residues 3–12) and midsequence (residues 8–27) epitopes and shows high affinity for fibrils. Early studies based on PET (positron emission tomography) demonstrated that gantenerumab significantly reduced brain amyloid plaques by recruiting microglial cells (Bohrmann et al. 2012). Similarly, SAR228810 is also a humanized antibody that recognizes particular conformational epitope that allows specific binding to protofibrils and fibrils.

14.5.1.5 Ponezumab

Ponezumab is a human IgG2 mAb that targets the Aβ C-terminus (residues 30–40) (La Porte et al. 2012). The IgG2 isotype has lower immune potential (Landen et al. 2013) limiting their clinical efficacy, and therefore ponezumab production was discontinued after few trials (Table 14.2).

14.5.1.6 BAN2401

BAN2401, a humanized IgG1 mAb, specifically binds to soluble Aβ protofibrils. This antibody was developed by E22G arctic mutation in the APP and has shown efficient reduction of Aβ protofibrils in the brain and CSF of tg-ArcSwe mice (Tucker et al. 2015). Multiple clinical trials are going on and BAN2401 has been reported to show no cases of ARIA, the major limitations of bapineuzumab (Lannfelt et al. 2014).

14.5.1.7 Aducanumab

Aducanumab, a human IgG1 mAb, selectively targets Aβ aggregates which includes soluble oligomers as well as insoluble fibrils (Sevigny et al. 2016). It targets the N-terminus (residues 3–6), which forms a conformational epitope absent in monomers. Analogues of this mAb exhibited the potential to cross BBB, bind parenchymal Aβ, and reduce soluble as well as insoluble Aβ (Sevigny et al. 2016).

So far, several clinical trials have been conducted using mAbs but these mAbs persist limitations regarding their production, lower tissue penetration, and adverse effects associated with inflammatory reactions and CAA-associated microhemorrhage (Pfeifer et al. 2002; Racke et al. 2005). The lack of specificity to the toxic pathological Aβ oligomers is the major limitation of these passive therapies. Antibodies directed against monomers may target the normal, nontoxic soluble Aβ that may restrict its vital physiological roles such as neuroprotection, modulating long-term potentiation and innate immunity. Moreover, the cross-reactivity with the soluble Aβ also increases the risk of autoimmune complications (Giuffrida et al. 2009; Puzzo et al. 2008; Soscia et al. 2010; Wisniewski and Goñi 2012). Early start of the multiple doses of drug further accentuates the need for a better treatment regime. Therefore, it can be carefully contended that the passive therapeutic approaches have limited efficacy in symptomatic AD.

14.5.2 Antibody Fragments

Different formats of recombinant antibody fragments (Ecker et al. 2015), such as single-chain fragment variable (ScFv), fragment antigen binding (Fab), single-domain antibody fragments (VHH or sdAbs), bispecific antibodies (BsAb), gammabodies, and intrabodies, are currently being investigated as therapeutics for AD. These are preferred over conventional full-length mAbs owing to their great specificity, higher affinity, stability, solubility, and reduced immunogenicity (Manoutcharian et al. 2017).

14.5.2.1 ScFv

The ScFvs are the smallest antibody fragments (V_H and V_L linked with a linker) with specificity and protective properties comparable to the parental antibody. These antibody fragments show increased stability as well as affinity (Frenkel et al. 2000; Malone and Sullivan 1996). ScFv can be delivered and distributed all over the brain via intracerebral, intranasal, or virus-mediated routes (Campana et al. 2009; Donofrio et al. 2005; Federoff 2009; Filesi et al. 2007). They are packaged in small viral vectors like recombinant adeno-associated virus (rAAV) and injected into the CNS (Fig. 14.3a). ScFvs exhibit little immunogenicity as they are unable to fix complement. Furthermore, since they do not require glycosylation, they can be largely produced in bacterial systems (Verma et al. 1998). The first ScFv-based anti-Aβ antibody, 508F (Fv), was derived from monoclonal IgM 508 antibody. This fragment demonstrated efficient disaggregation of Aβ fibrils and also reduced toxicity in cultured PC-12 cells. Similarly, N- and C-terminal binding ScFvs were produced and selected using naive human ScFv phage library with Aβ1–28 and Aβ1–40, respectively (Liu et al. 2004; Robert et al. 2008). Only the ScFv against N-terminal could inhibit Aβ aggregation in vitro (Liu et al. 2004). Another engineered ScFv based on mAb WO-2 that recognizes Aβ2–8 could disaggregate Aβ fibrils, inhibit Aβ fibril formation, and reduce Aβ-mediated toxicity (Robert et al. 2008). Despite the clinical approval of rAAV vector-based gene therapy, certain limitations involving hepatic genotoxicity urge a demand for a safer vector-based delivery system (Chandler et al. 2015; Donsante et al. 2007; Gil-Farina et al. 2016). Consequently, naked DNA plasmids may act as an appropriate vector as they are not associated with a risk of genome integration (Flingai et al. 2015; Muthumani et al. 2013, 2016).

Other novel fragments called as catabodies have been developed. These catabodies are ScFv fragments generated by affinity maturation of the corresponding parent mAb with improved catalytic activity. They catalyze proteolysis of Aβ and reduce the accumulation of toxic amyloid in brain. The first catabody, Asec-1A, inhibited the aggregation of Aβ and reduced Aβ toxicity on human neuroblastoma cells (Kasturirangan et al. 2010). Similarly, a bispecific tandem ScFv produced by combining iBSEC1 and Asec-1A showed inhibition of amyloidogenic APP processing and enhanced Aβ proteolysis (Boddapati et al. 2012). The iBSEC1 ScFv selectively inhibits amyloidogenic APP processing and has demonstrated a decrease in both intracellular and extracellular Aβ in APP overexpressing Chinese hamster ovary cells (Boddapati et al. 2011). Recently, a catalytic IgV construct 2E6 composed of VL1 and VL2 has been shown to reduce Aβ deposits in the brain of 5xFAD mice after intravenous injection (Planque et al. 2015).

14.5.2.2 Fab Fragments

Fab fragments contain one heavy and one light chain with binding avidity lower than IgG but affinity parallel to it (Fig. 14.3b). These are small in size and also stable when compared to ScFvs. Tammer et al. (2002) produced a recombinant Fab

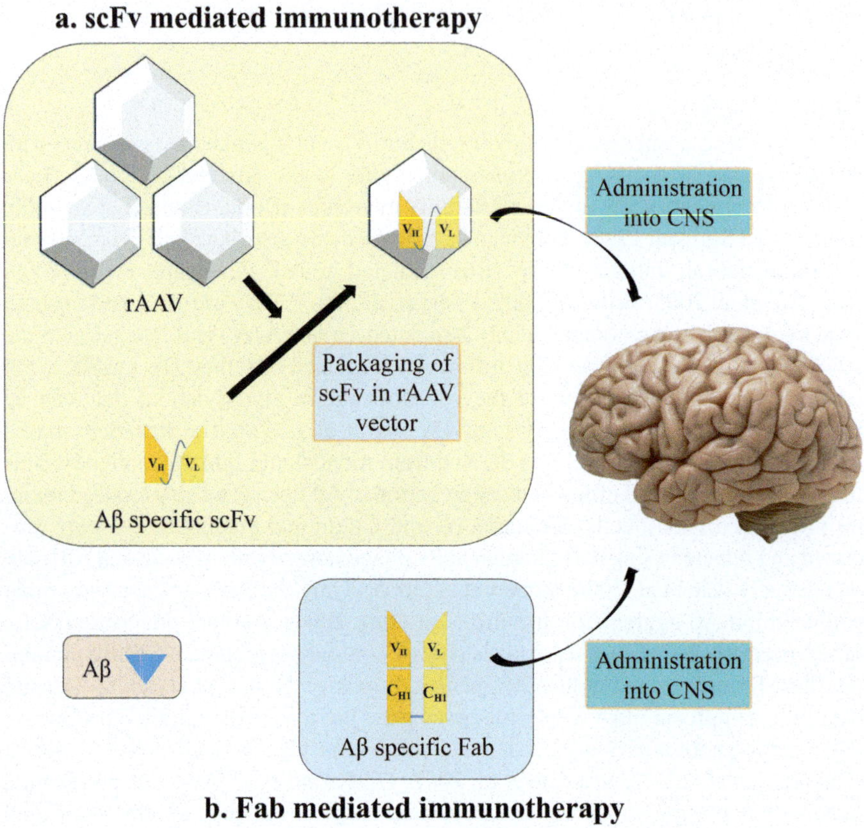

Fig. 14.3 *Immunotherapy by ScFv and Fab.* (**a**) Structure and rAAV-mediated delivery of ScFv. Small ScFv can be packaged in rAAV vector and efficiently delivered to CNS. (**b**) Fab fragments can be administered directly into CNS for therapeutic purpose

(rFab) against the central region of Aβ derived from the parent hybridoma 1E8. These rFab were efficient binder of amyloid plaques. Similarly, another rFab generated from hybridoma WO-2 inhibited fibril formation and dissolved the preformed fibrils in vitro (Robert et al. 2008). Also a humanized version of this rFab (hWO-2 rFab) was engineered by the same group. This rFab retained the ability to inhibit fibril formation and associated toxicity with a strong affinity (Kd-6 nm) for Aβ, determined by surface plasmon resonance measurements (Robert et al. 2010).

14.5.2.3 Bispecific Antibodies (BsAbs)

A highly innovative approach, the molecular Trojan horse technology (Pardridge 2008) has been employed to generate fusion proteins called BsAbs against AD. These BsAbs contained binding sites for Aβ peptide and for the insulin/transferrin receptor (IR/TfR), which are highly expressed on the BBB. A ScFv directed against Aβ is linked to the Fc region of a monoclonal antibody directed against IR/

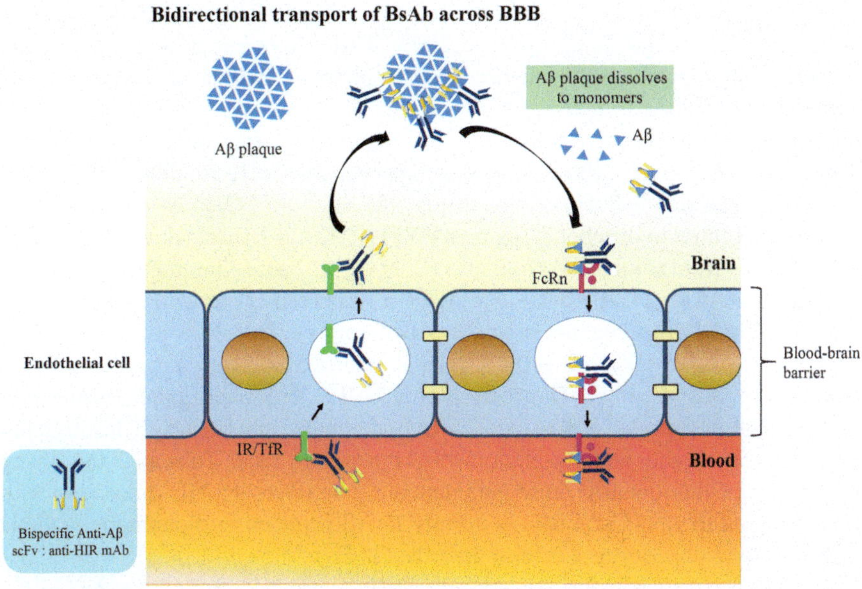

Fig. 14.4 *BsAb-mediated Aβ clearance from the brain*. The BsAbs are transported into the brain via IR/TfR where the ScFv binds to Aβ plaques. The Aβ bound BsAbs are then effluxed to circulation via FcRn-mediated transcytosis

TfR. These tetravalent BsAbs have two binding sites for IR/TfR present in mAb and two for Aβ from two ScFvs (Fig. 14.4). The binding conditions are optimized to minimize the steric hindrance along with retention of affinity for both the targets. The central part of the BsAb comprising CH2–CH3 domain of the mAb provides the binding site for FcRn receptor expressed on the BBB (Robert and Wark 2012). The antigen-antibody complex is then efficiently effluxed out from the brain via this FcRn-mediated transcytosis (Fig. 14.4). The low affinity anti-TfR antibodies demonstrated 5 times higher infiltration across BBB compared to their high affinity counterparts at the same dose owing to faster dissociation constant (Robert and Wark 2012; Yu et al. 2009),

In a study, above 50% of human IR mAb-ScFv fusion BsAbs were effluxed from the brain within minutes after administration and this was blocked by adding human Fc fragments (Boado et al. 2007). The renal clearance rate of this fusion protein (1.72 mL/min/Kg) was comparable to that of the IR mAb alone (0.22–1.0 mL/min/Kg). This format of antibody appears to raise a new hope for therapy against AD which can be translated to clinic. Although the anti-IR mAbs have been reported to be safe as therapeutics, the adverse effects of these antibodies on the endogenous insulin pathways have to be worked out.

14.5.2.4 Gammabodies

Gammabodies are grafted amyloid-motif antibodies. They are designed based on the principle that the amyloidogenic motifs of one Aβ peptide interact with identical motifs of adjacent Aβ monomers forming stacks of parallel β sheets (Lührs et al. 2005; Perchiacca et al. 2012; Yu et al. 2009). This homotypic interaction between peptide motifs induces Aβ fibrillation and this phenomenon has been exploited for antibody engineering to recognize specific Aβ oligomers and fibrils. A small amyloidogenic peptide (6–10 residues) from Aβ-42 is grafted into the complementarity determining regions (CDRs) of V_H domain that can recognize soluble Aβ oligomers and fibrils. The gammabodies which display the Aβ motif (^{18}VFFA^{21}) react with Aβ fibrils. The central hydrophobic segment ^{18}VFFA^{21} forms β sheets during the formation of fibrils from soluble Aβ oligomers (Lührs et al. 2005). Gammabodies specific to fragments Aβ12–21, Aβ15–24, and Aβ18–27 can readily recognize the Aβ18–21 motif present in β sheet but not in soluble Aβ oligomers. Therefore, the development of β sheet by Aβ18–21 motif is the crucial structural modification during fibril formation from Aβ oligomers (Perchiacca et al. 2012). However, gammabodies displaying C-terminal motif (^{34}LMVGGVVIA^{42}) recognize and bind both the oligomer and fibrils but weakly with Aβ monomers (Ahmed et al. 2010; Yu et al. 2009; Zhang et al. 2013). Consequently, these gammabodies neutralize toxicity associated with both the conformers. Gammabody affinity to Aβ peptides in monomeric, soluble oligomers and fibril forms has been described in Fig. 14.5.

14.5.2.5 Intrabodies

Intrabodies are the intracellularly expressed antibody fragments that recognize and bind molecules within the cell (Cardinale and Biocca 2008; Miller and Messer 2005). The fragments are expressed with a trafficking signal which facilitates efficient subcellular localization. Accumulation of Aβ inside the neuronal cells in AD opens up the possibility for intrabody-based therapy (LaFerla et al. 1995). The intrabody ScFv-β1, developed to recognize N-terminal region near the β-secretase cleavage site, demonstrated significant reduction in Aβ generation in APP overexpressing human embryonic kidney cells (Paganetti et al. 2005). This was further improved when the KDEL sequence was incorporated which facilitated expression specifically in the endoplasmic reticulum. ScFv-β1 acts by shielding the cleavage site for β-secretase which facilitates the inoffensive α-secretase induced cleavage of APP (Sudol et al. 2009).

In AD, the Fc region of antibody plays a significant role in eliciting adverse reactions which include meningoencephalitis and cerebral hemorrhages. Strategies that help lower the affinity of Fc to the Fc receptor (FcγR) present on immune cells and inhibit complement activation through c1q binding are being worked upon. Point mutation for deglycosylation of the asparagine at 297th position to alanine or glutamine, or replacing leucine with alanine in the lower hinge region (L234 and L235), helped lower the Fc-FcγR interactions (Alegre et al. 1994; Walker et al. 1989). IgG4 format is preferred over others due to lower affinity toward FcγR. But the Fab arm exchange property results in formation of bispecific antibody in vivo. However, this is being worked upon and can be avoided by introducing point mutation (S228P) in hinge region of IgG4 (Salfeld 2007).

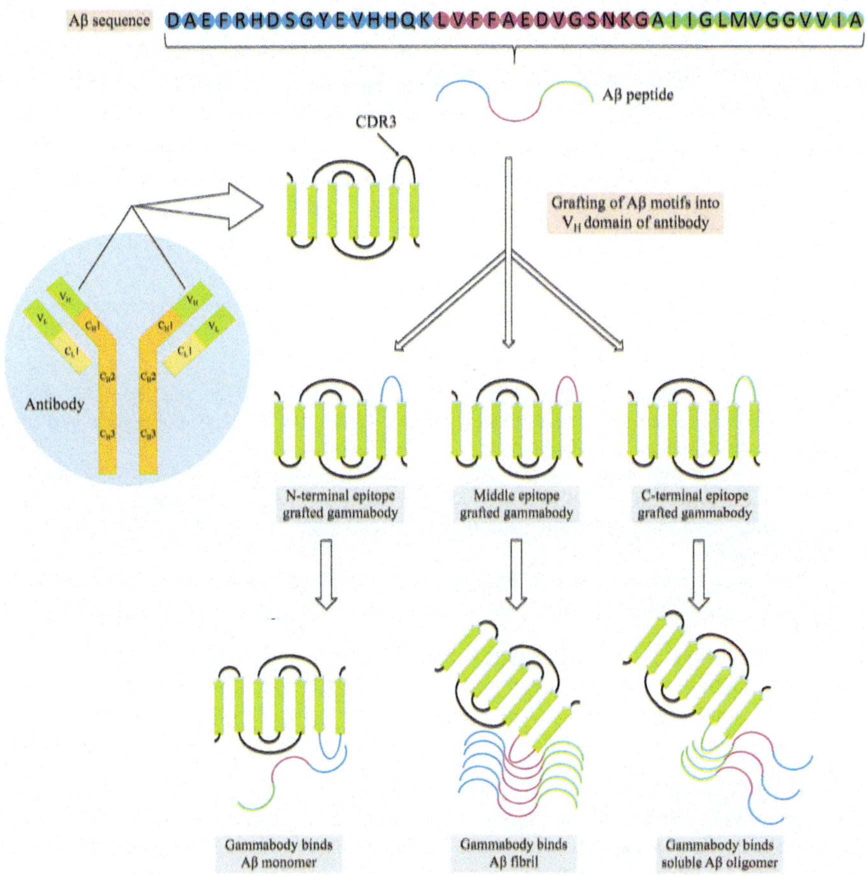

Fig. 14.5 *Gammabodies with affinity for Aβ peptide*. Aβ motifs from different regions of Aβ grafted in the CDR3 of V_H domain of antibody. The binding affinity of gammabodies differs based on the grafted motifs: gammabodies with N-terminal motif (blue) have affinity for Aβ monomers, gammabodies with midregion motif (purple) have affinity for fibrils, and gammabodies with C-terminal motif (green) have affinity for insoluble oligomers

14.6 Conclusion

The immunotherapy works in close collaboration with human immune system to counterbalance the Aβ aggregation. At present, it appears to be the best approach to lower the neurodegeneration and the cognitive decline in patients with AD. Nevertheless, further studies/trials are needed to develop effective vaccines with more specificity, less toxicity, and lowered autoimmune response. Similarly, for passive immunization, the efficiency of mAb and antibody fragments to cross the BBB needs to be enhanced, the cross-reactivity minimized, and the inflammatory alterations lowered. Most of the clinical trials have targeted and engaged primarily on Aβ. However, recent studies are widening the immunotherapeutic approaches targeting hyperphosphorylated, aggregated, and insoluble toxic Tau (Gerson et al. 2014; Nisbet et al. 2015). There needs to be a comprehensive research for developing therapeutics against the Tau aggregates involved in AD pathology (Iqbal et al. 2014, Iqbal et al. 2016). At this point of technical advancements, it can be contemplated that use of non-immunogenic compounds such as DNA and RNA aptamers with strong affinity to target amyloids for their clearance and neutralization in brain can help overcome the problems observed with immunization (Qu et al. 2017).

References

Abushouk AI, Elmaraezy A, Aglan A, Salama R, Fouda S, Fouda R, AlSafadi AM (2017) Bapineuzumab for mild to moderate Alzheimer's disease: a meta-analysis of randomized controlled trials. BMC Neurol 17:66

Adolfsson O, Pihlgren M, Toni N, Varisco Y, Buccarello AL, Antoniello K, Lohmann S, Piorkowska K, Gafner V, Atwal JK (2012) An effector-reduced anti-β-amyloid (Aβ) antibody with unique aβ binding properties promotes neuroprotection and glial engulfment of Aβ. J Neurosci 32:9677–9689

Ahmed M, Davis J, Aucoin D, Sato T, Ahuja S, Aimoto S, Elliott JI, Van Nostrand WE, Smith SO (2010) Structural conversion of neurotoxic amyloid-β 1–42 oligomers to fibrils. Nat Struct Mol Biol 17:561

Alegre M-L, Peterson LJ, Xu D, Sattar HA, Jeyarajah DR, Kowalkowski K, Thistlethwaite JR, Zivin RA, Jolliffe L, Bluestone JA (1994) A non-activating "humanized" anti-CD3 monoclonal antibody retains immunosuppressive properties in vivo. Transplantation 57:1537–1543

Amit AG, Mariuzza RA, Phillips SE, Poljak RJ (1986) Three-dimensional structure of an antigen-antibody complex at 2.8 A resolution. Science 233:747–753

Asuni A, Boutajangout A, Scholtzova H, Knudsen E, Li Y, Quartermain D, Frangione B, Wisniewski T, Sigurdsson EM (2006) Aβ derivative vaccination in alum adjuvant prevents amyloid deposition and does not cause brain microhemorrhages in Alzheimer's model mice. Eur J Neurosci 24:2530–2542

Bard F, Cannon C, Barbour R, Burke R-L, Games D, Grajeda H, Guido T, Hu K, Huang J, Johnson-Wood K (2000) Peripherally administered antibodies against amyloid β-peptide enter the central nervous system and reduce pathology in a mouse model of Alzheimer disease. Nat Med 6:916

Bayer AJ, Bullock R, Jones RW, Wilkinson D, Paterson KR, Jenkins L, Millais SB, Donoghue S (2005) Evaluation of the safety and immunogenicity of synthetic Aβ42 (AN1792) in patients with AD. Neurology 64:94–101

Bhat TN, Bentley GA, Fischmann TO, Boulot G, Poljak RJ (1990) Small rearrangements in structures of Fv and Fab fragments of antibody D 1.3 on antigen binding. Nature 347:483

Blond S, Goldberg M (1987) Partly native epitopes are already present on early intermediates in the folding of tryptophan synthase. Proc Natl Acad Sci 84:1147–1151

Boado RJ, Zhang Y, Zhang Y, Xia C-F, Pardridge WM (2007) Fusion antibody for Alzheimer's disease with bidirectional transport across the blood- brain barrier and Aβ fibril disaggregation. Bioconjug Chem 18:447–455

Boche D, Nicoll JA (2008) SYMPOSIUM: clearance of Aβ from the brain in Alzheimer'Disease: the role of the immune system in clearance of Aβ from the brain. Brain Pathol 18:267–278

Boddapati S, Levites Y, Sierks MR (2011) Inhibiting β-secretase activity in Alzheimer's disease cell models with single-chain antibodies specifically targeting APP. J Mol Biol 405:436–447

Boddapati S, Levites Y, Suryadi V, Kasturirangan S, Sierks MR (2012) Bispecific tandem single chain antibody simultaneously inhibits β-secretase and promotes α-secretase processing of AβPP. J Alzheimers Dis 28:961–969

Bohrmann B, Baumann K, Benz J, Gerber F, Huber W, Knoflach F, Messer J, Oroszlan K, Rauchenberger R, Richter WF (2012) Gantenerumab: a novel human anti-Aβ antibody demonstrates sustained cerebral amyloid-β binding and elicits cell-mediated removal of human amyloid-β. J Alzheimers Dis 28:49–69

Campana V, Zentilin L, Mirabile I, Kranjc A, Casanova P, Giacca M, Prusiner SB, Legname G, Zurzolo C (2009) Development of antibody fragments for immunotherapy of prion diseases. Biochem J 418:507–515

Cardinale A, Biocca S (2008) The potential of intracellular antibodies for therapeutic targeting of protein-misfolding diseases. Trends Mol Med 14:373–380

Carlson JD, Yarmush ML (1992) Antibody assisted protein refolding. Nat Biotechnol 10:86

Chandler RJ, LaFave MC, Varshney GK, Trivedi NS, Carrillo-Carrasco N, Senac JS, Wu W, Hoffmann V, Elkahloun AG, Burgess SM (2015) Vector design influences hepatic genotoxicity after adeno-associated virus gene therapy. J Clin Invest 125:870–880

D'andrea MR, Nagele RG, Wang H-Y, Peterson PA, Lee DHS (2001) Evidence that neurones accumulating amyloid can undergo lysis to form amyloid plaques in Alzheimer's disease. Histopathology 38:120–134

Davies DR, Cohen GH (1996) Interactions of protein antigens with antibodies. Proc Natl Acad Sci 93:7–12

Deane R, Bell RD, Sagare A, Zlokovic BV (2009) Clearance of amyloid-β peptide across the blood-brain barrier: implication for therapies in Alzheimer's disease. CNS Neurol Disord 8:16–30

Deane R, Sagare A, Hamm K, Parisi M, LaRue B, Guo H, Wu Z, Holtzman DM, Zlokovic BV (2005) IgG-assisted age-dependent clearance of Alzheimer's amyloid β peptide by the blood–brain barrier neonatal Fc receptor. J Neurosci 25:11495–11503

DeMattos RB, Bales KR, Cummins DJ, Dodart J-C, Paul SM, Holtzman DM (2001) Peripheral anti-Aβ antibody alters CNS and plasma Aβ clearance and decreases brain Aβ burden in a mouse model of Alzheimer's disease. Proc Natl Acad Sci 98:8850–8855

DeMattos RB, Bales KR, Cummins DJ, Paul SM, Holtzman DM (2002) Brain to plasma amyloid-β efflux: a measure of brain amyloid burden in a mouse model of Alzheimer's disease. Science 295:2264–2267

Dodel RC, Hampel H, Du Y (2003) Immunotherapy for Alzheimer's disease. Lancet Neurol 2:215–220

Donofrio G, Heppner FL, Polymenidou M, Musahl C, Aguzzi A (2005) Paracrine inhibition of prion propagation by anti-PrP single-chain Fv miniantibodies. J Virol 79:8330–8338

Donsante A, Miller DG, Li Y, Vogler C, Brunt EM, Russell DW, Sands MS (2007) AAV vector integration sites in mouse hepatocellular carcinoma. Science 317:477–477

Ecker DM, Jones SD, Levine HL (2015) The therapeutic monoclonal antibody market. In: MAbs. Taylor & Francis, pp 9–14

Farlow M, Arnold SE, Van Dyck CH, Aisen PS, Snider BJ, Porsteinsson AP, Friedrich S, Dean RA, Gonzales C, Sethuraman G (2012) Safety and biomarker effects of solanezumab in patients with Alzheimer's disease. Alzheimers Dement 8:261–271

Federoff HJ (2009) Development of vaccination approaches for the treatment of neurological diseases. J Comp Neurol 515:4–14

Filesi I, Cardinale A, Mattei S, Biocca S (2007) Selective re-routing of prion protein to proteasomes and alteration of its vesicular secretion prevent PrPSc formation. J Neurochem 101:1516–1526

Flingai S, Plummer EM, Patel A, Shresta S, Mendoza JM, Broderick KE, Sardesai NY, Muthumani K, Weiner DB (2015) Protection against dengue disease by synthetic nucleic acid antibody prophylaxis/immunotherapy. Sci Rep 5:12616

Frauenfelder H, Petsko GA, Tsernoglou D (1979) Temperature-dependent X-ray diffraction as a probe of protein structural dynamics. Nature 280:558

Frenkel D, Solomon B, Benhar I (2000) Modulation of Alzheimer's β-amyloid neurotoxicity by site-directed single-chain antibody. J Neuroimmunol 106:23–31

Galimberti D, Ghezzi L, Scarpini E (2013) Immunotherapy against amyloid pathology in Alzheimer's disease. J Neurol Sci 333:50–54

Gelinas DS, DaSilva K, Fenili D, George-Hyslop PS, McLaurin J (2004) Immunotherapy for Alzheimer's disease. Proc Natl Acad Sci 101:14657–14662

Gerson JE, Castillo-Carranza DL, Kayed R (2014) Advances in therapeutics for neurodegenerative tauopathies: moving toward the specific targeting of the most toxic tau species. ACS Chem Neurosci 5:752–769

Gil-Farina I, Fronza R, Kaeppel C, Lopez-Franco E, Ferreira V, D'avola D, Benito A, Prieto J, Petry H, Gonzalez-Aseguinolaza G (2016) Recombinant AAV integration is not associated with hepatic genotoxicity in nonhuman primates and patients. Mol Ther 24:1100–1105

Gilman S, Koller M, Black RS, Jenkins L, Griffith SG, Fox NC, Eisner L, Kirby L, Rovira MB, Forette F (2005) Clinical effects of Aβ immunization (AN1792) in patients with AD in an interrupted trial. Neurology 64:1553–1562

Gitter BD, Gannon KS, Cummins DJ, Brown-Augsburger PL, Bales KR, Bailey DL, Ballard DW, Brazelton AD, Czilli DL, Greene SJ (2002) Reduction in brain amyloid beta burden and reversal of memory impairment in APP V717F transgenic mice following chronic administration of the anti-amyloid beta antibody M266. 2. In: Neurobiology of aging. Elsevier Science Inc, New York, pp S105–S105

Giuffrida ML, Caraci F, Pignataro B, Cataldo S, De Bona P, Bruno V, Molinaro G, Pappalardo G, Messina A, Palmigiano A (2009) β-amyloid monomers are neuroprotective. J Neurosci 29:10582–10587

Gray AJ, Sakaguchi G, Shiratori C, Becker AG, LaFrancois J, Aisen PS, Duff K, Matsuoka Y (2007) Antibody against C-terminal Abeta selectively elevates plasma Abeta. Neuroreport 18:293–296

Greeneld JP, Haroutunian V, Buxbaum JD, Xu H, Greengard P, Relkin NR (2000) Intraneuronal Abeta42 accumulation in human brain. Am J Pathol 156:1520

Hashimoto M, Rockenstein E, Crews L, Masliah E (2003) Role of protein aggregation in mitochondrial dysfunction and neurodegeneration in Alzheimer's and Parkinson's diseases. NeuroMolecular Med 4:21–35

Hernandez F, Avila J (2007) Tauopathies. Cell Mol Life Sci 64:2219–2233

Hickman DT, López-Deber MP, Ndao DM, Silva AB, Nand D, Pihlgren M, Giriens V, Madani R, Pierre AS, Karastaneva H (2011) Sequence-independent control of peptide conformation in liposomal vaccines for targeting protein misfolding diseases. J Biol Chem 286:13966–13976

Hillen H, Barghorn S, Striebinger A, Labkovsky B, Müller R, Nimmrich V, Nolte MW, Perez-Cruz C, van der Auwera I, van Leuven F (2010) Generation and therapeutic efficacy of highly oligomer-specific β-amyloid antibodies. J Neurosci 30:10369–10379

Hooper NM (2005) Roles of proteolysis and lipid rafts in the processing of the amyloid precursor protein and prion protein. Portland Press Limited, London

Ida N, Hartmann T, Pantel J, Schröder J, Zerfass R, Förstl H, Sandbrink R, Masters CL, Beyreuther K (1996) Analysis of heterogeneous βA4 peptides in human cerebrospinal fluid and blood by a newly developed sensitive Western blot assay. J Biol Chem 271:22908–22914

Iqbal K, Gong C-X, Liu F (2014) Microtubule-associated protein tau as a therapeutic target in Alzheimer's disease. Expert Opin Ther Targets 18:307–318

Iqbal K, Liu F, Gong C-X (2016) Tau and neurodegenerative disease: the story so far. Nat Rev Neurol 12:15

Janus C, Pearson J, McLaurin J, Mathews PM, Jiang Y, Schmidt SD, Chishti MA, Horne P, Heslin D, French J (2000) Aβ peptide immunization reduces behavioural impairment and plaques in a model of Alzheimer's disease. Nature 408:979

Karlnoski RA, Rosenthal A, Alamed J, Ronan V, Gordon MN, Gottschall PE, Grimm J, Pons J, Morgan D (2008) Deglycosylated anti-Aβ antibody dose–response effects on pathology and memory in APP transgenic mice. J Neuroimmune Pharmacol 3:187–197

Karplus M, Petsko GA (1990) Molecular dynamics simulations in biology. Nature 347:631

Kasturirangan S, Boddapati S, Sierks MR (2010) Engineered proteolytic nanobodies reduce Aβ burden and ameliorate Aβ-induced cytotoxicity. Biochemistry 49:4501–4508

Katzav-Gozansky T, Hanan E, Solomon B (1996) Effect of monoclonal antibodies in preventing carboxypeptidase A aggregation. Biotechnol Appl Biochem 23:227–230

La Porte SL, Bollini SS, Lanz TA, Abdiche YN, Rusnak AS, Ho W-H, Kobayashi D, Harrabi O, Pappas D, Mina EW (2012) Structural basis of C-terminal β-amyloid peptide binding by the antibody ponezumab for the treatment of Alzheimer's disease. J Mol Biol 421:525–536

LaFerla FM, Tinkle BT, Bieberich CJ, Haudenschild CC, Jay G (1995) The Alzheimer's Aβ peptide induces neurodegeneration and apoptotic cell death in transgenic mice. Nat Genet 9:21

Lambert MP, Barlow AK, Chromy BA, Edwards C, Freed R, Liosatos M, Morgan TE, Rozovsky I, Trommer B, Viola KL (1998) Diffusible, nonfibrillar ligands derived from Aβ1–42 are potent central nervous system neurotoxins. Proc Natl Acad Sci 95:6448–6453

Landen JW, Zhao Q, Cohen S, Borrie M, Woodward M, Billing CB Jr, Bales K, Alvey C, McCush F, Yang J (2013) Safety and pharmacology of a single intravenous dose of ponezumab in subjects with mild-to-moderate Alzheimer disease: a phase I, randomized, placebo-controlled, double-blind, dose-escalation study. Clin Neuropharmacol 36:14–23

Lannfelt L, Möller C, Basun H, Osswald G, Sehlin D, Satlin A, Logovinsky V, Gellerfors P (2014) Perspectives on future Alzheimer therapies: amyloid-β protofibrils-a new target for immunotherapy with BAN2401 in Alzheimer's disease. Alzheimers Res Ther 6:16

Lemere CA (2013) Immunotherapy for Alzheimer's disease: hoops and hurdles. Mol Neurodegener 8:36

Lemere CA, Maron R, Selkoe DJ, Weiner HL (2001) Nasal vaccination with β-amyloid peptide for the treatment of Alzheimer's disease. DNA Cell Biol 20:705–711

Levine H III (1993) Thioflavine T interaction with synthetic Alzheimer's disease β-amyloid peptides: detection of amyloid aggregation in solution. Protein Sci 2:404–410

Lichtlen P, Mohajeri MH (2008) Antibody-based approaches in Alzheimer's research: safety, pharmacokinetics, metabolism, and analytical tools. J Neurochem 104:859–874

Liu R, Yuan B, Emadi S, Zameer A, Schulz P, McAllister C, Lyubchenko Y, Goud G, Sierks MR (2004) Single chain variable fragments against β-amyloid (Aβ) can inhibit Aβ aggregation and prevent Aβ-induced neurotoxicity. Biochemistry 43:6959–6967

Lührs T, Ritter C, Adrian M, Riek-Loher D, Bohrmann B, Döbeli H, Schubert D, Riek R (2005) 3D structure of Alzheimer's amyloid-β (1–42) fibrils. Proc Natl Acad Sci 102:17342–17347

Malone J, Sullivan MA (1996) Analysis of antibody selection by phage display utilizing anti-phenobarbital antibodies. J Mol Recognit 9:738–745

Manoutcharian K, Perez-Garmendia R, Gevorkian G (2017) Recombinant antibody fragments for neurodegenerative diseases. Curr Neuropharmacol 15:779–788

Miller TW, Messer A (2005) Intrabody applications in neurological disorders: progress and future prospects. Mol Ther 12:394–401

Morgan D (2011) Immunotherapy for Alzheimer's disease. J Intern Med 269:54–63

Muhs A, Hickman DT, Pihlgren M, Chuard N, Giriens V, Meerschman C, Van der Auwera I, van Leuven F, Sugawara M, Weingertner M-C (2007) Liposomal vaccines with conformation-specific amyloid peptide antigens define immune response and efficacy in APP transgenic mice. Proc Natl Acad Sci 104:9810–9815

Muthumani K, Block P, Flingai S, Muruganantham N, Chaaithanya IK, Tingey C, Wise M, Reuschel EL, Chung C, Muthumani A (2016) Rapid and long-term immunity elicited by DNA-encoded antibody prophylaxis and DNA vaccination against chikungunya virus. J Infect Dis 214:369–378

Muthumani K, Flingai S, Wise M, Tingey C, Ugen KE, Weiner DB (2013) Optimized and enhanced DNA plasmid vector based in vivo construction of a neutralizing anti-HIV-1 envelope glycoprotein Fab. Hum Vaccines Immunother 9:2253–2262

Nisbet RM, Polanco J-C, Ittner LM, Götz J (2015) Tau aggregation and its interplay with amyloid-β. Acta Neuropathol (Berl) 129:207–220

Ohnishi S, Takano K (2004) Amyloid fibrils from the viewpoint of protein folding. Cell Mol Life Sci 61:511–524

Paganetti P, Calanca V, Galli C, Stefani M, Molinari M (2005) β-site specific intrabodies to decrease and prevent generation of Alzheimer's Aβ peptide. J Cell Biol 168:863–868

Pardridge WM (2008) Re-engineering biopharmaceuticals for delivery to brain with molecular Trojan horses. Bioconjug Chem 19:1327–1338

Perchiacca JM, Ladiwala AR, Bhattacharya M, Tessier PM (2012) Structure-based design of conformation- and sequence-specific antibodies against amyloid β. Proc Natl Acad Sci U S A 109:84–89

Pfeifer M, Boncristiano S, Bondolfi L, Stalder A, Deller T, Staufenbiel M, Mathews PM, Jucker M (2002) Cerebral hemorrhage after passive anti-Aβ immunotherapy. Science 298:1379–1379

Planque SA, Nishiyama Y, Sonoda S, Lin Y, Taguchi H, Hara M, Kolodziej S, Mitsuda Y, Gonzalez V, Sait HB (2015) Specific amyloid β clearance by a catalytic antibody construct. J Biol Chem 290:10229–10241

Pride M, Seubert P, Grundman M, Hagen M, Eldridge J, Black RS (2008) Progress in the active immunotherapeutic approach to Alzheimer's disease: clinical investigations into AN1792-associated meningoencephalitis. Neurodegener Dis 5:194–196

Priller C, Bauer T, Mitteregger G, Krebs B, Kretzschmar HA, Herms J (2006) Synapse formation and function is modulated by the amyloid precursor protein. J Neurosci 26:7212–7221

Puzzo D, Privitera L, Leznik E, Fà M, Staniszewski A, Palmeri A, Arancio O (2008) Picomolar amyloid-β positively modulates synaptic plasticity and memory in hippocampus. J Neurosci 28:14537–14545

Qu J, Yu S, Zheng Y, Zheng Y, Yang H, Zhang J (2017) Aptamer and its applications in neurodegenerative diseases. Cell Mol Life Sci 74:683–695

Racke MM, Boone LI, Hepburn DL, Parsadainian M, Bryan MT, Ness DK, Piroozi KS, Jordan WH, Brown DD, Hoffman WP (2005) Exacerbation of cerebral amyloid angiopathy-associated microhemorrhage in amyloid precursor protein transgenic mice by immunotherapy is dependent on antibody recognition of deposited forms of amyloid β. J Neurosci 25:629–636

Robert R, Dolezal O, Waddington L, Hattarki MK, Cappai R, Masters CL, Hudson PJ, Wark KL (2008) Engineered antibody intervention strategies for Alzheimer's disease and related dementias by targeting amyloid and toxic oligomers. Protein Eng Des Sel 22:199–208

Robert R, Lefranc M-P, Ghochikyan A, Agadjanyan MG, Cribbs DH, Van Nostrand WE, Wark KL, Dolezal O (2010) Restricted V gene usage and VH/VL pairing of mouse humoral response against the N-terminal immunodominant epitope of the amyloid β peptide. Mol Immunol 48:59–72

Robert R, Wark KL (2012) Engineered antibody approaches for Alzheimer's disease immunotherapy. Arch Biochem Biophys 526:132–138

Ryan JM, Grundman M (2009) Anti-amyloid-β immunotherapy in Alzheimer's disease: ACC-001 clinical trials are ongoing. J Alzheimers Dis 17:243–243

Salfeld JG (2007) Isotype selection in antibody engineering. Nat Biotechnol 25:1369

Salloway S, Sperling R, Gilman S, Fox NC, Blennow K, Raskind M, Sabbagh M, Honig LS, Doody R, Van Dyck CH (2009) A phase 2 multiple ascending dose trial of bapineuzumab in mild to moderate Alzheimer disease. Neurology 73:2061–2070

Schenk D, Barbour R, Dunn W, Gordon G, Grajeda H, Guido T, Hu K, Huang J, Johnson-Wood K, Khan K (1999) Immunization with amyloid-β attenuates Alzheimer-disease-like pathology in the PDAPP mouse. Nature 400:173

Schneeberger A, Mandler M, Otava O, Zauner W, Mattner F, Schmidt W (2009) Development of AFFITOPE vaccines for Alzheimer's disease (AD)—from concept to clinical testing. J Nutr Health Aging 13:264–267

Sevigny J, Chiao P, Bussière T, Weinreb PH, Williams L, Maier M, Dunstan R, Salloway S, Chen T, Ling Y (2016) The antibody aducanumab reduces Aβ plaques in Alzheimer's disease. Nature 537:50–56

Siemers ER, Sundell KL, Carlson C, Case M, Sethuraman G, Liu-Seifert H, Dowsett SA, Pontecorvo MJ, Dean RA, Demattos R (2016) Phase 3 solanezumab trials: secondary outcomes in mild Alzheimer's disease patients. Alzheimers Dement 12:110–120

Sigurdsson EM, Knudsen E, Asuni A, Fitzer-Attas C, Sage D, Quartermain D, Goni F, Frangione B, Wisniewski T (2004) An attenuated immune response is sufficient to enhance cognition in an Alzheimer's disease mouse model immunized with amyloid-β derivatives. J Neurosci 24:6277–6282

Sigurdsson EM, Scholtzova H, Mehta PD, Frangione B, Wisniewski T (2001) Immunization with a nontoxic/nonfibrillar amyloid-β homologous peptide reduces Alzheimer's disease-associated pathology in transgenic mice. Am J Pathol 159:439–447

Solomon B, Koppel R, Frankel D, Hanan-Aharon E (1997) Disaggregation of Alzheimer β-amyloid by site-directed mAb. Proc Natl Acad Sci 94:4109–4112

Solomon B, Koppel R, Hanan E, Katzav T (1996) Monoclonal antibodies inhibit in vitro fibrillar aggregation of the Alzheimer beta-amyloid peptide. Proc Natl Acad Sci 93:452–455

Solomon B, Schwartz F (1995) Chaperone-like effect of monoclonal antibodies on refolding of heat-denatured carboxypeptidase A. J Mol Recognit 8:72–76

Soscia SJ, Kirby JE, Washicosky KJ, Tucker SM, Ingelsson M, Hyman B, Burton MA, Goldstein LE, Duong S, Tanzi RE (2010) The Alzheimer's disease-associated amyloid β-protein is an antimicrobial peptide. PLoS One 5:e9505

Sperling R, Salloway S, Brooks DJ, Tampieri D, Barakos J, Fox NC, Raskind M, Sabbagh M, Honig LS, Porsteinsson AP (2012) Amyloid-related imaging abnormalities in patients with Alzheimer's disease treated with bapineuzumab: a retrospective analysis. Lancet Neurol 11:241–249

Sudol KL, Mastrangelo MA, Narrow WC, Frazer ME, Levites YR, Golde TE, Federoff HJ, Bowers WJ (2009) Generating differentially targeted amyloid-β specific intrabodies as a passive vaccination strategy for Alzheimer's disease. Mol Ther 17:2031–2040

Tammer AH, Coia G, Cappai R, Fuller S, Masters CL, Hudson P, Underwood JR (2002) Generation of a recombinant Fab antibody reactive with the Alzheimer's disease-related Aβ peptide. Clin Exp Immunol 129:453–463

Tiraboschi P, Hansen LA, Thal LJ, Corey-Bloom J (2004) The importance of neuritic plaques and tangles to the development and evolution of AD. Neurology 62:1984–1989

Tucker S, Möller C, Tegerstedt K, Lord A, Laudon H, Sjödahl J, Söderberg L, Spens E, Sahlin C, Waara ER (2015) The murine version of BAN2401 (mAb158) selectively reduces amyloid-β protofibrils in brain and cerebrospinal fluid of tg-ArcSwe mice. J Alzheimers Dis 43:575–588

Turner PR, O'connor K, Tate WP, Abraham WC (2003) Roles of amyloid precursor protein and its fragments in regulating neural activity, plasticity and memory. Prog Neurobiol 70:1–32

Ultsch M, Li B, Maurer T, Mathieu M, Adolfsson O, Muhs A, Pfeifer A, Pihlgren M, Bainbridge TW, Reichelt M (2016) Structure of crenezumab complex with Aβ shows loss of β-hairpin. Sci Rep 6:39374

Verma R, Boleti E, George AJT (1998) Antibody engineering: comparison of bacterial, yeast, insect and mammalian expression systems. J Immunol Methods 216:165–181

Walker MR, Lund J, Thompson K, Jefferis R (1989) Aglycosylation of human IgG1 and IgG3 monoclonal antibodies can eliminate recognition by human cells expressing FcγRI and/or FcγRII receptors. Biochem J 259:347–353

Walsh DM, Klyubin I, Fadeeva JV, Cullen WK, Anwyl R, Wolfe MS, Rowan MJ, Selkoe DJ (2002) Naturally secreted oligomers of amyloid β protein potently inhibit hippocampal long-term potentiation in vivo. Nature 416:535

Wisniewski T, Frangione B (2005) Immunological and anti-chaperone therapeutic approaches for Alzheimer disease. Brain Pathol 15:72–77

Wisniewski T, Goñi F (2015) Immunotherapeutic approaches for Alzheimer's disease. Neuron 85:1162–1176

Wisniewski T, Goñi F (2014) Immunotherapy for Alzheimer's disease. Biochem Pharmacol 88:499–507

Wisniewski T, Goñi F (2012) Could immunomodulation be used to prevent prion diseases? Expert Rev Anti-Infect Ther 10:307–317

Yu L, Edalji R, Harlan JE, Holzman TF, Lopez AP, Labkovsky B, Hillen H, Barghorn S, Ebert U, Richardson PL (2009) Structural characterization of a soluble amyloid β-peptide oligomer. Biochemistry 48:1870–1877

Zhang C, Gao C, Mu J, Qiu Z, Li L (2013) Spectroscopic studies on unfolding processes of apo-neuroglobin induced by guanidine hydrochloride and urea. BioMed Res Int 2013:349542

The manufacturer's authorised representative in the EU is Springer Nature Customer Service Centre GmbH, Europaplatz 3, 69115 Heidelberg, Germany. If you have any concerns regarding our products, please contact ProductSafety@springernature.com

Printed and bound by CPI Group (UK) Ltd, Croydon, CR0 4YY
30/07/2026
02181675-0004